172
Advances in Polymer Science

Editorial Board:
A. Abe · A.-C. Albertsson · R. Duncan · K. Dušek · W. H. de Jeu
J. F. Joanny · H.-H. Kausch · S. Kobayashi · K.-S. Lee · L. Leibler
T. E. Long · I. Manners · M. Möller · O. Nuyken · E. M. Terentjev
B. Voit · G. Wegner

Advances in Polymer Science

Recently Published and Forthcoming Volumes

**Advanced Computer Simulation
Approaches for Soft Matter Sciences I**
Volume Editors: Holm, C., Kremer, K.
Vol. 173, 2004

Microlithography · Molecular Imprinting
Volume Editor: Kobayashi, T.
Vol. 172, 2005

Polymer Synthesis
Vol. 171, 2004

**NMR · Coordination Polymerization ·
Photopolymerization**
Vol. 170, 2004

Long-Term Properties of Polyolefins
Volume Editor: Albertsson, A.-C.
Vol. 169, 2004

Polymers and Light
Volume Editor: Lippert, T. K.
Vol. 168, 2004

New Synthetic Methods
Vol. 167, 2004

**Polyelectrolytes with Defined
Molecular Architecture II**
Volume Editor: Schmidt, M.
Vol. 166, 2004

**Polyelectrolytes with Defined
Molecular Architecture I**
Volume Editors: Schmidt, M.
Vol. 165, 2004

**Filler-Reinforeced Elastomers ·
Scanning Force Microscopy**
Vol. 164, 2003

**Liquid Chromatography ·
FTIR Microspectroscopy · Microwave
Assisted Synthesis**
Vol. 163, 2003

**Radiation Effects on Polymers
for Biological Use**
Volume Editor: Kausch, H.
Vol. 162, 2003

**Polymers for Photonics
Applications II**
Nonlinear Optical, Photorefractive and
Two-Photon Absorption Polymers
Volume Editor: Lee, K.-S.
Vol. 161, 2003

**Filled Elastomers · Drug Delivery
Systems**
Vol. 160, 2002

**Statistical, Gradient, Block
and Graft Copolymers by Controlled/
Living Radical Polymerizations**
Authors: Davis, K. A., Matyjaszewski, K.
Vol. 159, 2002

**Polymers for Photonics
Applications I**
Nonlinear Optical and
Electroluminescence Polymers
Volume Editor: Lee, K.-S.
Vol. 158, 2002

Degradable Aliphatic Polyesters
Volume Editor: Albertsson, A.-C.
Vol. 157, 2001

**Molecular Simulation · Fracture ·
Gel Theory**
Vol. 156, 2001

**Polymerization Techniques
and Synthetic Methodologies**
Vol. 155, 2001

Polymer Physics and Engineering
Vol. 154, 2001

Microlithography · Molecular Imprinting

With contributions by
H. Ito · J. D. Marty · M. Mauzak

 Springer

The series presents critical reviews of the present and future trends in polymer and biopolymer science including chemistry, physical chemistry, physics and material science. It is addressed to all scientists at universities and in industry who wish to keep abreast of advances in the topics covered.

As a rule, contributions are specially commissioned. The editors and publishers will, however, always be pleased to receive suggestions and supplementary information. Papers are accepted for "Advances in Polymer Science" in English.

In references Advances in Polymer Science is abbreviated Adv Polym Sci and is cited as a journal.

The electronic content of APS may be found at springerlink.com

Library of Congress Control Number: 2004115719

ISSN 0065-3195
ISBN 3-540-21862-9 **Springer Berlin Heidelberg New York**
DOI 10.1007/b14099

This work is subject to copyright. All rights are reserved, whether the whole or part of the material is concerned, specifically the rights of translation, re-printing, re-use of illustrations, recitation, broadcasting, reproduction on microfilms or in any other ways, and storage in data banks. Duplication of this publication or parts thereof is only permitted under the provisions of the German Copyright Law of September 9, 1965, in its current version, and permission for use must always be obtained from Springer-Verlag. Violations are liable to prosecution under the German Copyright Law.

Springer is a part of Springer Science+Business Media
springeronline.com
© Springer-Verlag Berlin Heidelberg 2005
Printed in The Netherlands

The use of registered names, trademarks, etc. in this publication does not imply, even in the absence of a specific statement, that such names are exempt from the relevant protective laws and regulations and therefore free for general use.

Cover: Künkellopka GmbH, Heidelberg; design & production GmbH, Heidelberg
Typesetting: Fotosatz-Service Köhler GmbH, Würzburg

Printed on acid-free paper 02/3141/xv – 5 4 3 2 1 0

Editorial Board

Prof. Akihiro Abe
Department of Industrial Chemistry
Tokyo Institute of Polytechnics
1583 Iiyama, Atsugi-shi 243-02, Japan
aabe@chem.t-kougei.ac.jp

Prof. A.-C. Albertsson
Department of Polymer Technology
The Royal Institute of Technology
S-10044 Stockholm, Sweden
aila@polymer.kth.se

Prof. Ruth Duncan
Welsh School of Pharmacy
Cardiff University
Redwood Building
King Edward VII Avenue
Cardiff CF 10 3XI, United Kingdom
duncan@cf.ac.uk

Prof. Karel Dušek
Institute of Macromolecular Chemistry,
Czech
Academy of Sciences of the
Czech Republic
Heyrovský Sq. 2
16206 Prague 6, Czech Republic
dusek@imc.cas.cz

Prof. Dr. W. H. de Jeu
FOM-Institute AMOLF
Kruislaan 407
1098 SJ Amsterdam, The Netherlands
dejeu@amolf.nl
and

Dutch Polymer Institute
Eindhoven University of Technology
PO Box 513
5600 MB Eindhoven, The Netherlands

Prof. Jean-François Joanny
Physicochimie Curie
Institut Curie section recherche
26 rue d'Ulm
75248 Paris cedex 05, France
jean-francois.joanny@curie.fr

Prof. Hans-Henning Kausch
c/o EPFL, Science de Base (SB-ISIC)
Station 6
CH-1015 Lausanne, Switzerland
kausch.cully@bluewin.ch

Prof. S. Kobayashi
Department of Materials Chemistry
Graduate School of Engineering
Kyoto University
Kyoto 615-8510, Japan
kobayasi@mat.polym.kyoto-u.ac.jp

Prof. Kwang-Sup Lee
Department of Polymer Science &
Engineering
Hannam University
133 Ojung-Dong
Daejeon 306-791, Korea
kslee@mail.hannam.ac.kr

Prof. L. Leibler
Matière Molle et Chimie
Ecole Supèrieure de Physique
et Chimie Industrielles (ESPCI)
10 rue Vauquelin
75231 Paris Cedex 05, France
ludwik.leibler@espci.fr

Prof. Timothy E. Long
Department of Chemistry
and Research Institute
Virginia Tech
2110 Hahn Hall (0344)
Blacksburg, VA 24061, USA
telong@vt.edu

Prof. Ian Manners
Department of Chemistry
University of Toronto
80 St. George St.
M5S 3H6 Ontario, Canada
imanners@chem.utoronto.ca

Prof. Dr. Martin Möller
Deutsches Wollforschungsinstitut
an der RWTH Aachen e.V.
Pauwelsstraße 8
52056 Aachen, Germany
moeller@dwi.rwth-aachen.de

Prof. Oskar Nuyken
Lehrstuhl für Makromolekulare Stoffe
TU München
Lichtenbergstr. 4
85747 Garching, Germany
oskar.nuyken@ch.tum.de

Dr. E. M. Terentjev
Cavendish Laboratory
Madingley Road
Cambridge CB 3 OHE
United Kingdom
emt1000@cam.ac.uk

Prof. Brigitte Voit
Institut für Polymerforschung Dresden
Hohe Straße 6
01069 Dresden, Germany
voit@ipfdd.de

Prof. Gerhard Wegner
Max-Planck-Institut
für Polymerforschung
Ackermannweg 10
Postfach 3148
55128 Mainz, Germany
wegner@mpip-mainz.mpg.de

Advances in Polymer Science
Also Available Electronically

For all customers who have a standing order to Advances in Polymer Science, we offer the electronic version via SpringerLink free of charge. Please contact your librarian who can receive a password for free access to the full articles by registering at:

springerlink.com

If you do not have a subscription, you can still view the tables of contents of the volumes and the abstract of each article by going to the SpringerLink Homepage, clicking on "Browse by Online Libraries", then "Chemical Sciences", and finally choose Advances in Polymer Science.

You will find information about the

– Editorial Board
– Aims and Scope
– Instructions for Authors
– Sample Contribution

at springeronline.com using the search function.

Contents

Molecular Imprinting: State of the Art and Perspectives
J. D. Marty · M. Mauzac 1

Chemical Amplification Resists for Microlithography
H. Ito 37

Author Index Volumes 101–172 247

Subject Index 267

Molecular Imprinting: State of the Art and Perspectives

Jean Daniel Marty (✉) · Monique Mauzac

Laboratoire IMRCP, UMR CNRS 5623, Université Paul Sabatier, 118 route de Narbonne, 31062 Toulouse Cedex 4, France
marty@chimie.ups-tlse.fr

1	Introduction	2
2	MIP Materials	4
2.1	The Different Stages Involved in the Preparation of the Materials	5
2.1.1	Preorganisation Stage	5
2.1.2	Step of Polymerisation-Crosslinking of the Monomers Around the Complex Formed in the First Step	10
2.1.3	Template Molecule Extraction Stage	12
2.2	Influence of the Solvent	13
3	Recognition Properties	14
3.1	Guest Capacity	14
3.2	Specificity of Recognition	14
3.3	Lifetime of the MIPs	16
4	Main Uses of Imprinted Polymers	17
4.1	Separation of Molecules	17
4.2	Preparation of Antibody Analogues	19
4.3	Sensors	20
4.4	Stereoselective Reactions and Catalysis	20
5	Summary and Outlook	22
5.1	Molecularly Imprinted Hydrogels	24
5.2	Two-Dimensional Molecular Imprinting	24
5.3	Liquid Crystalline Imprinted Materials	26
6	Conclusion	29
	References	30

Abstract Molecular recognition is central to how biological systems work. The molecular imprinting technique is a valuable polymerisation method for preparing synthetic materials able to mimic the molecular recognition phenomena present in living systems. A molecule that acts as a template is associated with functional monomers to form a complex by means of covalent linkages or noncovalent interactions. A polymerisation-crosslinking reaction is then performed around this complex. Upon removal of the template species, functionalised cavities, that have memorized the special features and bonding preferences of the template, are left inside the polymer network. A large number of potential applications for this class materials are being intensively developed, for example as chromatographic stationary phases or as stereospecific catalyst. To improve this technique, the challenge is now to rationalize the necessary stiffness of the network with the expansion of its capacity. From this perspective, the use of materials involving supramolecular organisation is of great interest to bring closer to the biological processes and so to improve the recognition properties.

Keywords Molecular recognition · Imprinting · Polymer · Chromatography · Catalysis

1
Introduction

Molecular recognition is central to how biological systems work, especially at the cellular level (example Fig. 1). The observation of the various systems where processes of recognition occur (enzyme substrate complexes, antibody-antigen systems, DNA replication, membrane receptors, and so on) has indicated a certain number of directions for the preparation of synthetic systems capable of molecular recognition.

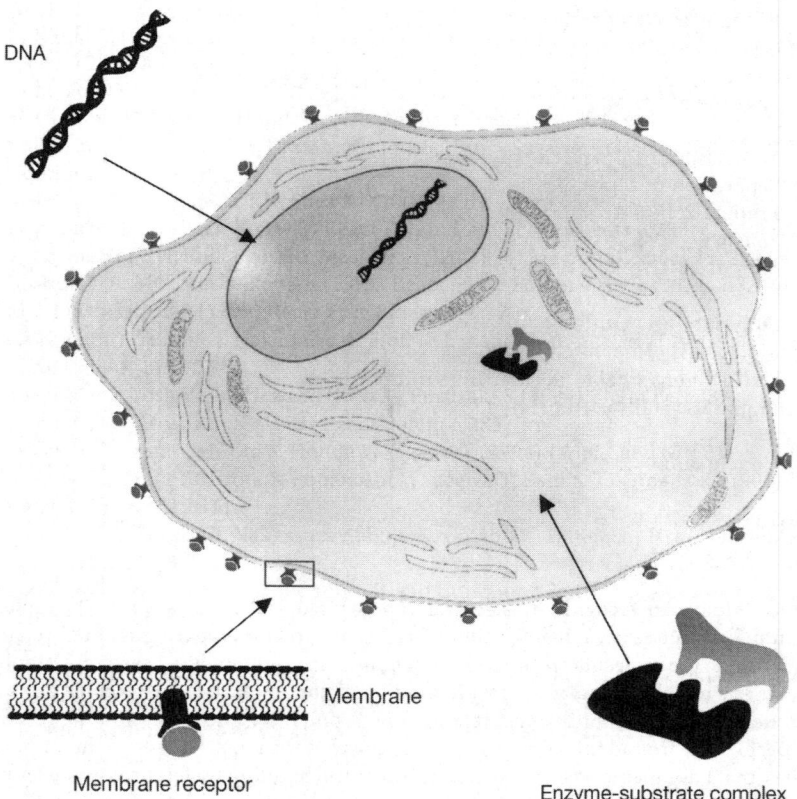

Fig. 1 Examples of molecular recognition systems occurring in the cell

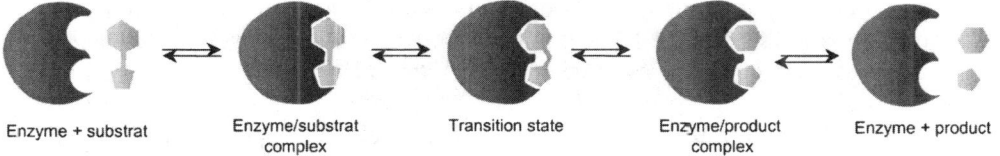

Fig. 2 Changes that occur in the shape of the enzyme when the substrate binds to the active site so that catalysis can take place

Let us consider for instance, the way enzyme-substrate type systems operate. In 1894, Emil Fisher developed the "lock-and-key" notion [1], which has become one of the most frequently mentioned concepts in Life Sciences. This model describes how proteins bear specific cavities into which a given substrate fits. The process was refined by Koshland [2] who considered that during the interaction phase, the geometry of the cavities became adapted to the substrate, optimising the interactions taking place (for example Fig. 2). In spite of its apparent complexity, the recognition process involves simple interactions between chemical groups – the overall exact location of the interactions leading to molecular recognition.

Considerable efforts have been made to synthesise systems mimicking the natural processes of molecular recognition [3, 4]. The first studies involved the preparation of small ring systems and 'cage' molecules that could become specifically associated to a given ion or to small molecules: crown ethers [5], cryptates [6, 7], cyclophanes [8], and cyclodextrins [9] (Fig. 3). Abzymes have been designed in an attempt to study recognition of more complex molecules [10]. However, they were found to be difficult to prepare and to use, and their cost is generally high.

Antibodies, receptors and enzymes are biomolecules frequently used in analytical chemistry and biochemistry in various applications aiming to detect

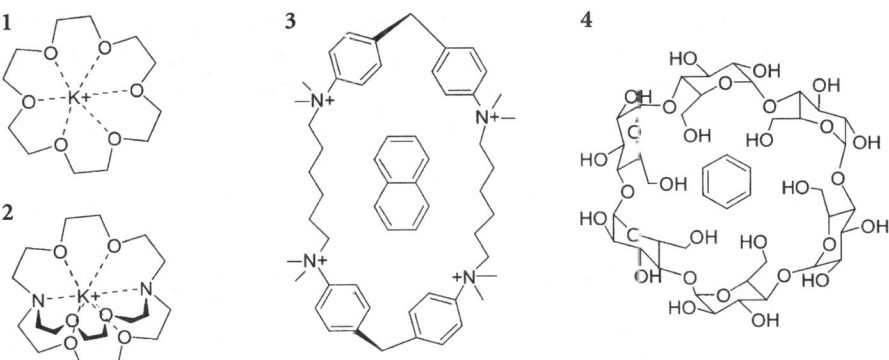

Fig. 3 Examples of synthetic receptors capable of molecular recognition: (1) crown ether; (2) cryptate; (3) cyclophane; (4) α-cyclodextrin (adapted from [5–9])

and quantify a host molecule (biosensors, affinity chromatography supports, immunoanalysis kits, etc.). These applications share a common first step where the guest molecule is recognised and selectively adsorbed by the receptor biomolecule. However, biomolecules can only be used in experimental conditions close to those found in nature, limiting their fields of application. Their replacement by more stable, cheaper materials with similar selectivity therefore became attractive.

The techniques for the synthesis of molecularly imprinted polymer (MIP) then developed. The principle consists of polymerising and crosslinking functional monomers, previously positioned by low-energy or covalent interactions around a molecule – the guest or template molecule – so as to "freeze" the imprint. Following extraction of the template molecule, the remaining three-dimensional network presents pores with a geometry and positioning of the functional groups complementary to those of the template. This enables it to specifically recognise the template molecule.

The conception of this technique was originally inspired by the works of Linus Pauling in the 1940s outlining the possible mode of action of antibodies [11]. One of his students, F.H. Dickey, was the first to prepare selective adsorbents by precipitating silica gel in the presence of dyes (methyl orange and its derivatives) [12]. When washed, the materials presented an enhanced affinity for the dyes. Similarly, materials used for chiral recognition [13] or for pesticide separation were studied. However, owing to the low separating powers obtained and the short useful life-time of the materials, this line of research was to be finally abandoned [14]. It was only with the works of G. Wulff from 1972 [15], followed by those of Mosbach in 1984 [16], that the principles of the technique of molecular imprinting were laid down.

The simplicity of use, the relatively low cost and the broad range of possible guest molecules (small organic molecules, ions but also biological macromolecules) have since led to the important development of this technique, as illustrated by the increasing numbers of publications over recent years [17–25]. The fields of application of these imprinted polymer networks are very diverse. We can mention chromatographic supports (particularly for the separation of enantiomers) recognition elements in the preparation of specific sensors, catalysts, systems for stereospecific synthesis, and selective adsorbents.

2
MIP Materials

The principle of molecularly imprinted polymer (MIP) production is schematised in Fig. 4.

The preparation of the materials starts by positioning the functional monomers around a template molecule. The monomers interact with sites on the template via interactions that can be reversible covalent or non-covalent (hydrogen, ionic, Van der Waals, π-π, etc.). They are then polymerised and cross-

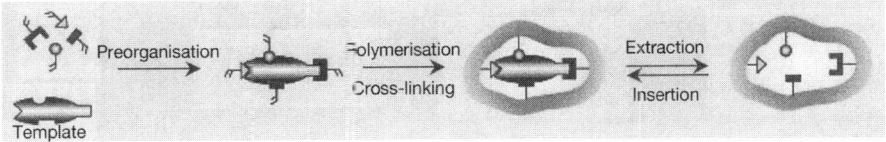

Fig. 4 Principle of the molecular imprinting technique

linked around the template in order to fix their position and to "freeze" the geometry of the pores in the network. The template molecule is then extracted, leaving a polymer with functional sites capable of molecular recognition.

2.1
The Different Stages Involved in the Preparation of the Materials

2.1.1
Preorganisation Stage

The development of an imprinted network starts with the choice of the template molecule. It is this molecule that the material will keep in memory. The choice remains limited, however, by the functional monomers available and susceptible to interact. Depending on the domain studied, diverse molecules have been used (Fig. 5): amino acids [26], sugar derivatives [27], steroids [28], nucleotides [29], pesticides [30], dyes [12], drugs [31–36], metal ions [37–39], more inert molecules (like anthracene, benzene and its derivatives) [40–42], and even more complex molecules such as proteins or enzymes [43, 44].

Note that with chiral guest molecules for which it is difficult to obtain a pure enantiomer, the use of a chiral functional monomer, interacting preferentially with one of the two enantiomers in a racemic mixture, can be an interesting approach [45].

Several types of monomers have been used for this (Fig. 6), especially easily polymerisable molecules. Acrylate monomers [46–51] and acrylamides have been used the most [52–56]. Systems with polystyrene [47, 57–63] and polysiloxane [64] backbones have also been synthesised. Recently, other types of polymer such as polyurethanes [65], polypyrroles [66–68], and polyimidazoles [69, 70] have been used in special applications. Renewed interest has also been shown in silicas and sol-gel glasses [71].

In the first step, the contact occurring between the functionalised monomers and the template molecule leads to the formation of a complex. Its structure and stability will then determine the behaviour of the future MIP. The interactions involved must be sufficiently strong to remain intact during the polymerisation stage but sufficiently labile to enable template extraction and reinsertion of guest molecules in the later stages. These interactions must therefore occur rapidly and be reversible. It is crucial to optimise the choice of the different components of the system at this point.

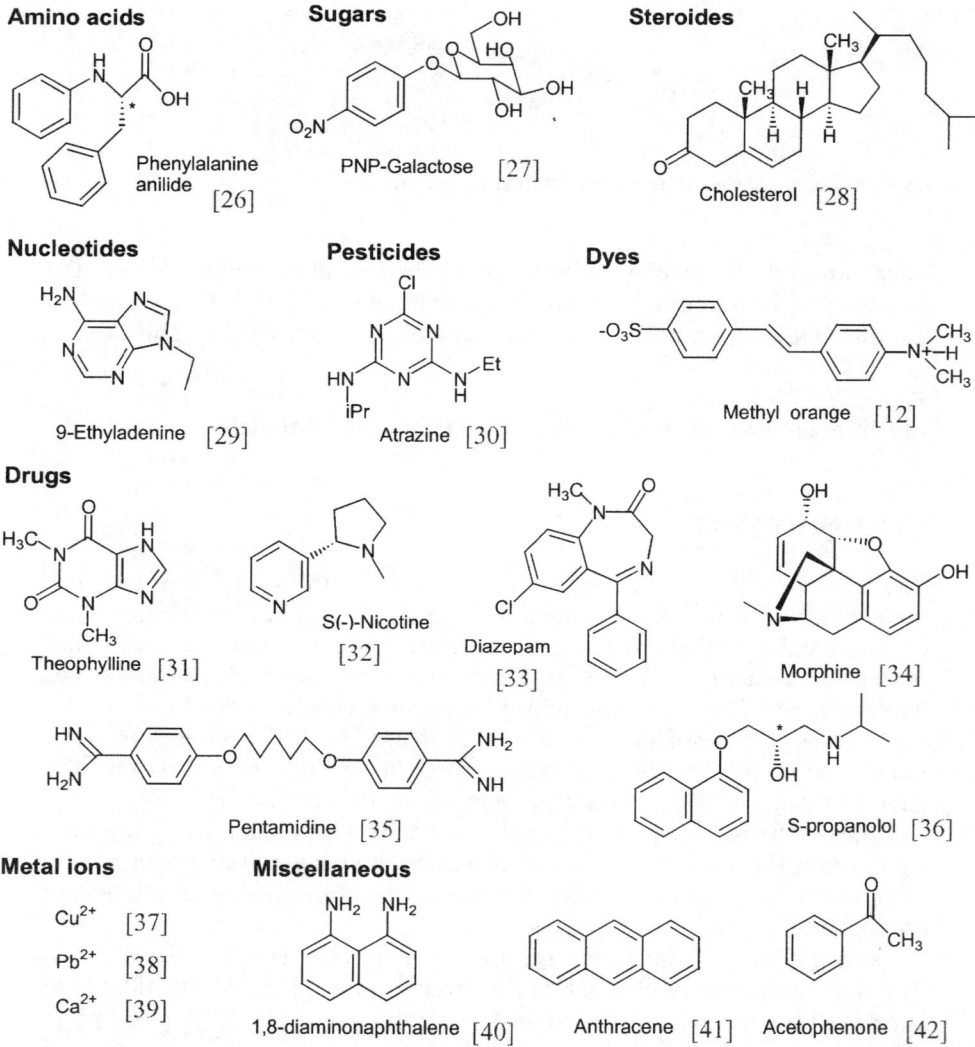

Fig. 5 Some examples of molecules used as templates in MIP

Different types of strategies can be distinguished depending on whether the bonding between template and host is covalent or not:

1. Covalent interactions (Fig. 7): this approach involves the formation of an easily cleavable functionalised monomer-template complex [61, 18]. The first example of an imprinted polymer network was prepared by Wulff's group and used the reversible formation of an ester bond between a diol and 4-vinylphenylboronic acid [15]. The reaction rates to reach equilibrium between the ester and boronic acid are comparable to those obtained with

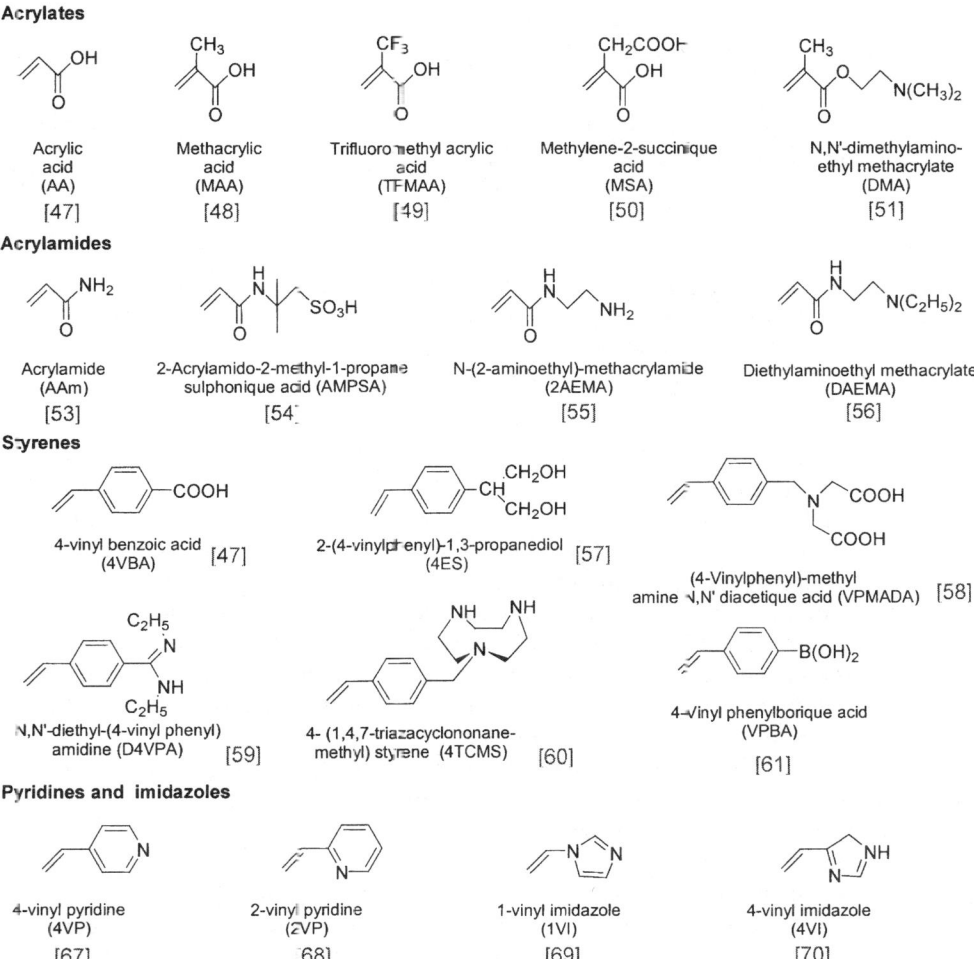

Fig. 6 The monomers most commonly used in MIP

non-covalent systems [72]. It can be noted that systems using two different boronic acids to interact with the template lead to MIPs [73] that ensure higher resolution of racemic mixtures than those consisting of a single boronic acid [74].

Covalent bonds involving Schiff bases [75], esters [76], amides [77], and ketals [57, 78] have also been used. However, their much slower monomer-template interaction kinetics have excluded them from several types of application.

Although this covalent bonding strategy provides well-defined cavities, the limited choice of functional monomers and hence of useable template molecules has restricted its use.

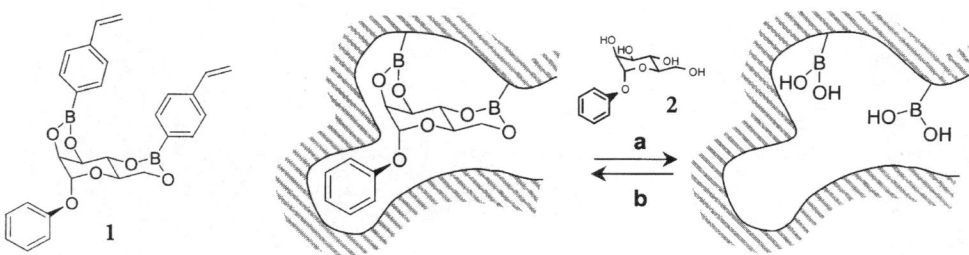

Fig. 7 Schematic representation of a cavity obtained by polymerisation of **1**. The interactions occurring between the network and the guest molecule **2** are covalent. The guest is extracted in the presence of water and methanol (**a**), and its reinsertion leads to the reoccupation of the imprinted cavity (**b**) (adapted from [61])

2. Non-covalent interactions: this approach is simpler to use and can be adapted to a larger range of template molecules. Here, the guest interacts with one or several functional monomers [79–82] to form a complex. The stability of the complex depends on the different components present in the mixture. The forces involved can include ionic interactions [83], hydrogen bonding [54], π-π interactions [84] and hydrophobic interactions [85]. Owing to the lower stability of the complex that is formed, systems based on these types of interactions often lead to MIP that are less specific than those that use covalent interactions. On the other hand, the host-guest exchange rates are more rapid and the MIP are suitable for use in chromatography.

Among the most frequently used monomers we can mention carboxylic acids (e.g. acrylic, methacrylic and vinyl benzoic acids), sulfonic acids and heterocyclic bases (e.g. vinylpyridines, vinylimidazoles).

Currently, great efforts are being made to broaden the range of templates that can be used and to improve the specificity of 'non-covalent' MIP. The first systems developed were not very selective [79]. It was only once the experimental conditions had been optimised (conditions of synthesis, host/guest ratios, eluent, etc.) that Mosbach et al. achieved materials with selectivities similar to those obtained with the covalent approach [48]. Owing to the weakness of the interactions occurring, a large excess of functional monomer must be added to shift the equilibrium towards formation of the complex (typical monomer:guest ratios are 4:1). This leads to the formation, inside the imprinted network, of sites with different affinities for the molecule to be recognised (polyclonality) and lowers the specificity of the MIP. In chromatographic applications, for instance, this causes broadening of the peaks and decreased resolution. To overcome this problem, new monomers interacting more specifically with the template at several points have been developed for addition in stoichiometric quantities [59, 86] (Fig. 8).

Template molecules with moieties capable of binding to a metal can also be incorporated in a MIP, if the metal presents polymerisable ligands. The strength of the metal-template bond can be easily modulated by adjusting

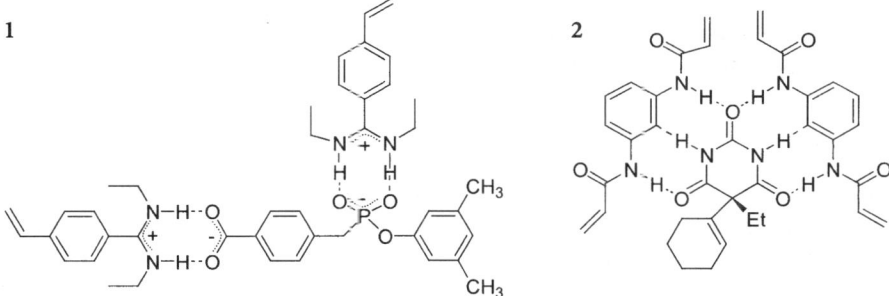

Fig. 8 Functionalised monomer/template molecule complexes based on strong non-covalent interactions (the various molecules are added in stoichiometric proportions): (1) adapted from [59]; (2) adapted from [86]

the experimental conditions. This approach was recently evaluated and gives very good results, in some cases equivalent to those obtained with enzyme systems [58, 60, 87].

3. Finally, there is an alternative strategy which uses a double approach: covalent plus non-covalent (Fig. 9). With the example of cholesterol taken as template molecule, M.J. Whitcombe and co-workers esterified the hydroxyl function, to enable incorporation of the cholesterol within the polymer network. After breaking the ester bond, the template only acted through non-covalent interactions [28]. This strategy, which was also used by A. Sahran [88, 89] does present the disadvantage of potentially modifying the geometry of the imprinted site following hydrolysis.

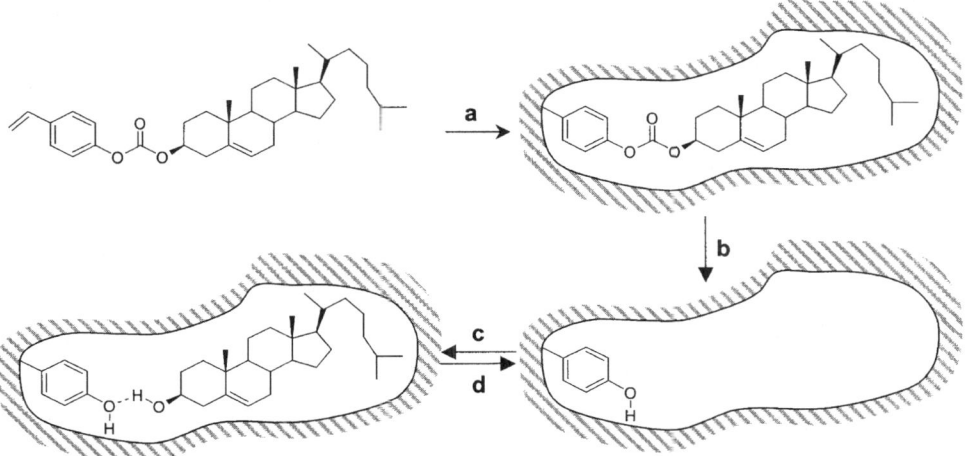

Fig. 9 Preparation of a MIP using covalent interactions during the polymerisation-crosslinking step (a) and the extraction of the template molecule (b). During the specificity tests, (c) and (d), the interactions occurring are of the hydrogen bonding type (adapted from [28])

2.1.2
Step of Polymerisation-Crosslinking of the Monomers Around the Complex Formed in the First Step

Polymerisation mode [90]. Owing to its ease of use, radical polymerisation is the most frequent. The crucial question is to determine how to carry out this polymerisation-crosslinking step with minimum disturbance of the complex already in place. Choices must be made, for instance in radical polymerisation, the radicals can be generated at 60 °C (α,α'-azoisobutyronitrile – AIBN) or 45 °C (with azobis valeronitrile – ABDV) which could cause heat destabilisation of the complex, or at 4 °C with low-temperature photochemical radical production (AIBN, 360 nm). Comparative studies on recognition specificity have shown that the photochemical approach gives the most specific materials [91].

Crosslinking. To date, only a small number of crosslinkers have been tested as their poor miscibility with the monomers considerably limits the choice (Fig. 10) [92–99]. The most frequently used crosslinkers in acrylate systems are acrylate esters of diols and triols such as ethylene glycol dimethacrylate and trimethylolpropane trimethacrylate [92–94]. In styrene-based systems, isomers of divinylbenzene are usually used [61a, 62].

Wulff et al. studied the influence of the degree of crosslinking on the suitability of a MIP used as a chromatographic support to separate two enantiomers – one of which had been used as the template [94]. The results obtained are reported in Fig. 11, selectivity being represented by values of separation factor α greater than 1. Irrespective of the crosslinker used, to obtain significant recognition of the guest molecule, the network has to be strongly crosslinked to limit chain relaxation [94]. For similar proportions of crosslinker, the use of a short molecule (ethylene glycol dimethacrylate rather than tetramethylene

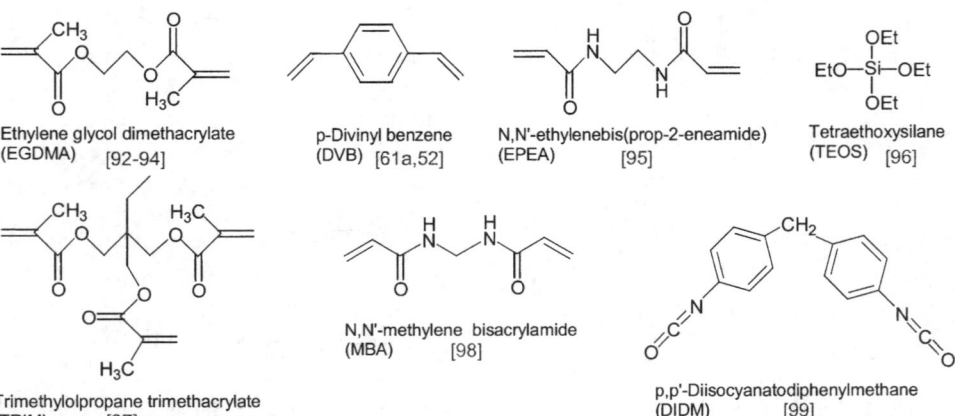

Fig. 10 Some of the crosslinkers used to prepare MIP

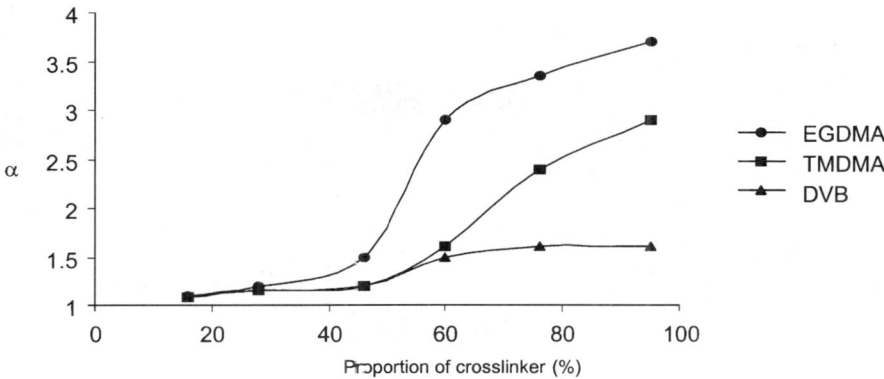

Fig. 11 Influence of the proportion of crosslinker on the recognition specificity of MIPs. The crosslinkers represented are ethylene glycol dimethacrylate (EGDMA), tetramethylene dimethacrylate (TMDMA) and divinyl benzene (DVB) (adapted from [94])

dimethacrylate) but not too rigid (ethylene glycol dimethacrylate rather than divinyl benzene) leads to more specific recognition. So, a high degree of crosslinking appears necessary and in most cases over 70 mol.% crosslinker is required.

While a high degree of crosslinking enhances the mechanical stability of the MIP, resulting in reduced deformation of the imprinted sites, it does rigidify the network, deteriorating the extraction and insertion kinetics of the guest molecule and reducing the accessibility of the imprinted sites in the network. So, in networks based on non-covalent interactions, at best 10% of the sites formed are active. The result is that the high proportions of crosslinker bring about a decrease in the capacity of the network meaning that the MIPs can only be basically used for analytical purposes. In an attempt to alleviate this shortcoming, it has been suggested that tri- and even tetravalent crosslinkers be used [100, 101]. Experiments have also been carried out with crosslinkers able to interact in the same way as functionalised monomers [86].

Shaping the MIPs. The use intended for the MIP will determine its shape. The most widely used technique consists of preparing a bulk polymer which is then ground and graded. The particles obtained (Fig. 12(1)) can then be used in various applications [68, 102, 103].

To use MIPs in chromatography, other techniques have been developed. Polymers have been prepared in situ in chromatographic columns [104] and in capillary systems for electrophoresis [105]. Their performance does not, however, come up to expectations and efforts have been made to obtain more homogeneous materials. This was done in two ways: i) grafting or covering preformed particles by the MIP. The particles covered can be silica or trimethylolpropane trimethacrylate [52, 106–108]. Sveç and Fréchet used a process they call two-step swelling. The first step consists of making polystyrene support beads and the

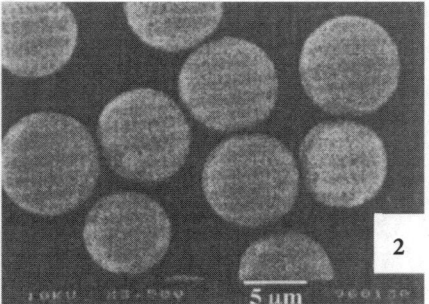

Fig. 12 Scanning Electron Microscopy images of MIP used as stationary phase in chromatography: (1) MIP particles obtained by grinding, grading and sedimentation; (2) beads prepared by an emulsion process (from [114])

second is the polymerisation around them of the MIP, leading to the formation of perfectly spherical beads [109] and ii) the preparation of polymer beads via processes in emulsion or in suspension [110–114]. Spherical imprinted particles with a very narrow size distribution are then obtained, and can be used as high performance chromatography stationary phase (Fig. 12(2)).

The formation of thin layers and membranes, for analytical applications or as sensors, has also been studied. Here, either the MIP is grown directly on a surface (for instance an electrode) [115], or particles of MIP are held together by a binder to generate a material similar to those used for thin layer chromatography [116]. The first method avoids the grinding step and thus reduces the risk of deforming the sites.

A surface imprinting technique has also been investigated [117–122]: after the template molecule has been dissolved in the presence of functionalised monomer, the preformed complex is grafted onto an activated surface (silica or a glass surface). This approach can prove to be particularly interesting for the recognition of macromolecules such as proteins which present much greater problems of insertion and extraction than small molecules [123]. Using similar techniques, metal adsorption sites can be integrated onto surfaces [124].

Finally, a new method for obtaining permeable MIPs has recently been developed: first, the guest molecule is grafted onto a very porous material (support network), then polymerisation is carried out around it. The support network is then destroyed leaving a highly porous MIP suitable for chromatographic applications [125].

2.1.3
Template Molecule Extraction Stage

The proportion of extraction of the template molecules interacting with the MIP via easily hydrolysable covalent bonds or non-covalent linkages is estimated to be about 90% [17]. The remaining molecules are trapped in highly crosslinked zones. This problem is exacerbated with macromolecules where steric hindrance lowers the efficiency of extraction.

In addition, the synthesis of MIPs requires large quantities of guest molecule (50 to 500 µmoles per gram of polymer). So, when pure template molecule is difficult or expensive to obtain, reaching quantitative template extraction yields can be primordial. Extraction conditions must then be optimised to obtain yields of over 99% [126, 127].

The extraction step uses an appropriate solvent. It often proves to be long and the actual process involved is dependent on the system in question. So, automation of the washing steps for industrial applications still remains problematic.

Extraction of the template leaves a three-dimensional material in which the cavity shapes and functional group locations are complementary to the guest molecule.

2.2
Influence of the Solvent

Although this parameter has been little studied, the choice of solvent is also particularly important in optimising the properties of MIPs:

- Particularly for non-covalent systems, in the steps of pre-organisation and polymerisation, the solvent plays an essential role in that it influences the type and the strength of the interactions occurring The interactions are strongly dependent on the polarity and the dissociating power of the solvent. In the example of a MIP composed of ethylene glycol dimethacrylate (EGDMA) and methacrylic acid (MAA) polymerised around atrazine, the best performance was obtained with fairly apolar solvents such as toluene and dichloromethane [128]. MIPs based on covalent interactions, on the other hand, are only slightly affected by the type of solvent used [129].
- In addition, the solvent generates pores and, in conjunction with other parameters such as temperature, influences the morphology of the material (size, shape and size distribution of the cavities). For chromatographic applications, macroporous networks are preferable. As the pores are more accessible, recognition is enhanced and retention times reduced. For instance, the use of acetonitrile as solvent in acrylate networks leads to a more macroporous structure than chloroform [130].
- When testing the properties of the MIP, the solvent modifies the morphology of the network through swelling processes. For instance, in acrylate systems, more swelling occurs in chlorinated solvents such as chloroform and dichloromethane than in tetrahydrofuran or acetonitrile. The swelling can decrease the recognition capacity. It is therefore often judicious to use an analytical solvent with a structure as close as possible to that of the synthesis solvent so as to create a micro-environment identical to that predominating when the network was originally prepared [131].

Most of the results obtained to date concern MIPs synthesised in organic media. They should now be transposed to aqueous systems which have a much broader field of application (protein recognition) and perspectives of development [132, 133]. This poses a certain number of problems since the interactions occurring in the two types of system are fundamentally different (Van der Waals interactions for organic systems and hydrophobic interactions and complexation in aqueous media [134]).

3
Recognition Properties

3.1
Guest Capacity

As mentioned above, a certain number of sites in the MIP remain inaccessible. This can be due to the presence of residual template molecules, excessive cross-linking limiting diffusion, or deformation of the network. In covalent systems, 80 to 90% of the sites vacated can be occupied again; in non-covalent systems, the proportion of reinsertion falls to 10 to 15% [135]. Non-covalent systems therefore provide low-capacity materials. This difference of behaviour is apparent from Fig. 13 [136b,c].

After extraction of the template, any deformation of the recognition sites, due to swelling of the MIP by solvent or relaxation of the polymer chains can hinder reinsertion. If the guest molecule is bound by covalent interactions (Fig. 13(1)), the sites it can interact with are located exclusively inside the 'imprinted' cavities which swell preferentially facilitating insertion. For non-covalent systems, on the other hand, (Fig. 13(2)), owing to the introduction of an excess of functionalised monomer, three-quarters of the sites are located outside cavities. Swelling occurs throughout the whole polymer and leads to irreversible deformation of the recognition sites. This process stresses the importance of using functionalised monomers that can interact in stoichiometric proportions with the template molecule. The capacity of MIPs for the guest is generally low (typical values are 0.1 mg per gram of MIP). The development of new polymeric systems involving for instance the use of new crosslinkers (trifunctional or more) has led to significant increases in capacity: some systems are now able to resolve up to 1 mg of a racemic mixture of peptides using one gram of MIP [101]. These values are of the same order of magnitude as those obtained with "classic" chiral stationary phases.

3.2
Specificity of Recognition

On reinsertion of the guest molecule or of molecules with a similar structure, how can the specificity of the imprinted sites for the guest molecule be deter-

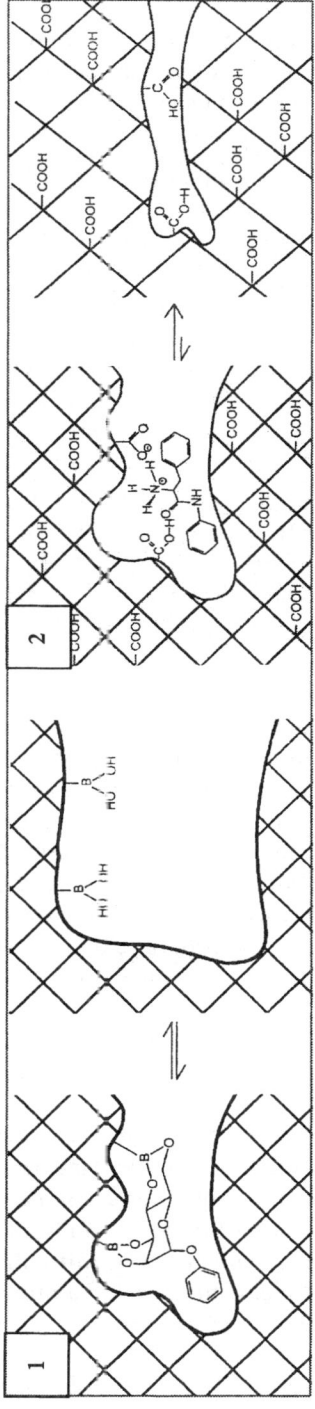

Fig. 13 Schematic representation of template molecule extraction and the resulting swelling of the network; (**1**) covalent system, 90–95% of the cavities are suitable for reinsertion; (**2**) non-covalent system, only 15% of the sites can be reused (adapted from [135])

mined on a macroscopic scale? In other words, how can the specificity of the recognition generate an easily identifiable signal? Owing in particular to the low capacity of the networks, this question is probably the most limiting for MIP applications. Two approaches have been considered to resolve this problem:

- Direct measurement of the interaction between the network and the guest molecule, when it is possible to quantify the variations of a property depending on these interactions. For instance, the introduction of fluorescent probes, either via the functionalised monomer or via the guest would allow accurate assessment of any change in the immediate environment of the probes [136]. The use of MIPs as recognition elements in sensors led to the development of this approach (see below).
- Indirect assessment of the interactions by observing the modifications brought about by the presence of the MIP on an element external to the network. For instance, using a MIP as a chromatographic stationary phase can provide a large amount of information about the network (capacity, specificity, association constants, etc.) by analysing the composition of the solution at the chromatograph outlet over time [114]. Similarly, for tests run in static conditions (batch rebinding) where the MIP is placed in a solution of known composition, the appearance or disappearance of molecules in the supernatant can be evaluated by isotopic labelling [31a] or spectroscopic analysis [30b].

Irrespective of how the problem is resolved, to evaluate accurately the contribution made by the molecular imprinting technique, the results obtained with the MIP should be viewed with respect to results obtained using a network with the same chemical composition synthesised in the same conditions but in the absence of the template molecule.

Remarks on the origin of the recognition process: Studies have shown in some instances that the position of the functions in the cavity was the major factor in the recognition process, the shape of the cavity as a whole simply improving selectivity [78a]. In a chiral context, experiments have shown that the chirality of the recognition sites caused, on a larger scale, chirality in the MIP polymer structure [137].

3.3
Lifetime of the MIPs

MIPs generally present good physical stability (mechanical resistance to high pressures and temperatures) [138] and good chemical stability (resistance to bases, acids, organic solvents and metal ions) [67, 138]. They can be reused over a hundred times and be stored at room temperature without losing their recognition specificity [50]. These elements give MIPs numerous advantages over their protein counterparts.

4
Main Uses of Imprinted Polymers

MIPs find uses in four main types of application: separation of molecules, preparation of antibody analogues, recognition elements for sensors, stereoselective reactions and catalysis.

4.1
Separation of Molecules

Imprinted materials can be used as stationary phase in affinity chromatography, especially for the separation of enantiomers. Compared to standard chiral stationary phases, these materials present the advantage of being synthesised "to order" for a given enantiomerically pure molecule. These chromatographic materials have allowed the separation of numerous compounds such as naproxene (anti-inflammatory) [139], timolol (beta-blocker) [50] and nicotine [140].

Separation performance can be expressed by the separation factor α, which is only dependent on the retention times of the two enantiomers, or by the resolution factor R_s which also takes into account the breadth of the chromatographic peaks. The higher these factors are, the better separation is. Remarkable selectivities have been obtained, for instance with an enantiomeric mixture of the dipeptide N-acetyl-Trp-Phe-O-Me (non-covalent system, R_s=17.8) [141] and

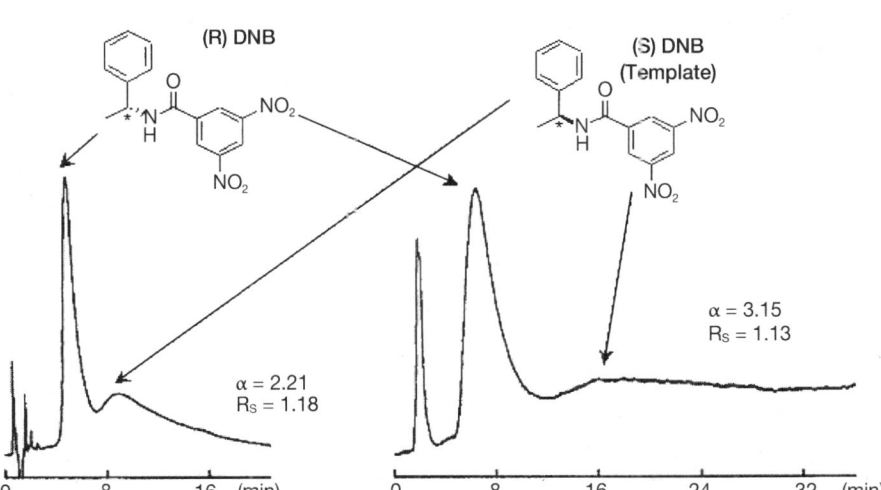

Fig. 14 Chromatographic separation of the two enantiomers of N-(3,5-dinitrobenzoyl)-α-methylbenzylamine (DNB) on networks imprinted around (S)-DNB. Effect of the mode of preparation on the chromatographic properties: (1) imprinted beads of uniform size (Fig. 12(2)); (2) ground graded MIP (Fig. 12(1)) (from [114])

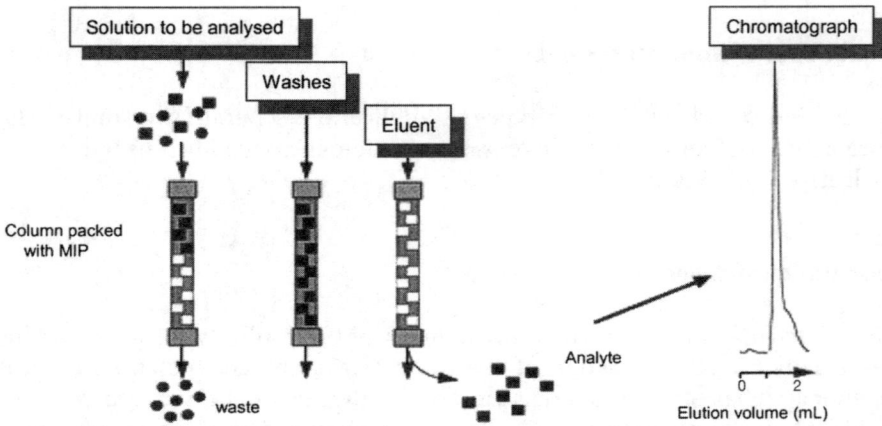

Fig. 15 Imprinted networks used for solid extraction. Explanatory schema

for a racemic mixture of phenyl-α-mannopyranoside (covalent system R_s=4.3) [142]. For example, Fig. 14 reports chromatograms obtained for the resolution of a racemic mixture of N-(3,5-dinitrobenzoyl)-α-methylbenzylamine obtained using the two materials of different morphologies presented in Fig. 12 [114].

The higher value of the separation factor α obtained with particles prepared by bulk polymerization can be explained by the weak template-MIP interactions involved (hydrogen bonding) when water is used as solvent, which decreases the strength of the interactions. In the same conditions of synthesis and use, the materials with a controlled structure ('beads') give a clearly higher resolution factor R_s.

These MIP were also successfully used for thin-layer chromatography [116] and capillary electrophoresis [105]. Extraction in the solid phase (Fig. 15) has also been extensively developed.

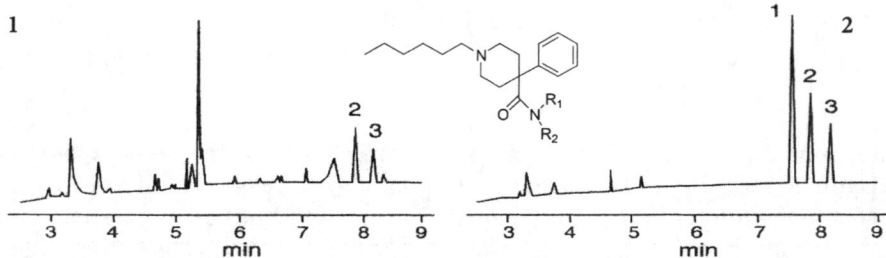

Fig. 16 Gas-phase chromatograms of extracts of human plasma; (1) after standard liquid-liquid extraction; (2) after solid-phase extraction with an imprinted polymer. Peak 1: guest molecule, R1=R2=CH_3, Peak 2: sameridine R1=CH_3, R2=C_2H_5, Peak 3: internal standard R1=R2=C_2H_5 (from [143])

In numerous fields, there is a great need for affinity matrices able to selectively extract and enrich analytes such as medical analysis, the food and drug industries and environmental applications. It has been demonstrated on many occasions that extraction on imprinted polymer can give better results than standard techniques such as liquid-liquid extraction, or extraction on C_{18} phase. Figure 16 gives the example of the extraction of sameridine, an analgesic, from human serum [143].

4.2
Preparation of Antibody Analogues

Recent studies have shown the efficiency of antibodies and artificial receptors prepared by the molecular imprinting technique and demonstrated the possibility of their use in therapeutic trials [102, 144]. Materials imprinted around diazepam (tranquilliser) and theophylline (broncho-dilator) templates have been found to have selectivities comparable to those of monoclonal antibodies and almost nil cross-reactions with related substances (Fig. 17) [102].

Receptors for morphine and for leu-enkephaline, efficient not only in organic media but also in aqueous solution, have also been synthesised [92]. Compared to their biological counterparts, imprinted networks present the advantage of being much more stable (chemically and physically), easy to synthesise and having a larger choice of template molecules.

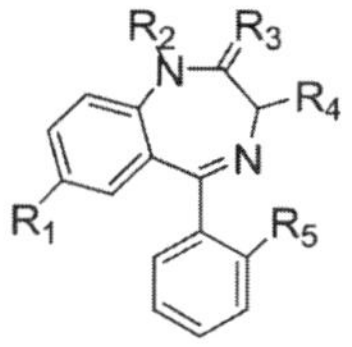

	R_1	R_2	R_3	R_4	R_5
Diazepam	Cl	Me	O	H	H
Desmethyldiazepam	Cl	H	O	H	H
Clonazepam	NO_2	H	O	H	Cl
Lorazepam	Cl	H	O	OH	Cl
Alprazolam	Cl	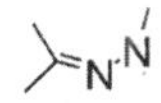 =N-N/		H	H

Ligand in competition	Cross-reactivity (%)	
	MIP	Antibody
Diazepam	100	100
Desmethyldiazepam	27	32
Clonazepam	9	5
Lorazepam	4	1
Alprazolam	2	1

Fig. 17 Cross-reactivities of various benzodiazepines for the adsorption of ^{3}H-diazepam on the imprinted network. These reactivities are expressed as the molar ratio diazepam/competing drug inhibiting 50% of the ^{3}H-diazepam binding (adapted from [102])

4.3
Sensors

One of the most promising applications of imprinted networks is their use as recognition elements in sensors [145]. A reminder of the principle of action of these sensors is given in Fig. 18.

The MIP is in contact with a transducer which converts the chemical or physical signal obtained on adsorption of the analyte into an easily quantifiable signal. Various principles of transduction have been used: we can mention ellipsometry [146], electric capacity [115], conductimetry [147], piezoelectric microgravimetry [148], evanescent wave IR [149], fluorescence [136], amperometry [150], voltammetry [151] and pH [152]. The imprinted materials allow the highly specific detection of even very dilute molecules in mixtures that can be complex and in extreme conditions (e.g. high temperature). They are therefore used or have been considered for use on ions, molecules with therapeutic properties, combat gases (sarin and soman) [153], etc. Here again, their great stability gives them a real advantage over the biomolecules commonly used.

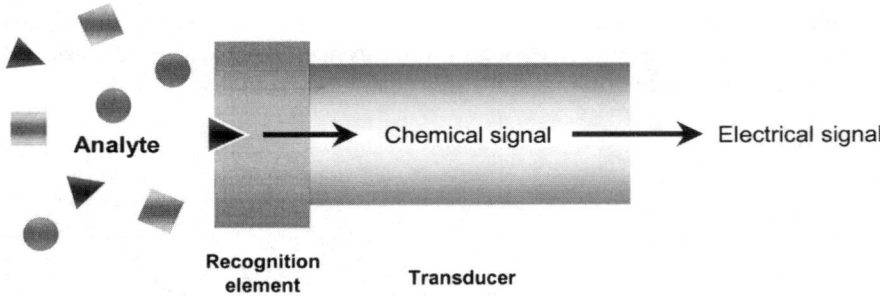

Fig. 18 Schematic representation of a sensor

4.4
Stereoselective Reactions and Catalysis

Reactions catalysed by enzymes are stereo- and regioselective. Abzymes and catalytic antibodies [154] constitute a first approach in mimicking such systems, but they are difficult to use.

Numerous research groups have attempted to make use of the selectivity of MIPs to prepare enzyme analogues with a catalytic activity [71, 155–170]. One strategy was to prepare a MIP around a template molecule with a structure similar to that of the substrate. The functional groups that play a catalytic role in the imprinted site are judiciously placed, by interaction with the functional groups on the guest molecule. An example of this strategy, represented in Fig. 19(2) [171, 172] concerns the catalysis of the dehydrofluorination of an

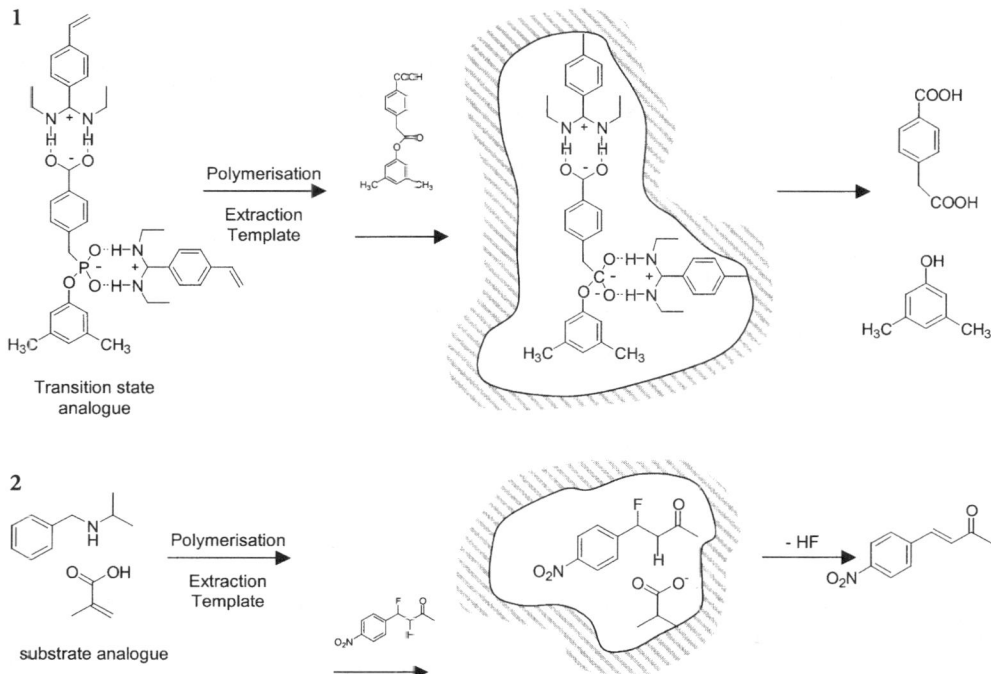

Fig. 19 MIPs used in catalysis: (1) ester hydrolysis (adapted from [156]); (2) dehydrofluorination (adapted from [171])

α-fluoroketone. The carboxylic catalytic group, placed in a favourable position by interaction with an amine function, catalyses the reaction very efficiently: the substrate, which is introduced afterwards, reacts 600 times faster than when the reaction occurs in solution. This value is close to that obtained with catalytic antibodies (1600 times faster) [171].

The most widely used strategy involves the synthesis of the network around a structural analogue of the transition state of the reaction. The imprinted sites then correspond to the conformation of the substrates in the transition state. For ester hydrolysis this state can, for instance, be simulated by a phosphonate derivative as template [156, 167]. An imprinted network with an esterase-type catalytic activity can then be obtained. For the MIP represented in Fig. 19(1), the reaction rate is increased 100-fold with respect to the reaction without catalyst and kinetics of the Michaelis-Menten type, as well as inhibition by an analogue of the transition state are observed [156].

Other reactions catalysed by MIPs have been described, including Diels-Alder type reactions [168, 170], aldol condensations [173] and isomerisation of benzisoxazoles [174], etc.

Apart from having a simple catalytic role, MIPs can be used in stereoselective [175, 182] and/or regioselective reactions [183]. In the example presented

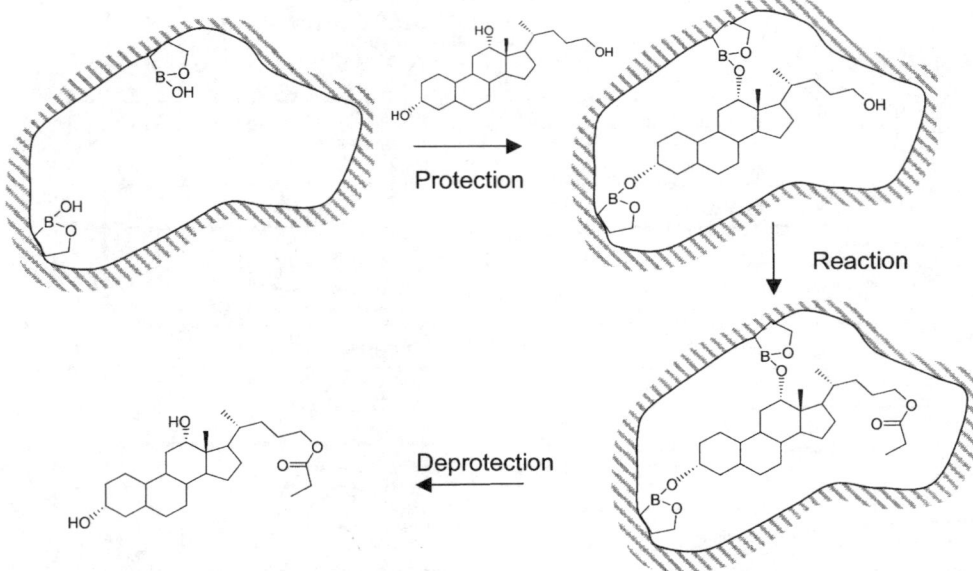

Fig. 20 Example of a regioselective reaction: the MIP enables acetylation on a specific hydroxyl group of the steroid while the other free hydroxyls are masked (adapted from [184])

in Fig. 20, a steroid with three free hydroxyl groups can be selectively acetylated on one of the hydroxyls, the others being masked by interactions with the imprinted network [184].

Finally, note that it has been proposed to modify enzymes using the principles of the molecular imprinting technique to modulate their action [185].

The catalytic activity of molecularly imprinted networks is still well below that of enzyme systems. However, even though imprinted materials cannot as yet compete with their biological counterparts, their high chemical stability and the possibility to use them in organic phases give them a promising future.

5
Summary and Outlook

The technique of molecular imprinting covers a wide range of applications. Compared to alternative techniques (involving biomolecules, abzymes, etc.), it presents a number of advantages: cost effectiveness, mechanical, thermal and chemical stability, long lifetime. As illustrated by the numerous patent applications, the first industrial uses for these materials have already been considered [186]. However, a certain number of factors limit the development of these materials.

In systems based on non-covalent interactions, a large excess of funtionalised monomer is introduced to favour the formation of the complex. The result is a random distribution of moieties liable to interact in addition to the active sites. This leads to a very heterogeneous population of recognition sites, lowering the performance of the imprinted networks (low capacity, reduced specificity, etc.). To overcome this, new monomers with a higher affinity for the guest molecule are being developed [59, 86].

In addition, the transposition of MIPs to aqueous media poses a certain number of problems related to the different types of interactions involved in this type of medium [134]. Similarly, for use in catalysis, much progress is yet to be made before industrial application can be considered [155].

Most of the drawbacks in MIPs have been linked to the fact that a large amount of crosslinker is needed (usually around 80–90%) to restrict distortion of the polymer backbone [17, 157]. The resulting stiffness of the network hinders the extraction and reinsertion of the template in the imprinted cavities and drastically decreases the capacity of the material [17]. Various 'surface imprinted' materials [117–122] have been reported to solve some of these problems but their capacities are very low.

Some promising materials have been proposed that have a certain rigidity maintaining the integrity of the recognition sites while remaining sufficiently flexible to enhance transfer of molecules and optimise the host-guest interactions.

These apparently antagonistic properties can however co-exist as shown by the way in which cell membranes operate (Fig. 1). The membranes occur in the form of a closed surface composed of a liquid film bathing in another immiscible liquid. They are essentially composed of a phospholipid bilayer integrating cholesterol, proteins and polysaccharides.

The hydrophobic alkyl chains are directed towards the interior of the film, whereas the hydrophilic poles cover the two faces of the film. Phospholipids tend to orient their chains perpendicularly to the film forming an ordered 2-D liquid. The proteins that are inserted into the bilayer themselves have regions that are more hydrophobic or more hydrophilic depending on the amino acids of the polypeptide chain. The hydrophobic parts are generally buried inside the bilayer while the hydrophilic parts protrude outwards. This membrane structure, although highly dynamic, remains coherent: the various molecules that make it up are in motion but the interactions that exist between them are sufficient to maintain the integrity of the structure. The supramolecular organisation is sufficient to maintain the recognition sites (especially on the proteins) in optimal positions to ensure that recognition is specific.

The use of systems based on supramolecular organisation therefore appears to be an interesting alternative in the technique of molecular imprinting. The integrity of the structure would not only be ensured, like in standard networks, by covalent chemical bonding but by the so-called "weak interactions" between the components of the network. It is in this light that polymer gels, 2-D films, and materials with a liquid crystal organisation have started to be developed.

5.1
Molecularly Imprinted Hydrogels

Hydrogels are cross-linked polymer networks that have the ability to absorb significant amounts of water. Their swelling behaviour can be modulated by external stimuli such as pH, temperature, ionic strength, electric field or concentration gradients. Tanaka and colleagues first obtained lightly crosslinked imprinted gels (less than 3 mol% crosslinker per mole monomer) that memorised their molecular conformation upon swelling and shrinking [188]. So, in a recent paper [189], Byrne and colleagues discussed the possibility of using a molecular imprinting technique in hydrogels to control the swelling behaviour of the material and so modulate their analyte binding ability (Fig. 21).

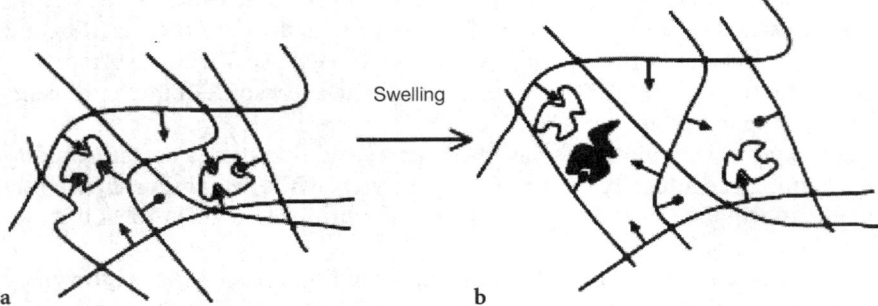

Fig. 21 Less crosslinked imprinted hydrogels: **a** rebinding of template; **b** swelling of the network leads to sites with varying affinity (adapted from [189])

5.2
Two-Dimensional Molecular Imprinting

Use has been made of lyotropic phases (such as micelles or vesicles) or Langmuir-Blodgett structures based on local 2-D organisation of molecules to form functionalised surfaces with a sharply defined molecular geometry. Two types of strategy have been applied to reach this aim:

- The simultaneous deposition of surfactant and template molecules, bound covalently or non-covalently, leading to the formation of a monolayer. [190, 28c]. This involves the formation of a deposit on a metal electrode (Fig. 22). The extraction of the template leaves a monolayer bearing sites allowing specific access to the electrode. The technique is well suited to the design of sensors.
- The formation of 2-D structures including functionalised parts. Mobility within the 2-D surfaces generated can optimise the positioning of the functionalised molecules with respect to the template. These structures can then be "frozen" by polymerising the vinylic surfactants used to form the layers [191]. In this way, Arnold et al. [87] prepared polymeric material function-

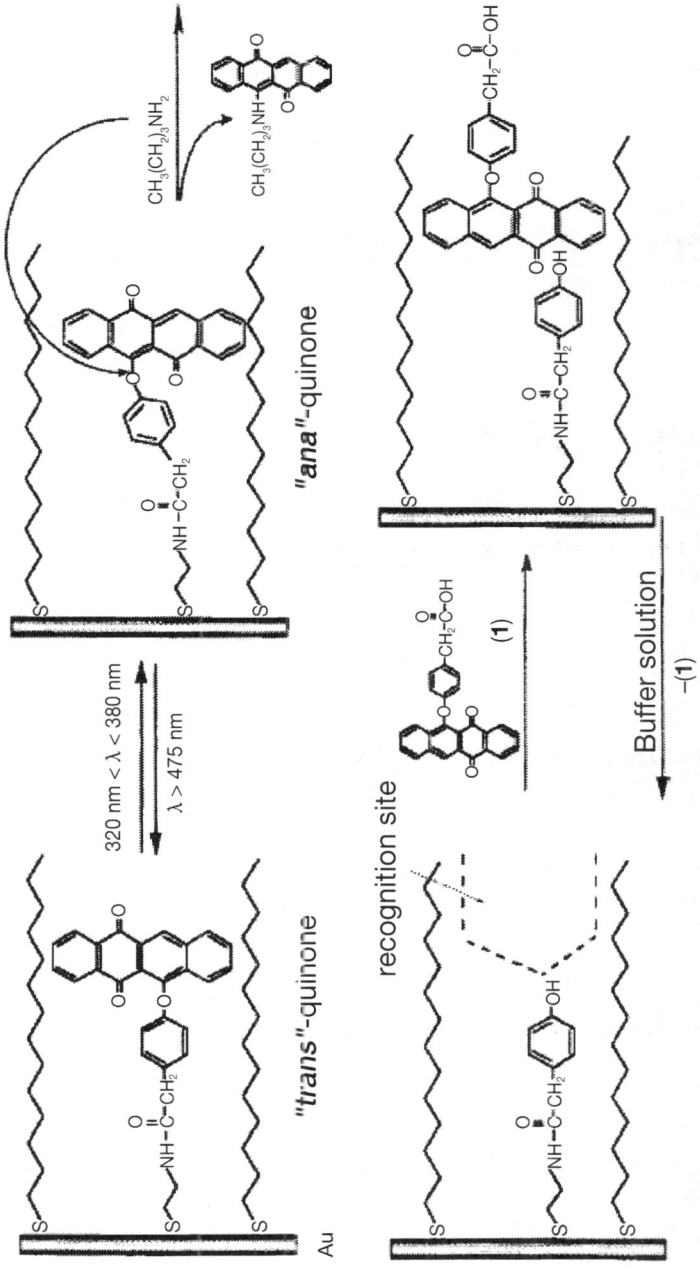

Fig. 22 Preparation of 'trans'-quinone/$C_{14}H_{29}SH$ mixed monolayer electrode and the photoinduced formation of molecular recognition sites in the array. Uptake and release of 'trans'-quinone to and from the sites (from [190])

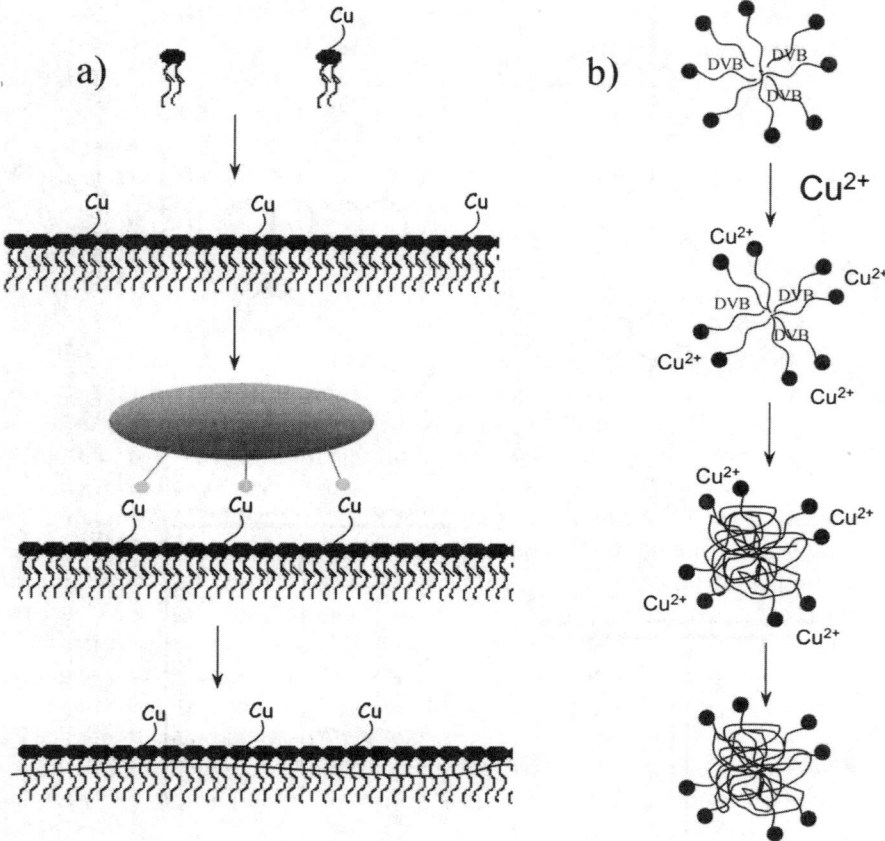

Fig. 23 a Schematic illustration of molecular imprinting of proteins on the surface of a metal-coordinating bilayer assembly (adapted from [87]). b Synthetic procedure for Cu(II) imprinted microspheres (adapted from [37, 124])

alised with metal ions to interact specifically with proteins (Fig. 23a). Likewise, superficially imprinted polymer beads were obtained from micellar structures [37, 124]. They presented good selectivity for the guest, which was the cupric ion (Fig. 23b).

In all cases, the imprinted surfaces generated in this way presented a higher affinity towards the template with respect to non-imprinted surfaces.

5.3
Liquid Crystalline Imprinted Materials

In order to soften the network while preserving the memory of the template, the use of liquid-crystal networks should be a useful tool. In such systems, the interactions that develop between mesogenic substituents confers a stiffness to the network formed from non-covalent reversible interactions [192]. Moreover, any manifestations of the interaction between the polymer backbone and the mesogenic side-chains could be transmitted to the macroscopic level provided that chemical cross-links are introduced between the polymer backbones to form liquid crystal elastomers. Such behaviour was predicted by de Gennes [193] and subsequently a number of phenomena have been observed experimentally, including electrically-induced shape changes [194], strain-induced switching [195], transfer of chirality [196] and memory effects [197, 198]. The effect of cross-linking biases the structure towards the backbone configuration present at the time of network formation; any distortion of this configuration is opposed by the elasticity of the network. Consequently, such materials display a memory of both backbone anisotropy and (by virtue of coupling) side-chain orientation. As a result, in such liquid crystalline materials used as MIPs, the template can be extracted without losing the imprinted information even with low crosslinking ratios [197–200]. Moreover, the template would be easily extracted by use of a solvent that swells the network or by heating the network above the liquid-crystal/isotropic transition.

Liquid crystalline imprinted materials have been synthesized (Fig. 24) from polysiloxanes with mesogenic side-chains. In some of them, acetophenone was chosen as template and was covalently linked to the mesomorphous network via a ketal link [199]. In the others, the templates (carbobenzoxy-L-phenylalanine, 1,8-diaminonaphthalene or theophylline) were in interaction with the mesomorphous polymer via hydrogen bonding [200]. All the imprinted networks were obtained as dense membranes.

It was observed that a high proportion of template (10%) can be introduced without losing the mesomorphic order [199]. Moreover after template extraction, the polymorphism was quite different from that of the non-imprinted material, the variations depending on the structure of the template used and on its concentration [200]. The last point is the manifestation of a significant memory effect of the template, imprinted inside the mesomorphic structure. It arises from the interactions between template and the other parts of the network which can induce conformational constraints inside the networks during cross-linking. It occurs even though the amount of crosslinker is low (5%). Moreover, this point constitutes a means for the direct study of the imprint left in the network by the template.

Concerning molecular recognition properties, for all the materials synthesised, imprinted networks exhibit a much higher affinity towards the template than non-imprinted networks. For instance, in the case of 1,8-diaminonaphthalene used as the template [199], the amount of this molecule rebound by the materials in the mesomorphic state is reported in Fig. 25. It is obvious that the non-imprinted network exhibits significantly lower template uptake compared to the imprinted network. These results indicate that, in addition to the hydrogen bonding or electrostatic interactions between the functional groups of the polymers and template, microcavities corresponding to the shape of the

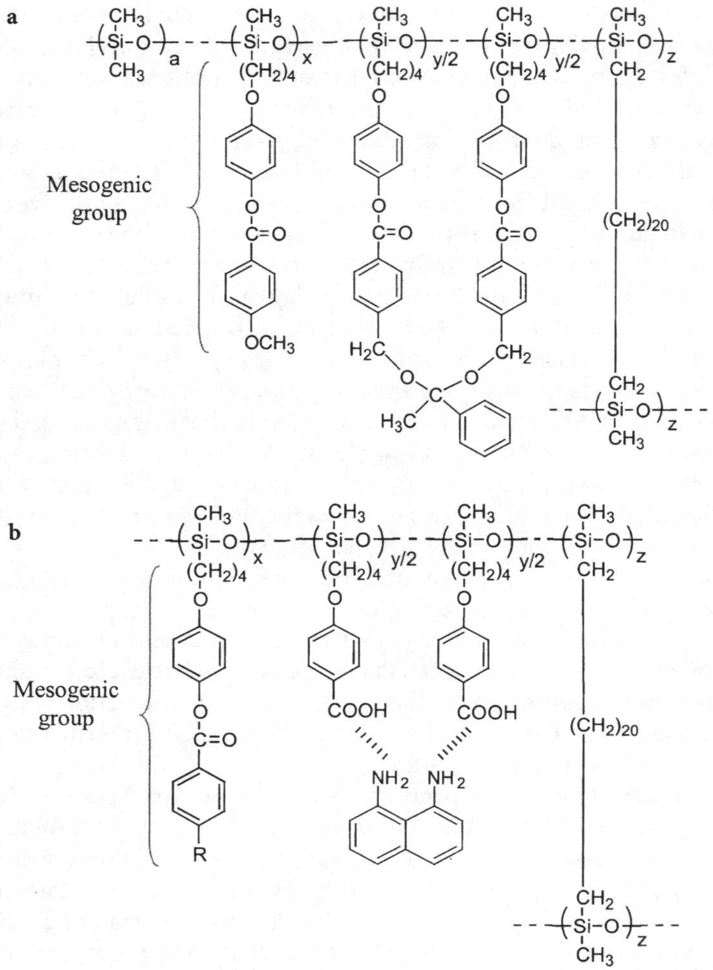

Fig. 24 Liquid crystalline materials imprinted via: **a** covalent linkages with acetophenone as template (adapted from [199]); **b** hydrogen bonding with 1,8-diaminonaphtalene as template (adapted from [200])

template are necessary for effective binding. Consequently, the polymer gains affinity for the template through the optimisation of molecular imprinting technology. Moreover, the network showed higher selectivity for the template than for a closely related compound (Fig. 25). Other experiments, obtained when theophylline or carbobenzoxy-L-phenylalanine were used as templates, have shown similar results [201].

On the other hand, the molecular trapping capacity of the networks (150 µmol/g of polymer) was shown to be much greater than that of most of the previously studied non-mesomorphous systems [197].

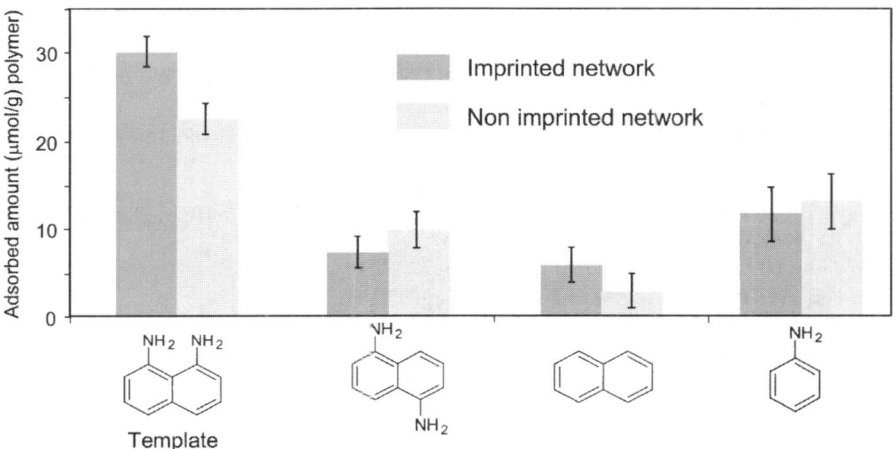

Fig. 25 Molecular recognition properties of a mesomorphous network imprinted around 1,8-diaminonaphtalene compared to a non-imprinted network (adapted from [201])

It is also possible to envisage exploiting other characteristics of liquid crystals to improve the recognition properties: the use of chiral mesophases to optimise the stereospecificity or modulate interactions between the recognition sites and the guest molecule by passing through the isotropic state, for example. These mesomorphic networks could also adapt to the geometry of the substrate to optimise the interactions occurring and thus enhance catalytic reactions.

6
Conclusion

The technique of molecular imprinting is currently in full development both from the point of view of design and study of the materials and of their actual domains of application. Widely used for the separation of molecules, it is now playing an increasingly active role in synthesis, in particular to generate stereospecific microreactors. The appearance of new materials involving supramolecular organisation appears promising in the improvement of recognition properties. Adapting the geometry of the cavities to the substrate and modulating it during the various phases of the process (formation, extraction, insertion), would then bring us another step closer to mimicking biological processes.

Acknowledgements The authors are greatly indebted to A. Lattes for fruitful discussions and P. Winterton for preparation of the English version of the manuscript.

References

1. a) Fischer E (1894) Ber Dtsch Chem Ges 27:2985; b) Lichtenthaler FW (1994) Angew Chem Int Ed Engl 33:2364
2. a) Koshland DE (1965) Angew Chem Int Ed Engl 33:2375; b) Yankeelov JA, Koshland DE J Biol Chem 240:1593
3. Lichtenthaler FW (1994) Angew Chem Int Ed Engl 33:2364
4. a) Vögtle F (ed) (1991) Supramolecular chemistry. Wiley (London); b) Hoss R, Vögtle F (1994) Angew Chem Int Ed Engl 33:375
5. Cram DJ (1988) Angew Chem Int Ed Engl 27:1009
6. Pedersen P (1988) Angew Chem Int Ed Engl 27:1021
7. Lehn JM (1988) Angew Chem Int Ed Engl 27:89
8. Schneider HJ (1991) Angew Chem Int Ed Engl 30:1417
9. Wenz G (1994) Angew Chem Int Ed Engl 33:803
10. Lerner RA, Benkovic SJ, Schulz RG (1991) Science 252:659
11. a) Pauling L (1940) J Am Chem Soc 60:2643; b) Pauling L (1949) Chem Eng News 27:913
12. a) Dickey FH (1949) Proc Natl Acad Sci USA 35:227; b) Dickey FH (1955) J Phys Chem 59:695
13. a) Curti R, Colombo U (1952) J Am Chem Soc 74:3961; b) Beckett AH, Anderson P (1957) Nature (London) 179:1074
14. Bernhard SB (1952) J Am Chem Soc 74:4946
15. a) Wulff G, Sarhan A (1972) Angew Chem Int Engl 11:341; b) Wulff G, Sarhan A, Zabrocki K (1973) Tetrahedron Lett 4329; c) Wulff G, Sarhan A (1978) US Patent 4727730
16. Andersson LI, Sellergren B, Mosbach K (1984) Tetrahedron Lett 25:5211
17. Wulff G (1995) Angew Chem Int Ed Engl 34:1812
18. Shea KJ (1994) Trends Polym Sci 2:166
19. Bartsch RA, Maeda M (eds) (1997) ACS Symposium Series, p 703
20. Mosbach K (1994) Trends Biochem 19:9
21. Anderson HS, Nicholls IA (1997) Recent Res Dev Pure Appl Chem 1:133
22. Takeuchi T, Matsui J (1996) Acta Polymer 47:471
23. a) Haupt K, Fradet A (2001) Actualité Chimique 4:23; b) Haupt K (2002) Nature Biotechnol 50:5; c) Haupt K (2003) Chem Commun 2:171; d) Haupt K, Mosbach, K (2000) Chem Rev 100:2495
24. Steinke J, Sherrington DC, Dunkin IR (1995) Adv Polym Sci 123:81
25. Brüggemann O (2002) In: Freitag R (ed) Synthetic polymers for biotechnology and medecine. Landes Bioscience (ISBN: 1-58706-027-2)
26. a) Lepistö M, Sellergren B (1989) J Org Chem 54:6010; b) Hamase K, Iwashita K, Zaitsu K (1999) Anal Sci 15:411; c) Whitcombe MJ, Vulfson EN (2001) Adv Mater 31:467
27. a) Fischer E (1894) Ber Dtsch Chem Ges 27:2985; b) Lichtenthaler FW (1994) Angew Chem Int Ed Engl 33:2364; c) Mayes AG, Andersson LI, Mosbach K (1994) Anal Biochem 222:483
28. a) Whitcombe MJ, Rodriguez ME, Villar P, Vulfson EN (1995) J Am Chem Soc 117:7105; b) Asanuma H, Kakazu M, Shibata M, Hishiya T, Komiyana M (1997) Chem Commun 1971; c) Piletsky SA, Piletskaya EV, Sergeyeva TA, Panasyuk TL, El'skaya AV (1999) Sens Actuators B Chem 60:216; d) Hwang CC, Lee WC (2002) J Chromatogr A962:69
29. a) Shea KJ, Spivak D, Sellergren B (1993) J Am Chem Soc 115:3368; b) Krotz JM, Shea KJ (1996) J Am Chem Soc 118:8154; c) Spivak D, Gilmore MA, Shea KJ (1997) J Am Chem Soc 119:4388; d) Matsui J, Tachibana Y, Takeuchi T (1998) Anal Commun 35:225
30. a) Matsui J, Miyoshi Y, Doblhoff-Dier O, Takeuchi T (1995) Anal Chem 67:4404; b) Muldoon MT, Stanker LH (1997) Anal Chem 69:803; c) Matsui J, Kubo H, Takeuchi T (1998) Anal Sci 14:699

31. a) Vlatakis G, Andersson LI, Müller R, Mosbach K (1993) Nature 361:645; b) Kabayashi T, Wang HY, Fujii N (1995) Chem Lett 927; c) Wang HY, Kabayashi T, Fujii N (1996) Langmuir 12:4850; d) Wang HY, Kabayashi T, Fukaya T, Fujii N (1997) Langmuir 13:5396; e) Kobayashi T, Wang HY, Fujii N (1998) Anal Chim Acta 365:81; f) Ye L, Cormack PAG, Mosbach K (1999) Anal Commun 36:35
32. Andersson HS, Koch-Schmidt AC, Ohlson S, Mosbach K (1996) J Mol Recognit 9:675
33. Vlatakis G, Andersson LI, Müller R, Mosbach K (1993) Nature 361:645
34. Andersson LI, Müller R, Vlatakis G, Mosbach K (1995) Proc Natl Acad Sci 92:4788
35. a) Sellergren B (1994) J Chromatogr 673:133; b) Sellergren B (1994) Anal Chem 66:1578
36. a) Haupt K, Noworyta K, Kutner W (1999) Anal Commun 36:391; b) Bowman MAE, Allender CJ, Brain KR, Heard CM (1998) Drug development assay approaches, including molecular imprinting and biomarkers. Reid R, Hill H, Wilson I (eds), Royal Society of Chemistry, 25:37; c) Andersson LI (1996) Anal Chem 68:111
37. Kido H, Miyajima T, Tsukagoshi K, Maeda M, Takagi M (1992) Anal Sci 8:749
38. Bae SY, Zeng X, Murray GM (1998) J Anal At Spectrom 13:1177
39. Rosatzen T, Andersson LI, Simon W, Mosbach K (1991) J Chem Soc Perkin Trans 2 1261
40. a) Matsui J, Kato T, Takeuchi T, Suzuki M, Yokoyama K, Tamiya E, Karube I (1993) Anal Chem 65:2223; b) Hosoya K, Yoshikako K, Tanaka N, Kimata K, Araki T, Haginaka J (1994) Chem Lett 1437
41. Dickert FL, Besenböck H, Tortschanoff M (1998) Adv Mater 10:149
42. a) Shea KJ, Dougherty TK (1986) J Am Chem Soc 108:1091; b) Shea KJ, Sasaki DY (1989) J Am Chem Soc 111:3442; c) Shea KJ, Sasaki DY (1991) J Am Chem Soc 113:4109; d) Shea KJ, Sasaki DY, Stoddard GJ (1989) Macromolecules 22:1722; e) Shea KJ, Dougherty TK (1986) J Am Chem Soc 108:1091
43. Ratner BD, Shi H (1999) Curr Opin Solid State Mater Sci 4:395
44. Rich JO, Mozhaev VV, Dordick JS, Clark DS, Khmelnitsky Y L (2002) J Am Chem Soc 124:5254
45. Hosoya K, Shirasu Y, Kimata K, Tanaka N (1998) Anal Chem 70:943
46. Arshady R, Mosbach K (1981) Makromol Chem 182:687
47. Andersson LI, Sellergren B, Mosbach K (1984) Tetrahedron Lett 25:5211
48. Sellergren B, Lepistö M, Mosbach K (1988) J Am Chem Soc 110:5853
49. Matsui J, Miyoshi Y, Takeuchi T (1995) Chem Lett 1007
50. Fischer L, Müller R, Ekberg B, Mosbach K (1991) J Am Chem Soc 113:9358
51. Kugimiya A, Takeuchi T (1999) Anal Chim Acta 395:251
52. Norrlöw O, Glad M, Mosbach K (1984) J Chromatogr 299:29
53. Ramström O, Ye L, Yu C, Gustavsson PE (1997) Molecular and ionic recognition with imprinted polymers. ACS Symposium 703, p 82
54. Dunkin IR, Lenfeld J, Sherrington DC (1993) Polymer 34:77
55. Beach JV, Shea KJ (1994) J Am Chem Soc 116:379
56. McNiven S, Kato M, Levi R, Yano K, Karube I (1998) Anal Chim Acta 365:69
57. Shea KJ, Dougherty TK (1986) J Am Chem Soc 108:1091
58. Dhal PK, Arnold FH (1991) J Am Chem Soc 113:7417
59. Wulff G, Gross T, Schönfeld R (1997) Angew Chem Int Ed Engl 36:1962
60. Chen H, Olmstead MM, Albright RL, Devenyi J, Fish RH (1997) Angew Chem Int Ed Engl 36:642
61. a) Wulff G (1986) Polym Reagents Catal 308:186; b) Wulff G (1993) Trends Biotechnol 11:85
62. Matsui J, Nicholls IA, Karube I, Mosbach K (1996) Anal Chim Acta 335:71
63. Sarhan A, Wulff G (1982) Makromol Chem 183:85
64. a) Glad M, Norrlöw O, Sellergren B, Siegbahn B, Mosbach K (1985) J Chromatogr 347:11; b) Venton DL, Gudipati E (1995) Biochim Biophys Acta 1250:126

65. Dickert FL, Forth P, Lieberzeit P, Tortschanoff M (1998) Fresenius J Anal Chem 360:759
66. Malitesta C, Losito I, Zambonin PG (1999) Anal Chem 71:285
67. Haupt K (1997) Molecular and ionic recognition with imprinted polymers. ACS Symposium 703, p 135
68. Kriz D, Ramström O, Svensson A, Mosbach K (1995) Anal Chem 67:2142
69. Kempe M, Fischer L, Mosbach K (1993) J Mol Recognit 6:25
70. Leonhardt A, Mosbach K (1987) Reactive Polym 6:285
71. Katz A, Davis ME (2000) Nature 403:286
72. a) Wulff G, Lauer M, Böhnke H (1984) Angew Chem Int Ed Engl 23:741; b) Wulff G, Stellbrink H (1990) Recl Trav Chim Pays Bas 109:216
73. a) Wulff G, Vesper W (1978) J Chromatogr 167:171; b) Wulff G, Vietmeier J, Poll HG (1987) Makromol Chem 188:731; c) Wulff G, Schauhoff S (1991) J Org Chem 56:395
74. a) Sarhan A, Wulff G (1982) Makromol Chem 183:1603; b) Sarhan A, Wulff G (1982) Makromol Chem 183:2469
75. a) Wulff G (1982) Pure Appl Chem 54:2093; b) Wulff G, Best W, Akelah A (1984) React Polym Ion Exch Sorbents 2:167; c) Wulff G, Vietmeier J (1989) Makromol Chem 190:1727; d) Wulff G, Heide B, Helfmeier G (1986) J Am Chem Soc 108:1089; e) Wulff G, Vietmeier J (1989) Makromol Chem 190:1717
76. a) Sellergren B, Andersson LI (1990) J Org Chem 55:3381; b) Damen J, Neckers DC (1980) J Am Chem Soc 102:3265; c) Whitcombe MJ, Rodriguez ME, Villar P, Vulfson EN (1995) J Am Chem Soc 117:7105
77. Wulff G, Sarhan A, Zabrocki K (1973) Tetrahedron Lett 4329
78. a) Shea KJ, Sasaki DY (1989) J Am Chem Soc 111:3442; b) Shea KJ, Sasaki DY (1991) 113:4109
79. Arshady R, Mosbach K (1981) Makromol Chem 182:687
80. Ekberg B, Mosbach K (1989) Trends Biotechnol 7:92
81. Mosbach K (1994) Trends Biotechnol Sci 19:9
82. Zimmerman SC, Wendland MS, Rakow NA, Zharov I, Suslick KS (2002) Nature 418:399
83. Sellergren B, Ekberg B, Mosbach K (1985) J Chromatogr 347:1
84. Dunkin IR, Lenfeld J, Sherrington DC (1993) Polymer 34:77
85. Nicholls IA, Ramström O, Mosbach K (1995) J Chromatogr 691:349
86. Tanabe K, Takeuchi T, Matsui J, Ikebukuro K, Yano K, Karube I (1995) J Chem Soc Chem Comm 2303
87. Mallik S, Plunkett S, Dhal P, Johnson R, Pack D, Shnek D, Arnold F (1994) New J Chem 18:299
88. Sarhan A, El-Zahab MA (1987) Makromol Chem Rapid Comm 8:555
89. a) Whitcombe MJ, Rodriguez ME, Vulfson EN (1994) In: Separations for Biotechnology 3. Royal Society of Chemistry Special Publication, vol 158, p 565; b) Whitcombe MJ, Rodriguez ME, Villar ME, Vulfson EN (1995) J Am Chem Soc 117:7105
90. Piletsky SA, Piletska EV, Karim K, Freebairn KW, Legge CH, Turner APF (2002) Macromolecules 35:7499
91. a) O'Shannessy DJ, Ekberg B, Mosbach K (1989) Anal Biochem 177:144; b) Andersson LI, Mosbach K (1990) J Chromatogr 516:313
92. Kempe M, Mosbach K (1995) Tetrahedron Lett 36:3563
93. Andersson LI, Sellergren B, Mosbach K (1984) Tetrahedron Lett 25:5211
94. Wulff G, Kemmemer R, Vietmeier J, Poll H (1982) Nouv J Chim 6:681
95. Okkubo K, Funakoshi Y, Urata Y, Hirota S, Usui S, Sagawa T (1995) J Chem Soc Chem Comm 2143
96. Glad M, Norrlöw O, Sellergren B, Siegbahn N, Mosbach K (1985) J Chromatogr 347:11
97. Kempe M, Mosbach K (1995) Tetrahedron Lett 36:3563
98. Asanuma H, Kajiya K, Hishiya T, Komiyama M (1999) Chem Lett 665

99. Dickert FL, Tortschanoff M, Bulst WE, Fischerauer G (1999) Anal Chem 71:4559
100. Kempe M (1996) Anal Chem 68:1948
101. Kempe M, Mosbach K (1995) Tetrahedron Lett 36:3563
102. Vlatakis G, Andersson LI, Müller R, Mosbach K (1993) Nature 361:645
103. O'Shannessy DJ, Andersson LI, Mosbach K (1989) J Mol Recogn 2:1
104. Matsui J, Kato T, Takeuchi T, Suzuki M, Yokoyama K, Tamiya E, Karube I (1993) Anal Chem 65:2223
105. Nilsson K, Lindell J, Sellergren B, Norrlöw O, Mosbach K (1994) J Chromatogr 680:57
106. Yilmaz E, Ramstrom O, Moller P, Sanchez D, Mosbach K (2002) J Mater Chem 12:1577
107. Glad M, Reinholdsson P, Mosbach K (1995) React Polym 25:47
108. Dhal PK, Vidyasankar S, Arnold FH (1995) Chem Mater 7:154
109. Sveç F, Fréchet JM (1992) Anal Chem 64:820
110. Byström S, Börje A, Åkermak B (1993) J Am Chem Soc 115:2081
111. Hosoya K, Yoshizako K, Tanaka N, Kitama K, Araki T, Haginaka (1994) J Chem Lett 1437
112. Sellergren B (1994) J Chromatogr A 673:133
113. Mayes A, Mosbach K (1996) Anal Chem 68:3769
114. Hosoya K, Tanaka N (1997) Molecular and ionic recognition with imprinted polymers. ACS Symposium Series 703, p 143
115. Hedborg E, Winquist F, Andersson LI, Mosbach K (1993) Sens Actuators 37:796
116. Kriz D, Berggren Kriz C, Andersson LI, Mosbach K (1994) Anal Chem 66:2636
117. Norrlöw O, Månsson MO, Mosbach K (1987) J Chromatogr 396:374
118. Dhal PK, Arnold FH (1992) Macromolecules 25:7051
119. Shnek DR, Pack DW, Sasaki DY, Arnold FH (1994) Langmuir 10:2382
120. Morihara K, Takiguchi M, Shimada T (1994) Bull Chem Soc Jpn 67:1078
121. Tahmassebi DC, Sasaki T (1994) J Org Chem 59:579
122. Araki K, Goto M, Furusaki S (2002) Anal Chim Acta 469:173
123. Ratner BD, Shi H (1999) Curr Opin Solid State Mater Sci 4:395
124. Tsukagoshi K, Yu KY, Maeda M, Takagi M (1993) Bull Chem Soc Jpn 66:114
125. Yilmaz E, Haupt K, Mosbach K (2000) Angew Chem Int Ed Engl 39:2115
126. Müller R, Andersson LI, Mosbach K (1993) Makromol Chem Rapid Commun 14:637
127. Andersson LI, O'Shannessy DJ, Mosbach K (1990) J Chromatogr 516:167
128. Matsui J, Kubo H, Takeuchi T (1998) Anal Sci 14:699
129. Wulff G, Poll HG, Minarik M (1986) J Liq Chrom 385
130. a) Rosenberg JE, Flodin P (1986) Macromolecules 19:1543; b) Rosenberg JE, Flodin P (1987) Macromolecules 20:1518
131. a) O'Shannessy DJ, Ekberg B, Mosbach K (1989) Anal Biochem 177:144; b) Andersson LI, Mosbach K (1990) J Chromatogr 516:313
132. Carter SR, Rimmer S (2002) Adv Mater 14(9):667
133. Hart BR, Shea KJ (2002) Macromolecules 35:6192
134. Andersson LI, Müller R, Vlatakis G, Mosbach K (1995) Proc Natl Acad Sci USA 92:4788
135. a) Sellergren B (1989) Makromol Chem 190:2703; b) Sellergren B, Shea KJ (1993) J Chromatogr 635:31; c) Wulff G, Gross T, Schönfeld R, Schrader T, Kirsten C (1997) Molecular and ionic recognition with imprinted polymers. ACS Symposium Series 703, p 82
136. a) Kriz D, Ramström O, Svensson A, Mosbach K (1995) Anal Chem 67:2142; b) Al-Kindy S, Badia R, Diaz-Garcia ME (2002) Anal Lett 35:1763
137. a) Wulff G, Zabrocki K, Hohn J (1978) Angew Chem Int Ed Engl 17:535; b) Wulff G (1993) Makromol Chem Makromol Symp 70/71:285; c) Wulff G, Schmidt H, Witt H, Zentel R (1994) Angew Chem Int Ed Engl 33:188
138. Andersson LI, Ekberg B, Mosbach K (1993) Bioseparation and catalysis in molecularly imprinted polymers. In: Ngo TT (ed) Molecular interactions in bioseparations. New York, Plenum Press, p 383

139. Kempe M, Mosbach K (1994) J Chromatogr 664:276
140. Zander A, Findlay P, Renner T, Sellergren B (1998) Anal Chem 70:3304
141. Ramström O, Nicholls IA, Mosbach K (1994) Tetrahedron Assymetry 5:649
142. Wulff G, Minarik M (1990) J Liq Chromatogr 13:2987
143. Andersson LI, Paprica A, Arvidsson T (1997) Chromatographia 46:57
144. Ramström O, Ye L, Mosbach K (1996) Chem Biol 3(6):471
145. Dickert FL, Sikorski R (1999) Mater Sci Eng C 10:39
146. Andersson LI, Mandenius CF, Mosbach K (1988) Tetrahedron Lett 29:5437
147. Sergeyeva TA, Piletsky SA, Brovko AA, Slinchenko EA, Sergeeva LM, Panasyuk TL, Elskaya AV (1999) Analyst 124:331
148. Malitesta C, Losito I, Zambonin PG (1999) Anal Chem 71:1366
149. Jakusch M, Janotta M, Mizaikoff B, Mosbach K, Haupt K (1999) Anal Chem 71:4786
150. Kriz D, Mosbach K (1995) Anal Chim Acta 300:71
151. Kröger S, Turner APF, Mosbach K, Haupt K (1999) Anal Chem 71:3698
152. Chen GH, Guan ZB, Chen CT, Fu LT, Sundaresan V, Arnold FH (1997) Nature Biotechnol 15:354
153. Jenkins AL, Uy OM, Murray GM (1999) Anal Chem 71:373
154. Hasserodt J (1999) Synlett 12:2007
155. a) Whitcombe MJ, Alexander C, Vulfson EN (2000) Synlett 6:911; b) Ramström O, Mosbach K (1999) Curr Opin Chem Biol 3:759; c) Liu Q, Zhou Y, Liu Y Chin J (1999) Anal Chem 1341; d) Davis ME, Katz A, Ahmad WR (1996) Chem Mater 8:1820
156. a) Wulff G, Gross T, Schönfeld R (1997) Angew Chem Int Ed Engl 36:1962; b) Wulff G (2002) Chem Rev 102(1):1
157. Robinson DK, Mosbach K (1989) J Chem Soc Chem Commun 14:969
158. a) Ohkubo K, Urata Y, Honda Y, Nakashima Y, Yoshinaga K (1994) Polymer 35:5372; b) Ohkubo K, Funakoshi Y, Urata Y, Hiraota S, Usui S, Sagawa T (1995) J Chem Soc Chem Commun 2143
159. Sellergren B, Karmalkar RN, Shea KJ (2000) J Org Chem 65:4009
160. Leonhardt A, Mosbach K (1987) React Polym Ion Exch Sorbents 6:285
161. a) Karmalkar RN, Kulkarni MG, Mashelkar RA (1996) Macromolecules 29:1366; b) Lele BS, Kulkarni MG, Mashelkar RA (1999) React Funct Polym 40:215; c) Lele BS, Kulkarni MG, Mashelkar RA (1999) React Funct Polym 39:37
162. Kawanami Y, Yunoki T, Nakamura A, Fujii K, Umano K, Yamuachi H, Masuda K (1999) J Mol Catal A 145:107
163. Toorisaka E, Yoshida M, Uezu K, Goto M, Furusaki S (1999) Chem Lett 387
164. Leonhardt A, Mosbach K (1987) React Polym 6:285
165. Ohkubo K, Urata Y, Hirota S, Honda Y, Sagawa T (1994) J Mol Catal 87:L21
166. Ohkubo K, Urata Y, Hirota S, Funakoshi Y, Sagawa T, Usui S, Yoshinaga K (1995) J Mol Catal A Chem 101:L111
167. Kawanami Y, Yunoki T, Nakamura A, Fujii K, Umano K, Yamuachi H, Masuda K (1999) J Mol Cat A 145:107
168. Liu XC, Mosbach K (1997) Macromol Rapid Commun 18:609
169. Santora BP, Larsen AO, Gagné MR (1998) Organometallics 17:3138
170. Damen J, Neckers DC (1980) J Am Chem Soc 102:3265
171. Müller R, Andersson LI, Mosbach K (1993) Makromol Chem Rapid Commun 14:637
172. Beach JV, Shea KJ (1994) J Am Chem Soc 116:379
173. Matsui J, Nicholls IA, Karube I, Mosbach K (1996) J Org Chem 61:5414
174. Liu XC, Mosbach K (1998) Macromol Rapid Commun 19:671
175. Wulff G, Vietneier J (1989) Makromol Chem 190:1727
176. Ohkubo K, Urata Y, Hirota S, Funakoshi Y, Sagawa T, Usui S, Yoshinaga K (1995) J Mol Catal 101:L111

177. Sellergren B, Shea KJ (1994) Tetrahedron Assymetry 5:1403
178. a) Gamez P, Dunjic B, Pinel C, Lemaire M (1995) Tetrahedron Lett 36:8779; b) Locatelli F, Gamez P, Lemaire M (1998) J Mol Cat A 135:89
179. Morihara K, Kurokawa M, Kamata Y, Shinada T (1992) J Chem Soc Chem Commun 358
180. Byström SE, Börje A, Akermark B (1993) J Am Chem Soc 115 2081
181. Polborn K, Severin K (1999) Chem Commun 2481
182. Hamase K, Iwashita E, Zaitsu K (1999) Anal Sci 15:411
183. Byström SE, Boerge A, Akernah B (1993) J Am Chem Soc115:2081
184. Alexander C, Smith CR, Whitcombe MJ, Vulfson EN (1999) J Am Chem Soc 121:6640
185. Braco L, Dabulis K, Klibanov AM (1990) Proc Natl Acad Sci USA 87:274
186. a) Wulff G, Sarhan A (1978) US Patent 4,111,863; b) Wulff G, Sarhan A (1978) US Patent 4,127,730; c) Mosbach K, Nilsson KGI (1983) US Patent 4,415,665; d) Larsson PO, Mosbach K, Borchert A (1985) US Patent 4,532,232; e) Nilsson K, Mosbach K (1990) US Patent 4,935,365; f) Nilsson K, Mosbach K (1991) US Patent 5,015,576 ;g) Mosbach K (1992) US Patent 5,110,833; h) Arnold FH, Dhal P, Shnek D, Plunkett S (1994) US Patent 5,310,648; i) Douglas A, Shea KJ (1994) US Patent 5,321,102; j) Afeyan NB, Varady L, Regnier F (1994) US Patent 5,372,719; k) Domb AJ (1997) US Patent 5,630,978; l) Mosbach K, Cormack PAG, Ramström O, Haupt K (1999) US Patent 5,994,110
187. Ramström O, Ansell RJ (1998) Chirality 10:195
188. a) Enoki T, Tanaka K, Watanabe T, Oya T, Sakiyama T, Takeota Y, Ito K, Wang G, Annaka M, Hara K, Du R, Chuang J, Wasserman K, Grosberg AY, Masamune S, Tanaka T (2000) Phys Rev Lett 85:5000; b) Alvarez-Lorenzo C, Guney O, Oya T, Sakai Y, Kobayashi M, Enoki T, Takeoka Y, Ishibashi T, Kuroda K, Tanaka K, Wang G, Grosberg AY, Masamune S, Tanaka T (2001) J Chem Phys 114:2812; c) Alvarez-Lorenzo C, Hiratani H, Tanaka K, Stancil K, Grosberg AY, Tanaka T (2001) Langmuir 17:3616
189. Byrne ME, Park K, Peppas NA (2002) Adv Drug Deliv Rev 54:149
190. Lahav M, Katz E, Doron A, Patolsky F, Willner I (1999) J Am Chem Soc 121:862
191. Miyahara T, Kurihara K (2000) Chem Lett 1356
192. Davis FJ (1993) J Mater Chem 3:551
193. de Gennes PG (1971) Mol Cryst Liq Cryst 12:193
194. Zentel R (1986) Liq Cryst 1:589
195. Davis FJ, Mitchell GR (1987) Polymer Commun 28:8
196. Wulff G, Schmidt H, Witt H, Zentel R (1994) Angew Chem Int Ed Engl 33:188
197. Roberts PMS, Mitchell GR, Davis FJ (1997) Mol Cryst Liq Cryst 299:223
198. Brand HR, Finkelmann H (1998) Physical properties of liquid crystalline elastomers. In: Demus, Goodby, Gray, Spiess (eds) Handbook of liquid crystals VIII, vol. 3, chap V, p S277. Wiley-VCH, UK
199. Marty JD, Tizra M, Mauzac M, Rico-Lattes I, Lattes A (1999) Macromolecules 32:8674
200. Marty JD, Mauzac M, Fournier C, Rico-Lattes I, Lattes A (2002) Liq Cryst 29:529
201. Marty JD, Labadie L, Mauzac M, Fournier C, Rico-Lattes I, Lattes A (2004) Mol Cryst Liq Cryst 411:561

Editor: Jean-François Joanny
Received: December 2003

Chemical Amplification Resists for Microlithography

Hiroshi Ito (✉)

IBM Almaden Research Center, 650 Harry Road, San Jose, California 95120-6099, USA
hiroshi@almaden.ibm.com

1	Introduction	41
2	Chemical Amplification Concept	47
3	Photochemical Acid Generators	48
4	Chemically Amplified Imaging Mechanisms	53
4.1	Deprotection	54
4.1.1	Carbonates	55
4.1.2	Esters and Ethers	62
4.1.3	Hydrolysis	64
4.1.4	Polyhydroxystyrenes	65
4.1.5	Copolymers	70
4.1.6	Blends and Dissolution Inhibitors	80
4.1.7	Deprotection Involving Ring-Opening (Mass Persistent Resists)	85
4.1.8	Dendritic Polymers and Small Amorphous Materials	86
4.1.9	Evolution of 248 nm Positive Resists	88
4.1.10	Polymethacrylates and Norbornene Polymers for 193 nm Lithography	97
4.1.11	Fluoropolymers for 157 nm Lithography	121
4.1.12	Resists for Next Generation Lithography	137
4.2	Depolymerization	139
4.2.1	Thermodynamically-Driven Depolymerization	140
4.2.2	Repeated Catalytic Main Chain Scission	145
4.3	Rearrangement	146
4.3.1	Polarity Reversal	146
4.3.2	Claisen Rearrangement	148
4.3.3	Pinacol Rearrangement	149
4.4	Intramolecular Dehydration	151
4.5	Condensation/Intermolecular Dehydration	152
4.6	Esterification	161
4.7	Polymerization/Crosslinking	164
5	Environmentally Friendly Processes	166
5.1	Water-Processable Resists (Casting and Development)	166
5.2	CO_2-Processable Resists (Casting, Development, Rinse, and Strip)	172
6	Bilayer Lithography and Top-Surface Imaging	175
6.1	Bilayer Lithography with Organosilicon Resists	177
6.1.1	Semi-Dry Bilayer Lithography (Wet Development/O_2 RIE Pattern Transfer)	178
6.1.2	All-Dry Bilayer Lithography	184
6.2	Silylation of Organic Resists	188

6.2.1	Reaction-Controlled Silylation	188
6.2.2	Diffusion-Controlled Silylation	192
6.2.3	Bilayer Silylation	194
6.2.4	Silylation after Wet Development	195
6.2.5	Flood Silylation/Imagewise Desilylation	198
6.2.6	Surface Modification	198
6.3	Process Issues in Bilayer and Dry Lithographic Techniques	201
7	**Resist Characterization**	**202**
7.1	Molecular Weight Determination	203
7.2	Thermal Analysis	203
7.3	Spectroscopic Analysis	204
7.3.1	UV Spectroscopy	204
7.3.2	NMR Spectroscopy	205
7.3.3	IR Spectroscopy	206
7.3.4	Mass Spectroscopy	207
7.4	Surface Analysis	207
7.5	Neutron Scattering	208
7.6	Neutron and X-Ray Reflectometry	208
7.7	^{14}C Labeling/Scintillation	208
7.8	Laser Interferometry	208
7.9	Quartz Crystal Microbalance	209
7.10	Measurements of Optical Properties of Films	211
8	**Resist Performance Parameters**	**212**
8.1	Resist Sensitivity and Contrast	212
8.2	Linear Resolution	213
8.3	Depth of Focus (DOF) or Focus Latitude	214
8.4	Exposure Latitude (Dose Latitude)	215
8.5	Resist Image Profiles	216
8.6	Isolated-Nested Bias	216
8.7	PEB and PED Stability	217
8.8	Line Edge Roughness	217
8.9	Dry Etch Resistance	217
9	**Resolution Limit – Acid Diffusion/Image Blur**	**217**
10	**Closing Remarks and Future Perspectives**	**221**
References		**223**

Abstract This chapter describes polymers employed in formulation of chemically amplified resists for microlithography, which have become the workhorse in device manufacturing for the last few years and are continuing to be imaging materials of choice for a few more generations. The primary focus is placed on chemistries that are responsible for their lithographic imaging. Furthermore, innovations in the polymer design and processing that have supported the evolution of photolithography and advancement of the microelectronics technology are described. The topics covered in this chapter include the history, rapid development in the last 20 years, the current status, and the future perspective of chemical amplification resists.

List of Abbreviations

2D-NOESY	Two-dimensional nuclear Overhauser effect spectroscopy
AA	Acrylic acid
ACOST	Acetoxystyrene
AFM	Atomic force microscope
AIBN	2,2'-Azobis(isobutyronitrile)
APEX	Name of a positive deep UV resist
ARC	Antireflection coating
BPO	Benzoyl peroxide
CAMP	Name of 248 nm positive chemical amplification photoresists
CARL	Chemical amplification of resist lines
CASUAL	Chemically amplified Si-contained resist using silsesquioxane for ArF lithography
CD	Critical dimension
CKS	Chemical kinetics simulator
COBRA	Name of positive 193 nm resists based on polynorbornene
COG	Chrome-on-glass
COMA	Cycloolefin-maleic anhydride
DESIRE	Diffusion enhanced silylation resist
DMA	Dissolution modifying agent
DNQ	Diazonaphthoquinone
DOF	Depth of focus
DRAM	Dynamic random access memory
DSC	Differential scanning calorimetry
DTBIONf	Di-(*tert*-butylphenyl)iodonium nonaflate (perfluorobutanesulfonate)
EL	Exposure latitude
ESCA	Electron spectroscopy for chemical analysis
ESCAP	Environmentally stable chemical amplification photoresist
EUV	Extreme UV
GPC	Gel permeation chromatography
HFA	Hexafluoroisopropanol
IR	Infrared
KRS	Name of a positive electron beam resist (ketal resist system)
LAMMS	Laser ablation microprobe mass spectroscopy
LEEPL	Low energy electron beam projection lithography
LER	Line edge roughness
MA	Maleic anhydride
MAA	Methacrylic acid
MALDI-TOF	Matrix-assisted laser desorption ionization-time of flight
MCP	Methylcyclopentyl
MEMS	Micro-electromechanical systems
MMA	Methyl methacrylate
NA	Numerical aperture
NB	Norbornene
NBCA	Norbornene carboxylic acid
NBHFA	5-(2-Trifluoromethyl-1,1,1-trifluoro-2-hydroxypropyl)-2-norbornene
NBTBE	*tert*-Butyl 5-norbornene-2-carboxylate
NEXAFS	Near edge X-ray absorption fine structure
NGL	Next generation lithography
NMP	*N*-Methylpyrrolidone
NMR	Nuclear magnetic resonance

NSF	National Science Foundation
OD	Optical density
PAB	Postapply bake
PAC	Photoactive compound
PAG	Photochemical acid generator
PALS	Positron annihilation spectroscopy
PBOCST	Poly(*tert*-butoxycarbonyloxystyrene)
PDP	Plasma developable photoresist
PEB	Postexposure bake
PED	Postexposure delay
PEL	Projection electron beam lithography
PGMEA	Propylene glycol methyl ether acetate
PHOST	Polyhydroxystyrene
PMMA	Poly(methyl methacrylate)
PMTFMA	Poly(methyl 2-trifluoromethylacrylate)
PREVAIL	Projection reduction exposure with variable-axis immersion lenses
PSTHFA	Poly[4-(1,1,1,3,3,3-hexafluoro-2-hydroxylpropyl)styrene]
PTBMA	Poly(*tert*-butyl methacrylate)
QCM	Quartz crystal microbalance
RBS	Rutherford backscattering
RELACS	Resolution enhancement lithography assisted by chemical shrink
RET	Resolution enhancement technique
RIE	Reactive ion etching
R_{max}	Maximum dissolution rate
R_{min}	Minimum dissolution rate
ROMP	Ring opening metathesis polymerization
SABRE	Silicon-added bilayer resist
SAC-CI	Symmetry adapted cluster configuration interaction
SAFIER	Shrink assist film for enhanced resolution
SANS	Small angle neutron scattering
SCALPEL	Scattering angular-limited projection electron beam lithography
SCF	Supercritical fluid
SEC	Size exclusion chromatography
SEM	Scanning electron microscope
SILYAL	Silylation process after alkaline wet development
SIMS	Secondary ion mass spectrometry
SPM	Scanning probe microscope
SRC	Semiconductor Research Corporation
STHFA	4-(1,1,1,3,3,3-Hexafluoro-2-hydroxylpropyl)styrene
STM	Scanning tunneling microscope
STUPID	Simple transmission understanding and prediction by incremental dilution
SUCCESS	Sulfonium compounds containing expellable sophisticated side groups
SUPER	Sub-micron positive dry etch resist
SVM	Scanning viscoelasticity microscope
TAGA	Trace atmospheric gas analyzer
TBA	*tert*-Butyl acrylate
TBMA	*tert*-Butyl methacrylate
*t*BOC	*tert*-Butoxycarbonyl
TBTFMA	*tert*-Butyl 2-trifluoromethylacrylate
T_c	Ceiling temperature
TD-DFT	Time-dependent density functional theory

TEMPO	2,2,6,6-Tetramethyl-1-piperidinyloxy
TFE	Tetrafluoroethylene
TFMAA	2-Trifluoromethylacrylic acid
T_g	Glass transition temperature
TGA	Thermogravimetric analysis
THF	Tetrahydrofuran
THP	Tetrahydropyranyl
TMAH	Tetramethylammonium hydroxide
TMS	Trimethylsilyl
TSI	Top surface imaging
UV	Ultraviolet
VEMA	Vinyl ether-maleic anhydride
VOC	Volatile organic compound
VUV	Vacuum ultraviolet
XPS	X-ray photoelectron spectroscopy

1
Introduction

Computers, once considered to be expensive computation tools only for scientists and engineers, have become household electronic gadgets like domestic appliances. The advancement and evolution of microelectronic devices we have witnessed for the last two decades are astounding. The rapid improvement of the performance of semiconductor devices has been brought about by miniaturization – by reducing the minimum feature size on the chip. The rapid innovation cycles unique to the semiconductor technology have been often expressed by the famous "Moore's Law" [1]. Gordon E. Moore, a co-founder of Intel, made his famous observation in 1965 that circuit densities of semiconductors had and would continue to double on a regular basis, which has not only been validated but has since been dubbed "Moore's Law" and now carries with it enormous influence (Fig. 1).

Semiconductor devices, so-called computer "chips", are fabricated by a technology known as microlithography. As depicted in Fig. 2, the lithographic imaging process involves use of radiation-sensitive polymeric materials called "resists" to produce circuit patterns in substrates such as single crystals of silicon. The resist material is applied by spin-coating to form 1–0.1 µm-thick films on substrates (wafers). The wafers are then baked on hot plates to remove casting solvents (postapply bake (PAB), pre-exposure bake, or prebake). The resist films are next exposed in an imagewise fashion through a mask (photo- and X-ray lithography) or directly with finely focused electron beams. The exposed resist films are subsequently developed with a developer solvent to generate three-dimensional relief images. The resist materials are classified in general as positive or negative systems. In the positive-tone imaging, the exposure renders the resist film more soluble in the developer. Conversely, negative-tone

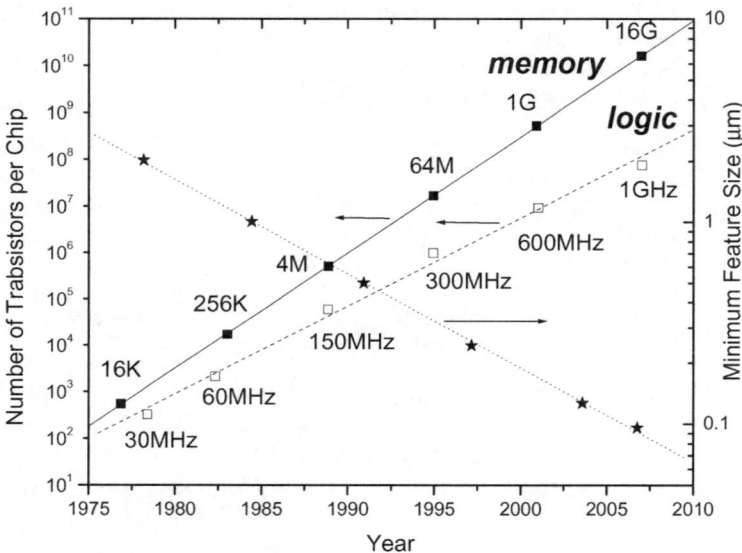

Fig. 1 Moore's law

images are produced when the exposed areas become less soluble in the developer. The resist film remaining after the development process functions as a protective stencil during etching of the substrate. The resist film must "resist" the etchant and protect the underlying substrate while the areas bared by development are being etched. The resist film remaining after etching is finally removed, leaving behind desired circuit patterns in the substrate. The process is repeated several times to fabricate complex semiconductor devices.

Resist materials must satisfy numerous demanding requirements, which have become and are becoming more and more stringent as the minimum feature size continues to shrink. Such requirements are:

- Good solubility in "safe" solvent
- Formation of uniform defect-free films
- Good adhesion to various substrates
- Ease of synthesis
- Good storage stability
- Low health and environmental hazards
- High thermal stability
- Adequate UV absorption (in photolithography)
- Good dry etch resistance
- High contrast
- High resolution
- High sensitivity
- Wide process latitudes
- Low price

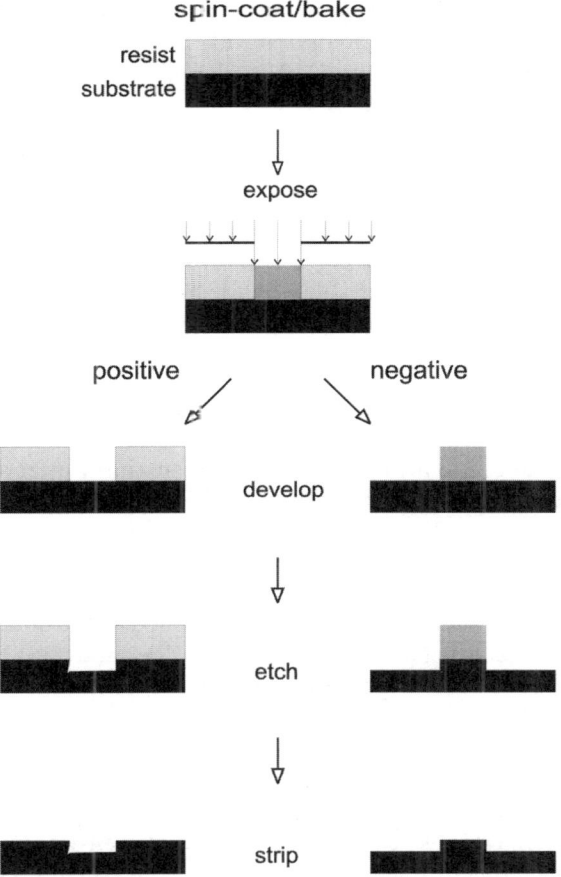

Fig. 2 Lithographic imaging process

Since the casting solvent constitutes the largest volume of resist formulations, the use of a so-called safe solvent is an important requirement. Although attempts to employ water and supercritical fluid (SCF) carbon dioxide as a casting solvent have emerged, the casting solvents mainly used in resist formulation today are propylene glycol methyl ether acetate (PGMEA), ethyl lactate, etc. Furthermore, the casting solvent can significantly affect the film and image qualities. However, resist polymers, which are the second largest component of resist formulations and the largest component in resist films after PAB, contribute to all aspects of resist characteristics and performance in a profound way. Thus, resist resins as functional polymers play a key role in microlithography and have provided exciting research opportunities to polymer scientists and engineers. Among the many requirements resist materials must satisfy, contrast and sensitivity are the most commonly employed parameters to describe resist

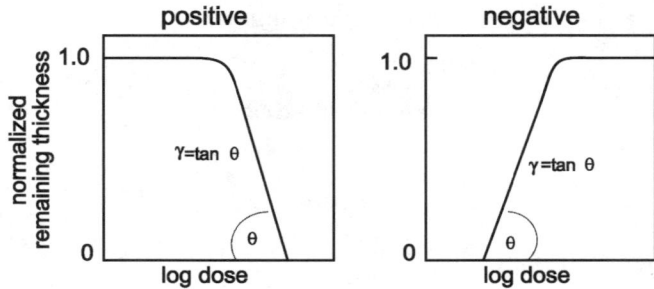

Fig. 3 Contrast (sensitivity) curves

performance. So-called "sensitivity (or contrast) curves" for positive and negative resists are schematically presented in Fig. 3. The thickness of the exposed resist film (normalized to the initial thickness) remaining after development is plotted as a function of logarithmic exposure dose (mJ/cm^2 for UV and X-ray resists and µC/cm^2 for e-beam resists). The sensitivity of positive resists is defined as the dose at which the exposed areas are cleanly developed to the substrate with a minimal thickness loss in the unexposed regions. In the case of negative systems, the sensitivity is generally defined as the dose at which 50% of the thickness is retained in the exposed areas. The contrast (γ) is determined from the slope of the linear portion of the curves. Higher contrasts lead to higher resolution in general and the resolution is the Holy Grail of microlithography.

In addition to the binary classification based on the imaging tone, resists can be divided on the basis of their design into 1) one-component and 2) multi-component systems (Fig. 4). One-component resists consist of pure radiation-sensitive polymers that must combine all the necessary attributes as mentioned above, and have long lost ground. The modern advanced lithography is ex-

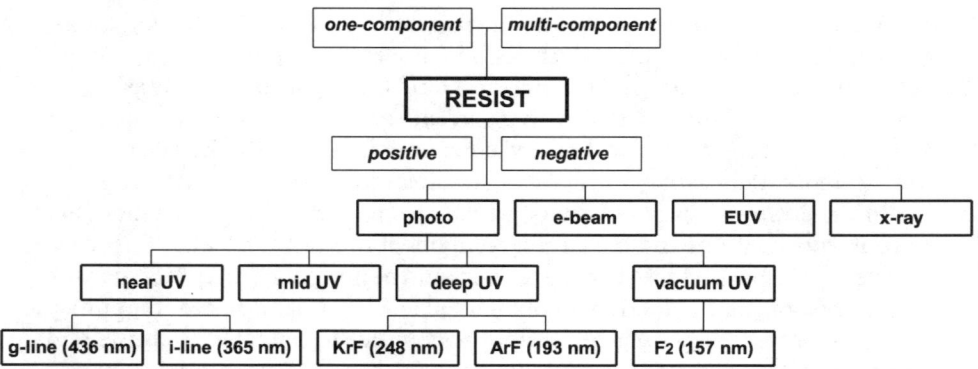

Fig. 4 Resist classification

clusively based on the two-component design, in which resist functions are provided by two separate components. In the two-component systems, the polymers are typically inert to radiation and function as an etch resistant binder. However, in the other important two-component systems, the polymers, which are radiation-insensitive, participate in reactions induced by irradiation of a radiation-sensitive component.

Another commonly employed resist classification is based on radiation sources: 1) UV or photo-resists; 2) extreme UV (EUV) resists; 3) X-ray resists; and 4) electron beam (e-beam) resists (Fig. 4). Photolithography employing ultraviolet (UV) light has been the dominant technology in logic and memory device manufacture for many years and is likely to maintain its dominance in the foreseeable future. X-ray lithography, primarily based on storage ring synchrotron radiation and capable of generating high resolution, high aspect ratio (height/width) images, has been pushed back due to advancement of photolithography. Electron beam lithography is primarily employed in fabrication of masks for photo- and X-ray imaging and may be employed in device manufacture in a projection mode in the future. Photolithography can be further divided into 1) near UV (350–450 nm), 2) mid-UV (300–350 nm), 3) deep UV (250–190 nm), and 4) vacuum UV (157 nm) technologies, depending on the wavelength of exposure (Fig. 4). EUV is actually soft X-ray at 13.4 nm and is not considered to be an extension of photolithography.

The above-mentioned resist classification based on radiation sources/exposure wavelength is the reflection of the industry's quest for higher resolution. The lithographic resolution (R) is proportional to the exposing wavelength (λ) and inversely proportional to the numerical aperture (NA) of the lens (Rayleigh's equation):

$$R = k_1 \lambda/NA$$

Thus, the migration has occurred from g-line (436 nm) to i-line (365 nm) of Hg discharge lamps, followed by an incremental increase in NA of the i-line lens, to achieve higher resolution. Improvements of resist materials and processes as well as optical resolution enhancement techniques (RET) decrease k_1. The astounding advancement of integrated circuit devices has been accomplished by increasing the number of components per chip; by continually reducing the minimum feature size on the chip. The quest for higher and higher resolution dictates the microlithographic technology.

The resist system which supported the i-line technology for many years in an exclusive manner was so-called diazonaphthoquinone (DNQ)/novolac resists (Fig. 5). This type of resists originally invented for printing by Süss [2] is a two-component system consisting of a novolac resin and a photoactive compound (PAC), diazonaphthoquinone. The novolac resin is soluble in aqueous base in virtue of the acidic phenolic OH functionality. However, the lipophilic diazonaphthoquinone dispersed in the phenolic matrix inhibits the dissolution of the resin film in an aqueous base developer. UV irradiation of the photoactive compound results in formation of a highly reactive carbene, accompanied

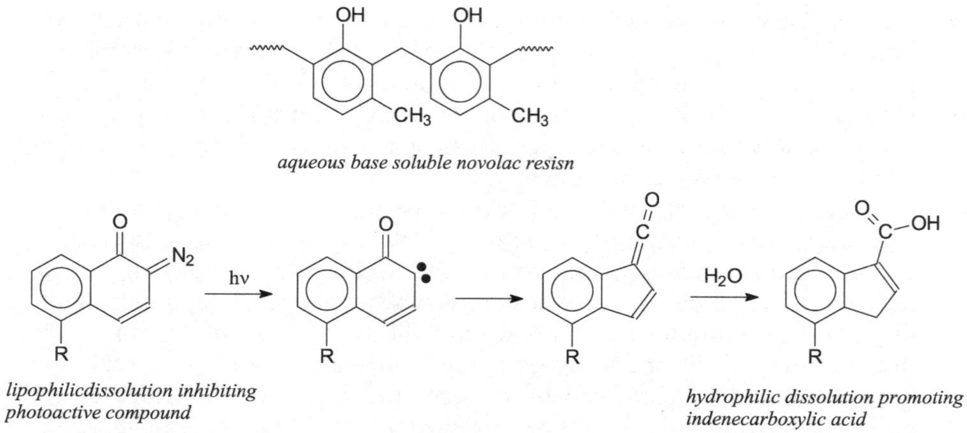

aqueous base soluble novolac resisn

lipophilicdissolution inhibiting photoactive compound

hydrophilic dissolution promoting indenecarboxylic acid

Fig. 5 Diazonaphthoquinone/novolac positive resist

by release of nitrogen, with a quantum yield of 0.2–0.3. The carbene intermediate undergoes the Wolff rearrangement to ketene, which rapidly reacts with ambient water in the film to produce indenecarboxylic acid [3] (Fig. 5). The net result is a photochemical transformation of dissolution inhibiting quinonediazide to base-soluble indenecarboxylic acid. In consequence, the exposed areas of the resist film dissolve much faster than the unexposed in an aqueous base developer. Thus, positive-tone relief images result.

The role of the novolac resin is not as minor as it may seem as a base-soluble binder. While the aromatic nature of the resin provides high dry etch resistance, the novolac structures and properties such as the ratio of o-cresol to m-cresol, the ratio of *ortho* to *para* backbone linkages, the molecular weight, and molecular weight distribution all affect the dissolution behavior, thermal flow resistance, and lithographic performance. The optimization of the novolac and diazonaphthoquinone properties in conjunction with the improvement of the i-line step-and-repeat exposure tools has pushed the resolution limit of photolithography to a sub-0.5-µm regime [4].

Miniaturization has continued. The drive to higher resolution has necessitated a further shift from i-line to deep UV employing the 254 nm emission from Xe-Hg lamps or the 248 nm emission from krypton fluoride (KrF) excimer lasers. Attempts to utilize the novolac/diazoquinone resist failed because of its insufficient sensitivity and poor imaging quality. The novolac resin strongly absorbs in the 250 nm region and the diazonaphthoquinone which bleaches cleanly upon irradiation in the near UV range exhibits a significant residual absorption in the deep UV region, thus limiting penetration of the light to the surface of the resist film, which results in formation of a sloped resist wall profile that is useless in the subsequent process steps. Efforts to design more transparent and bleachable photoactive dissolution inhibitors and search for more transparent base-soluble polymers were only partially successful. Use of transparent polymethacrylate-based one-component resists that undergo main chain scission upon deep UV irradiation did not bear a fruit either due to their low sensitivity. Sensitivity enhancement was the major research subject in the resist science and technology for many years but was very much incremental and marginal. The Hg discharge lamp had an extremely small output at 254 nm, demanding a couple of magnitude higher resist sensitivity for acceptable wafer throughput. The advent of the KrF excimer laser technology was expected to ease the sensitivity requirement. However, insertion of many optical elements between the laser source and the wafer plane dramatically reduced the radiation output available to the resist film and thus the high sensitivity still remained as an important issue. The shift to deep UV lithography required a completely new resist concept.

The concept of chemical amplification was proposed in the early 1980s to overcome the sensitivity limitation imposed by the imaging mechanism requiring several photons to produce one useful photoproduct. The chemical amplification concept has become the exclusive foundation for advanced high resolution resist systems and played a pivotal role in realization of the deep UV lithographic technology and in extension of photolithography to a higher dimension. The semiconductor industry implemented deep UV (248 nm) lithography first in manufacture of 256 Mbit DRAM and related logic devices with the minimum feature size of 0.25 µm and then extended it to a 130 nm node. The 248 nm lithography employing chemical amplification positive resists is the current workhorse of the device manufacturing. A deep UV technology employing ArF excimer lasers (193 nm) is currently being implemented for manufacture of devices with a critical dimension of <130 nm. Furthermore a new photolithographic technology has emerged, which employs F_2 excimer lasers emitting at 157 nm. All these photolithographic technologies operating at a wavelength below 248 nm and the next generation lithographic (NGL) technologies such as EUV, e-beam, and X-ray require chemical amplification resists as imaging materials. A number of reviews on various aspects of resist chemistry and processing are available [5–22].

2
Chemical Amplification Concept

The resist sensitivity is a very important parameter to be considered as it is directly related to wafer throughput and therefore device manufacturing cost. Thus, sensitivity enhancement was a primary research activity in the field of microlithography in the 1970s and early 1980s. However, the enhancement of sensitivity achieved at that time was too incremental and marginal. Quantum yields, expressed as the number of molecules transformed per photon absorbed, characterize the efficiency of photochemical events. Typical diazonaphthoquinone has a quantum yield of 0.2–0.3, which means that three to five photons are needed to convert a single molecule of the photoactive compound. An in-

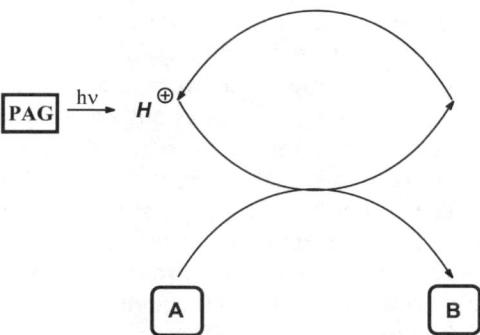

Fig. 6 Schematic illustration of chemical amplification

crease in the quantum yield to its theoretical upper limit of unity would result in only a three- to fivefold increase in sensitivity. However, orders of magnitude of sensitivity enhancement were required to make deep UV, e-beam, and X-ray lithography economically feasible in terms of unit area exposed per unit time. It was apparent that a gain mechanism in resist chemistry or processing was needed to meet the sensitivity requirement.

The concept of *chemical amplification* was proposed by Ito, Willson, and Fréchet in 1982 [23, 24]. In the chemical amplification scheme, a single photochemical event induces a cascade of subsequent chemical transformations in a resist film; irradiation produces active species that catalyze numerous chemical reactions (Fig. 6). In general, these chemical transformations are accomplished by heating the exposed resist film (postexposure bake (PEB) or postbake). Although the active species could be either ionic or radical in principle, use of photochemical acid generators (PAGs) which was proposed in the original chemical amplification concept [25] has become the primary and almost exclusive foundation for an entire family of advanced resist systems. The successful application of acid catalysis to the resist design has prompted research efforts in photochemical base generators [26–29]. However, since base catalysis has not been widely accepted as a viable resist design, this article deals with only acid-catalyzed chemical amplification resists.

3
Photochemical Acid Generators

Various PAGs have been synthesized specifically for use in chemical amplification resists, reflecting their important impact on lithographic performance (Fig. 7) [13, 30]. The choice of PAG depends on a number of factors such as the nature of radiation, quantum efficiency of acid generation, solubility, miscibility with resin, thermal and hydrolytic stability, plasticization effect, toxicity, strength and size of generated acid, impact on dissolution rates, cost, etc. In

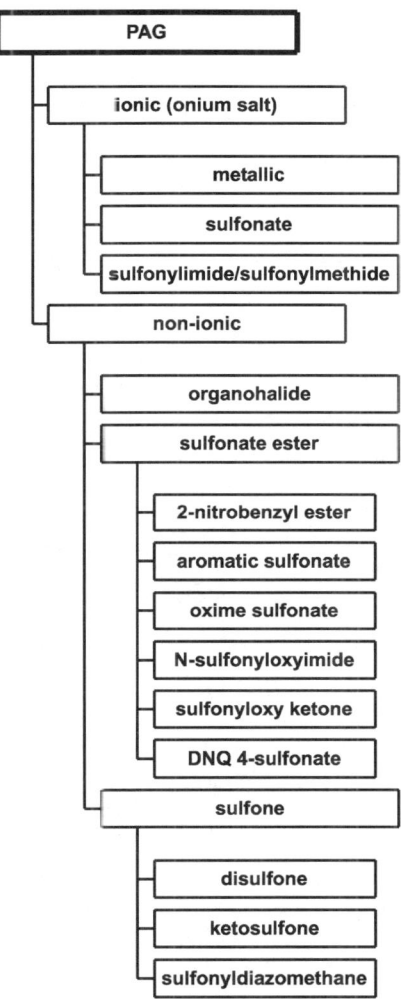

Fig. 7 Classification of photochemical acid generators

Fig. 8 are presented representative acid generators. In addition to the thermal stability of PAG by itself, its hydrolytic stability is important as PAG tends to decompose at lower temperatures in an acidic matrix polymer [31, 32].

Acid generators can be divided into two groups; ionic and non-ionic. So-called onium salt acid generators such as triarylsulfonium and diaryliodonium hexafluoroarsenates and antimonates (ionic and metallic) were initially developed for photochemical curing of epoxy resins by Crivello [33] and generate the strongest acids with excellent quantum yields. The mechanism of acid generation has been extensively studied [34–40] and is shown in Fig. 9 for triphenylsulfonium salts. In a viscous medium a cage effect dominates the distribution

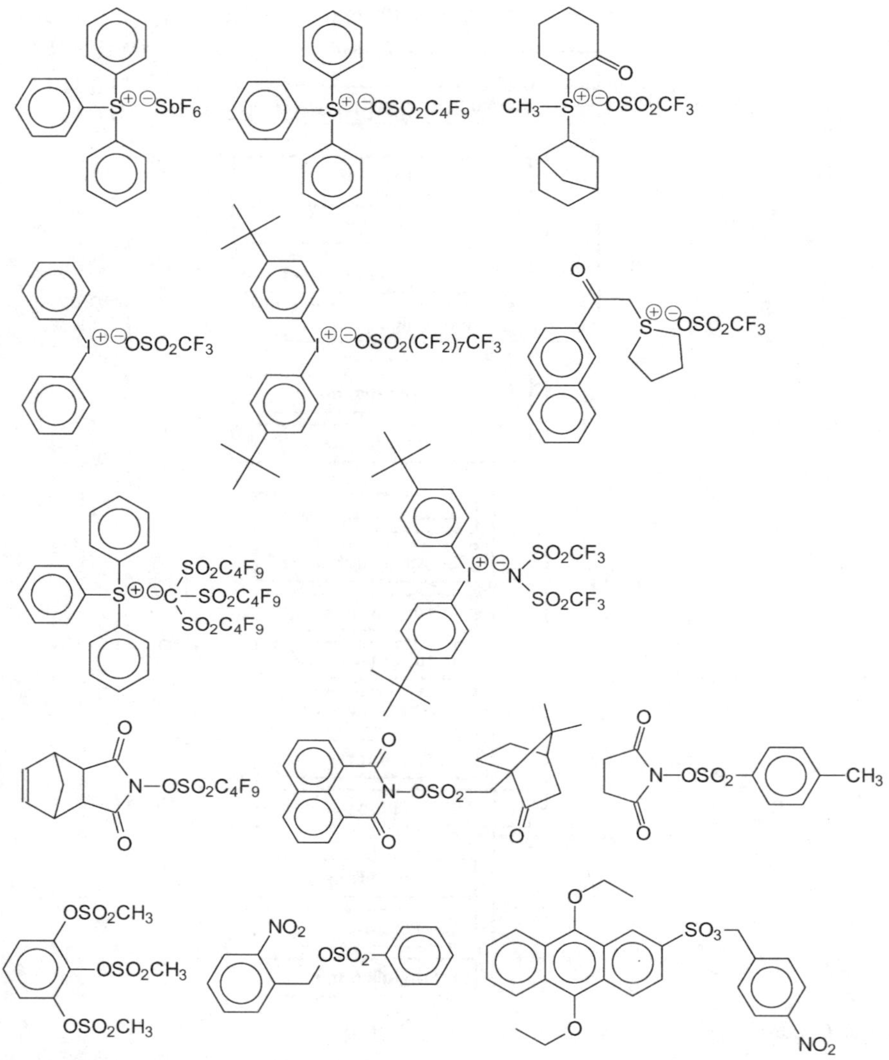

Fig. 8 Representative acid generators

of the photolysis products [39]. Quantum chemical calculation has been also performed to investigate the photolysis mechanism [41]. In the case of e-beam exposure, electron transfer from a matrix polymer to a sulfonium salt plays an important role [42]. Although triphenylsulfonium hexafluoroantimonate was employed in the very first chemical amplification resist which manufactured DRAMs in the mid-1980s [43], the lithography community has moved away from the metallic acid generators. Triphenylsulfonium trifluoromethanesulfonate (triflate) which generates the strongest organic acid was a logical alter-

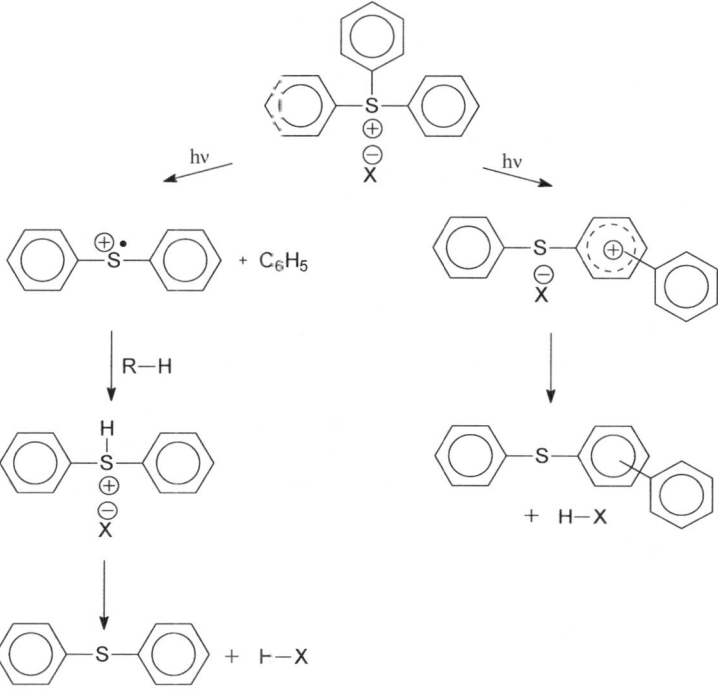

Fig. 9 Photolysis mechanism of triphenylsulfonium salts

native. Triarylsulfonium salts are the most thermally stable acid generators with the onset of decomposition observed at ca. 350 °C. The onium salts are typically sensitive to deep UV, e-beam, and X-ray irradiation and can be red-shifted for near UV use by incorporating a chromophore in the structure [44] or by adding a sensitizing molecule [36, 38, 45–47]. Diaryliodonium sulfonates such as di(*tert*-butylphenyl)iodonium perfluorobutanesulfonate (nonaflate) (DTBPIONf) and perfluorooctanesulfonate are heavily used in resist formulation.

The non-ionic acid generators have been primarily developed for use with chemical amplification resists. Organohalides such as tetrabromobisphenol A and 4,6-bis(trichloromethyl)-1,3,5-triazine generate hydrogen halide upon irradiation and were employed in early negative-tone chemical amplification resists [48]. Their high volatility and limited acidolysis capability are the primary reasons for their poor acceptance. The majority of efforts has been placed on non-ionic organic acid generators which produce sulfonic acids when irradiated. There are two groups; sulfonate esters and sulfonyl compounds. In some cases, sulfinic acid is generated [49, 50]. Polymerizable and polymeric acid generators have been also reported [51–55].

Because of the health hazard associated with fluorinated sulfonates, which was discovered recently, a new class of ionic acid generators has been proposed from 3M (imides and methides in Fig. 8) [56].

Quantum yields of acid generation and the concentration of acid photochemically generated in resist film are important factors in resist designing, processing, and modeling. Several experimental methods to quantify acid concentration in exposed resist film have been proposed [48, 57–61]. The most popular method is spectrophotometric detection of photochemically generated acid using an acid sensitive dye such as families of sulfonephthaleins (tetrabromophenol blue sodium salt, for example) [48, 59], merocyanines [57], monoazines [60a], xanthenes (Rhodamine B, for example) [60b], and benzothiazoles [60b]. Each exposed resist film on a Si wafer is extracted with a solvent, to which is added a dye solution. Absorbance of each resist extract is measured on a spectrophotometer and compared with a calibration curve to determine acid concentration. In situ spectrophotometric methods involving doping a resist with an acid sensitive dye have been also developed [62–65]. An ion conductivity method for acid quantification has been also reported [58]. Photochemical PAG decomposition in resist film was investigated by dissolving exposed resist film in a NMR solvent such as acetone-d_6 and performing quantitative inverse gate ^{13}C NMR analysis [66].

Photochemically generated acid must diffuse in resist film to catalyze desired reactions and to provide a gain mechanism for amplification. However, excessive diffusion (into the unexposed areas) destroys the linewidth control and eventually the resolution. Thus, as the minimum feature size becomes smaller and smaller, the control of acid diffusion plays a more important and difficult role. Therefore, investigation of acid diffusion in chemical amplification resist film is one of the most active areas of research today. A number of experimental procedures to measure acid diffusion length have been reported [67–88]:

- Hard contact X-ray lithography [72, 81]
- Conductivity measurements [69, 75, 78, 82]
- Integrated array electrodes [73, 81]
- Scanning tunneling microscopy [74]
- Laser confocal microscopy [86]
- Bilayer system consisting of an acid-containing layer and an acid-labile polymer layer [68, 75, 79, 84, 87]
- Irradiation of opaque resist films to study downward diffusion [88]

Although the primary function of acid generators is to produce acid upon irradiation for subsequent acid-catalyzed reactions, these photoactive compounds affect the lithographic performance of the resulting resist in a profound and complex way. For example, certain acid generators strongly inhibit dissolution of phenolic and other acidic resins in aqueous base developer [44, 89]. Photolysis of dissolution inhibiting PAG in a novolac resin results in faster dissolution of the exposed areas in aqueous base and thus positive images are generated upon base development (without PEB and without involving acid-catalyzed deprotection). In this case PAG functions like diazonaphthoquinone.

Distribution of PAG in spin-cast resist film can affect image profiles. Thus, efforts have been made to tailor the PAG structure for adequate polarity and to

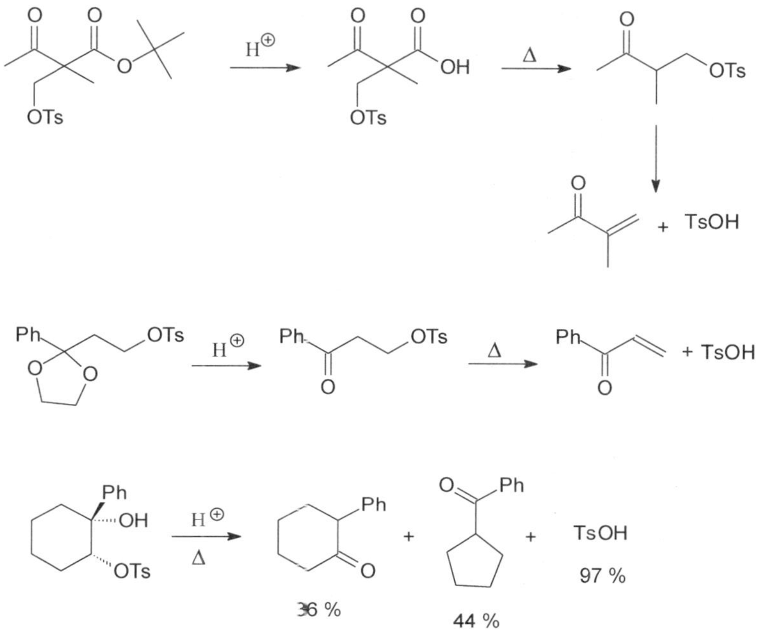

Fig. 10 Acid amplifiers

determine experimentally the depth distribution of PAG in spin cast films by, for example, Rutherford backscattering spectrometry (RBS) and dynamic secondary ion mass spectrometry (SIMS) [90, 91].

Use of alkylsulfonium iodides in conjunction with PAG has been reported to enhance contrast and to control acid diffusion in e-beam imaging [92].

A new interesting concept to further increase the sensitivity of chemical amplification resists has been proposed primarily for 193 nm lithography. An "acid amplifier" in resist film, inert to radiation, produces a large number of acid molecules by its catalytic reaction with a photochemically generated acid (Fig. 10) [93, 94]. In many cases only weak sulfonic acids such as p-toluenesulfonic acid (tosic acid, TsOH) are generated by this mechanism and an aromatic structure incorporated into acid amplifiers to promote acid cleavage increases 193 nm absorption.

4
Chemically Amplified Imaging Mechanisms

The original chemical amplification scheme reported in the early 1980s contained three mechanisms [25]: depolymerization for positive imaging, deprotection for dual tone imaging, and polymerization for negative imaging. Since then numerous imaging mechanisms based on photochemically-induced acid

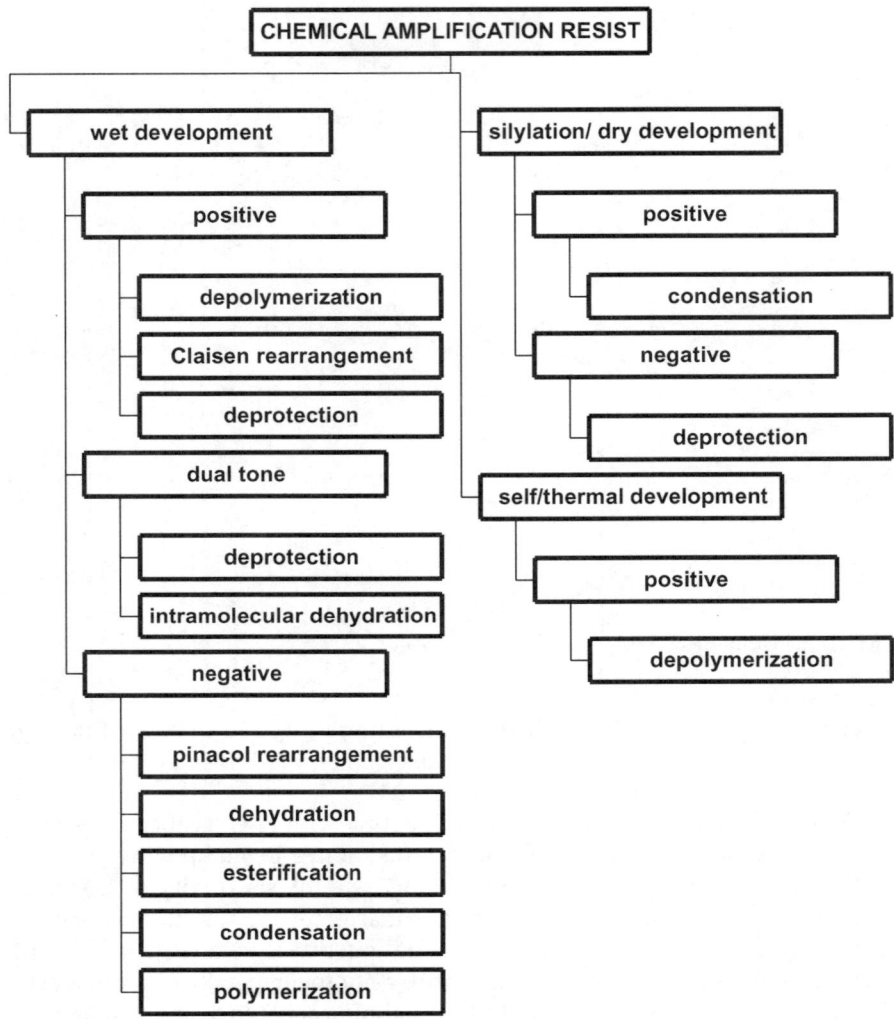

Fig. 11 Chemical amplification imaging mechanisms

catalysis have been reported as illustrated as a family tree in Fig. 11, which are described in this chapter.

4.1
Deprotection

Acid-catalyzed cleavage of pendant protecting groups to generate base-soluble acidic functionalities such as phenol and carboxylic acid was one of the three initial mechanisms of chemical amplification [23, 25]. This imaging mechanism has drawn a great deal of attention because it provides the basis for designing aqueous base developable positive resists for replacement of diazoquinone/novolac resists in short wavelength lithographic technologies. In fact, today's advanced positive resists are exclusively built on this deprotection mechanism.

4.1.1
Carbonates

The very first resist designed on the basis of acid-catalyzed deprotection was IBM's *t*BOC resist, which was a two-component system consisting of poly(4-*tert*-butoxycarbonyloxystyrene) (PBOCST) and an acid generator [23, 25]. Figure 12 illustrates the imaging chemistry of the classical *t*BOC resist. The phenolic functionality of poly(4-hydroxystyrene) (PHOST) is protected with an acid-labile *tert*-butoxycarbonyl (*t*BOC) group. This lipophilic carbonate polymer is converted upon PEB at ~100 °C for a couple of minutes to PHOST by reaction with a photochemically generated acid, releasing carbon dioxide, isobutene, and a proton, while the protecting group is stable thermally to 190 °C in the absence of acid. This deprotection reaction is truly catalytic A_{AL}-1 acidolysis which does not require a stoichiometric amount of water and thus the photochemically generated acid is not consumed by one reaction but regenerated to carry out a number of deprotection reactions. A catalytic chain length of the *t*BOC resist has been estimated to be ca. 1000 under normal processing conditions [57]. The acid-catalyzed deprotection reaction converts a lipophilic polymer to a hydrophilic polymer, providing a large change in the polarity and hence solubility of the polymer. This polarity change can be exploited in dual-tone imaging simply by changing the polarity of the developer solvent. Polar solvents such as alcohol and aqueous base, which are non-solvent for the lipophilic PBOCST remaining in the unexposed regions, dissolve only the polar PHOST produced in the exposed regions, thus providing positive-tone images (Fig. 13 top). Conversely, use of a nonpolar organic solvent such as anisole produces negative-tone images because such a solvent dissolves PBOCST readily but not PHOST (Fig. 13 bottom). Whereas negative resist systems based on crosslinking suffer from swelling-induced image distortion during organic development which

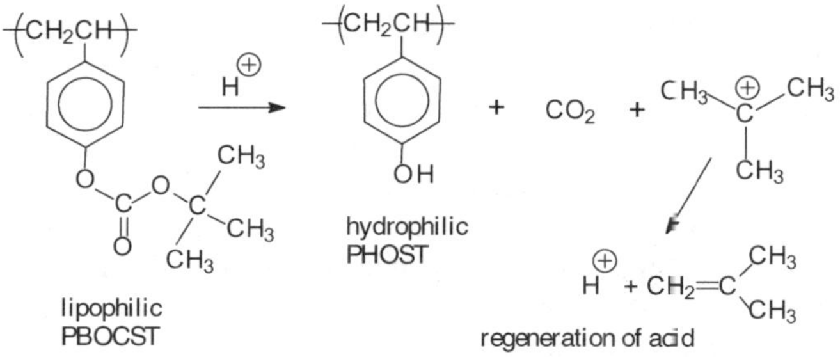

Fig. 12 Acid-catalyzed deprotection for polarity change (*t*BOC resist)

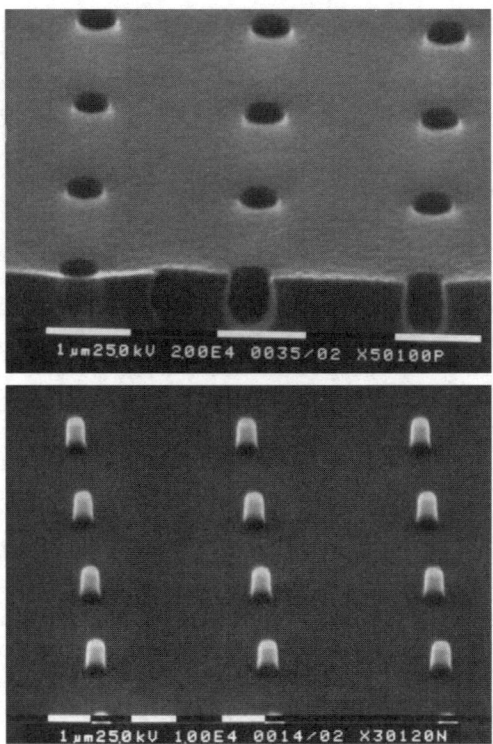

Fig. 13 Scanning electron micrographs of positive (*top*) and negative (*bottom*) images delineated in tBOC resist by X-ray irradiation [15, 16]

limits the resolution, the polarity change mechanism is less prone to such a problem even in negative imaging because the developer is a non-solvent for the exposed regions.

The tBOC resist containing 4.75 wt% of triphenylsulfonium hexafluoroantimonate was the very first chemical amplification resist employed in manufacture of semiconductor devices, producing millions of 1 Mbit dynamic random access memory (DRAM) devices by deep UV lithography on Perkin Elmer Micralign 500 mirror projection scanners (NA=0.17) at IBM in the mid-1980s (Fig. 14) [43], while the rest of the industry was using near UV lithography and a DNQ/novolac resist in device production. Although the deprotection mechanism has become the paradigm for positive resist systems, the tBOC resist was used in its negative mode with anisole as the developer in the manufacturing.

The conversion of PBOCST to PHOST in the solid state can be conveniently monitored by IR spectroscopy because the tBOC carbonyl absorption at 1755 cm^{-1} shrinks in intensity, accompanied by appearance of a phenolic hydroxyl absorption at ~3500 cm^{-1} as the acid-catalyzed deprotection proceeds (Fig. 15). Because this is the most classical and simplest form of the chemical

Chemical Amplification Resists for Microlithography 57

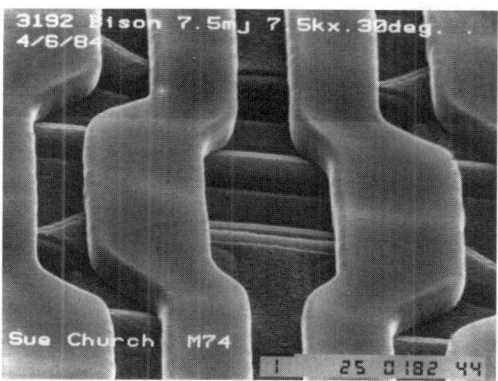

Fig. 14 One-μm space patterns on device topography printed in tBOC in a negative mode by deep UV lithography for 1 Mbit DRAM manufacturing [21]

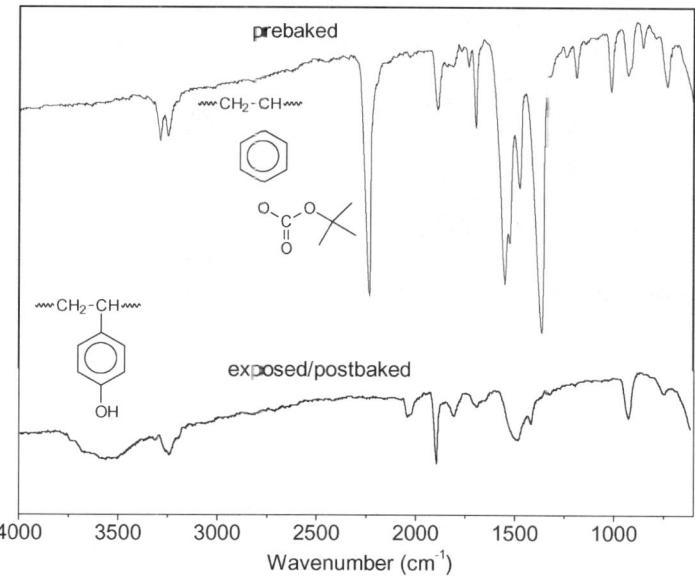

Fig. 15 IR spectra of tBOC resist before and after exposure/bake [25]

amplification resists, many fundamental studies are still being performed on this system. For example, the thermal and acid-catalyzed deprotection mechanisms of the tBOC resist have been investigated recently in detail by quantum mechanical calculation [95], by quantitative analysis of the resist film by inverse-gated ^{13}C NMR [66], and by using a trace atmospheric gas analyzer (TAGA) [96]. The ^{13}C NMR study has clearly indicated that the *tert*-butyl cation generated by acidolysis does not completely leave the film in the form of isobutene as shown

Fig. 16 Side reactions in *t*BOC deprotection

in Fig. 12 but reacts with the phenol through *O*-alkylation and *C*-alkylation upon PEB (Fig. 16). These side reactions could reduce the dissolution rate of the exposed film and therefore lower the contrast. Thermolysis at 180 °C in the absence of acid results in formation of 4% *tert*-butyl ether. While the semiempirical molecular orbital calculations postulated that protonation occurs on the carbonyl oxygen, followed by elimination of a *tert*-butyl cation to form an acid carbonate [95], the detailed mass spectroscopic studies of volatile products from thermal and acid-catalyzed deprotection have clearly indicated that there are two reaction pathways as illustrated in Fig. 17 [96]. Scission of CO-O to produce *tert*-butoxy intermediate generates acetone, *tert*-butyl alcohol, and methyl isopropenyl ether and scission of O-C to form *tert*-butyl intermediate generates isobutene as a gaseous product. In addition, numerous bimolecular products have been detected by the TAGA analysis.

PBOCST can be conveniently prepared by subjecting the corresponding monomer to radical polymerization or cationic polymerization in liquid sulfur

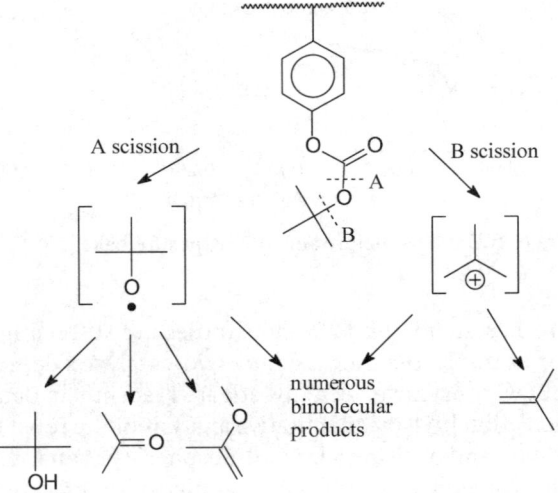

Fig. 17 *t*BOC deprotection pathways

dioxide [97] or alternatively by reacting PHOST with di-*tert*-butyl dicarbonate in the presence of base [98]. Narrow polydispersity PBOCST has been synthesized by living anionic polymerization of 4-*tert*-butyl(dimethyl)silyloxystyrene followed by desilylation with HCl to form PHOST and protection with di-*tert*-butyl dicarbonate [99]. PBOCST prepared by radical polymerization has a glass transition temperature (T_g) of ca. 135 °C and undergoes spontaneous autocatalytic thermolysis at ca. 190 °C to produce PHOST, carbon dioxide, and isobutene, with the weight loss amounting to 45.5 wt%. However, as mentioned above, the thermolysis is accompanied by about 4% of the C-alkylated side product [66]. PBOCST exhibits excellent UV transmission in the 250 nm region with its absorption of <0.1/µm and is highly suitable for deep UV imaging. The synthesis of PHOST from PBOCST has proven that "pure" PHOST is highly transparent in the 250 nm region with its absorption of ca. 0.2/µm while PHOST commercially available in the early 1980s was only slightly better than novolac resins in terms of the deep UV absorption. While the CN group attached to the polymer end slows down the acid-catalyzed deprotection reaction when 2,2'-azobis(isobutyronitrile) (AIBN) is used as the initiator, the benzoyl end group derived from benzoyl peroxide (BPO) increases a UV absorption in the 250 nm region especially when the molecular weights are low [100]. Anionically prepared polymers do not have absorbing or poisoning end groups [100].

The *t*BOC group has been later employed in protection of other aqueous base soluble polymers such as poly[styrene-*co*-N-(4-hydroxyphenyl)maleimide] [101], poly(styrene-*co*-maleimide) [102, 103], poly(4-hydroxystyrene sulfone) [104], and poly(4-hydroxy-α-methylstyrene) [105, 106] for lithographic applications. Novolac resins have been also protected with the *t*BOC group [107, 108], which has resulted in significant reduction of the deep UV absorption [108]. Poly(4-*tert*-butoxycarbonyloxy-α-methylstyrene) [105] and poly(4-*tert*-butoxycarbonyloxystyrene sulfone) [104] were thought to undergo main chain scission as well upon deep UV irradiation in the presence or absence of PAG, which was expected to corroborate with deprotection in positive imaging. Cationically-prepared poly(4-hydroxy-α-methylstyrene) and its *t*BOC-protected derivative undergo efficient acid-catalyzed depolymerization [109, 110]. Poly(4-*tert*-butoxycarbonyloxystyrene sulfone) can be synthesized by radical copolymerization of the substituted styrene with sulfur dioxide. This polymer has been reported to produce sulfinic or sulfonic acid through scission of the backbone C-S linkage by X-ray irradiation, though its sensitivity to deep UV radiation is poor [111]. Since the polymer itself generates acid that deprotects the *t*BOC group, this system can constitute a rare example of one-component chemical amplification resists (Fig. 18). The CAMP-series resists promoted and commercialized as deep UV resists by AT&T (now Lucent Technology) Bell Laboratories in collaboration with Olin-Ciba Geigy (currently ARCH Microelectronics) consist of the poly(styrene sulfone) and PAG [104]. In order to reduce the excessive shrinkage in the exposed areas (due to the loss of carbon dioxide and isobutene), which induces stress in the resist film and unacceptable thinning during reactive ion etching (RIE) employing a positive-tone image as

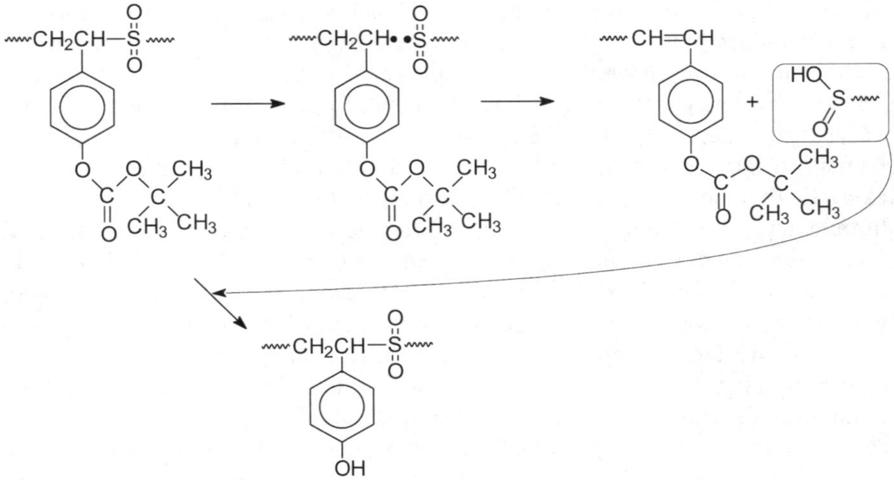

Fig. 18 One-component X-ray positive resist based on *t*BOC deprotection

a mask, 4-acetoxystyrene has been introduced as a third comonomer into the poly(*t*BOC-styrene sulfone) for CAMP6 [112]. The acetate group reportedly undergoes hydrolysis during aqueous base development only in the presence of a phenolic functionality (only in the exposed regions) and the sulfone structure (Fig. 19) while it is inert to acidolysis. Poly(4-hydroxystyrene sulfone) was shown to undergo main chain scission in aqueous base by ^{1}H NMR analysis [113]. The sulfone polymer was decomposed quantitatively upon dissolving it in a deuterium oxide solution of sodium deuteroxide to give *trans*-2-(4-hydroxyphenyl)ethenesulfinic acid and HOST in a molar ratio of about 2/3 [113].

A new class of aqueous-base soluble polymer, poly[4-(2-hydroxyhexafluoroisopropyl)styrene] (Fig. 20), has been reported as a replacement of PHOST for 248 nm lithography, which was also protected with the *t*BOC group [114]. The *t*BOC-protected polymer can be synthesized by radical polymerization of

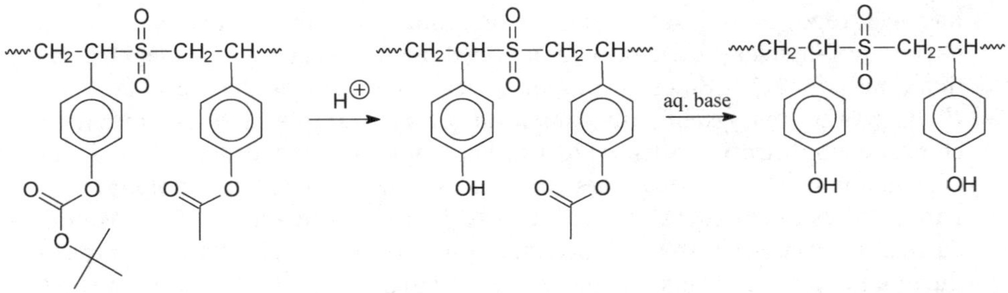

Fig. 19 Photochemically induced acid-catalyzed deprotection and subsequent base-catalyzed deacetylation during development

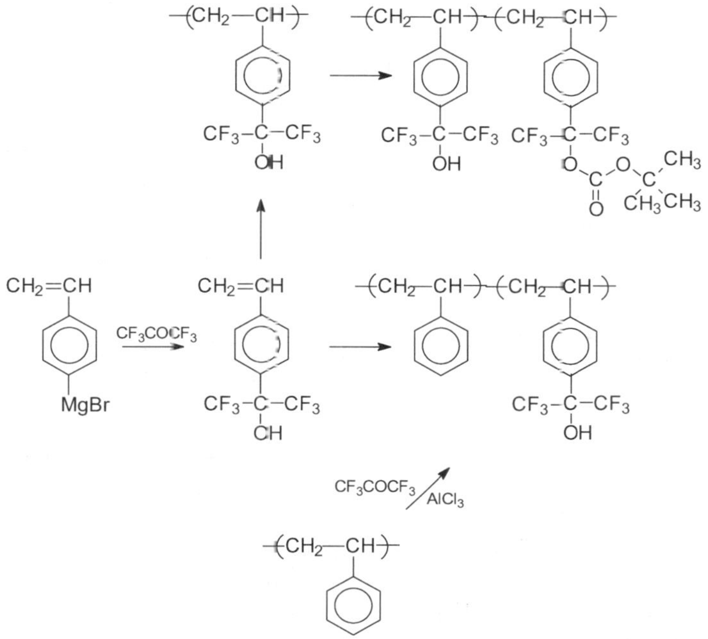

Chemical Amplification Resists for Microlithography 61

Fig. 20 Base-soluble polystyrene with a pendant hexafluoroisopropanol

a protected monomer or by reacting the fluoroalcohol polymer with di-*tert*-butyl dicarbonate. The base-soluble fluoroalcohol polymer may be prepared by radical polymerization of the corresponding unprotected monomer or by treating polystyrene with hexafluoroacetone in the presence of a Lewis acid catalyst (Fig. 20). The hexafluoroisopropanol group was attached to norbornene by Diels-Alder reaction of 1,1,1-trifluoro-2-(trifluoromethyl)pent-4-en-2-ol [115], which in turn had been prepared by reacting allylmagnesium chloride with hexafluoroacetone [116]. A *t*BOC-protected monomer was copolymerized with sulfur dioxide and maleic anhydride for 193 nm (ArF) lithography (Fig. 21) [115]. Since the hexafluoroalcohol group has become the acid group of choice for the emerging 157 nm (F_2 excimer lasers) and maturing 193 nm lithographic technologies, the resist systems based on the hexafluoroisopropyl group will be discussed in detail later.

Poly(hydroxyphenyl methacrylate)s, poly(*N*-hydroxyphenylmethacrylamide)s, and related copolymers have been masked with the *t*BOC group [117]. It has been reported that hydroxyphenyl methacrylate undergoes radical polymerization much more readily than 4-hydroxystyrene without protecting the phenolic hydroxyl group. These unprotected homopolymers are, however, unexpectedly opaque in the deep UV region, though the optical density (OD) of poly(4-hydroxyphenyl methacrylate) is somewhat smaller (0.48/µm) than that of a novolac resin. Masking with the *t*BOC group reduces the UV absorption of

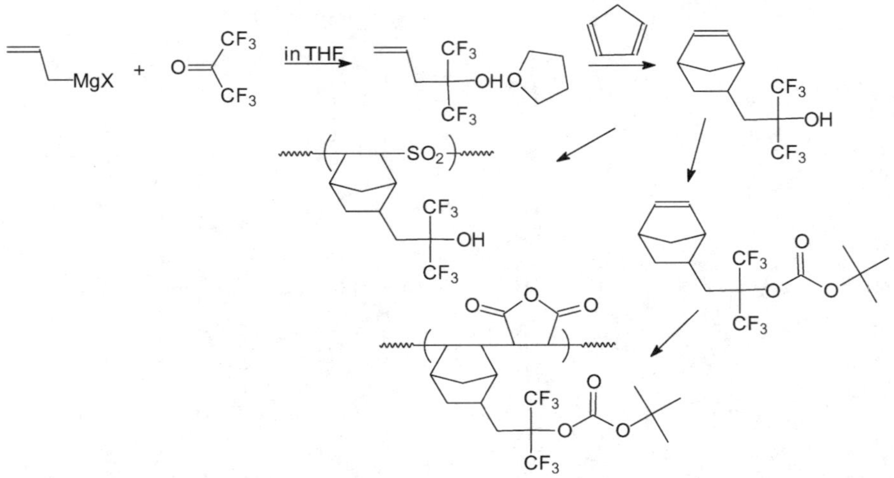

Fig. 21 Norbornene copolymers bearing hexafluoroisopropanol

poly(hydroxyphenyl methacrylate)s like novolac resins to 0.18–0.29/μm but does not improve the deep UV absorption characteristics of the corresponding polymethacrylamides.

Other carbonates, thermally more stable, have been also investigated; the *tert*-butyl group of *t*BOC has been replaced with isopropyl, substituted α-methylbenzyl, and 1-(2-tetrahydrofurfuryl)ethyl groups [118]. Attempts to increase the thermal stability could result in unacceptably low sensitivity toward acidolysis.

4.1.2
Esters and Ethers

Certain esters and ethers are also useful in the design of dual tone and positive chemical amplification resists. In fact, acid-cleavable *tertiary* esters are the basis for the advanced 248 nm and the majority of 193 and 157 nm positive resist systems (see later). Some examples are presented in Fig. 22, which are A_{AL}-1 acidolysis without a need for a stoichiometric amount of water. Poly(4-vinylbenzoate)s [119, 120] and poly(meth)acrylates [121, 122] are converted to poly-(4-vinylbenzoic acid) and poly[(meth)acrylic acid], respectively, by reaction with a photochemically generated acid. This polarity change from a nonpolar to polar state allows dual tone imaging, depending on a polarity of the developer solvent. The ester groups in these polymers must be selected to generate, upon heterolysis of the C-O bond, stable carbocations which undergo spontaneous β-proton elimination to form olefins. The thermal and acidolytic stability can be tuned by changing the ester structure. Thermogravimetric analysis (TGA) curves of select polymethacrylates are presented in Fig. 23 [121]. Once an ester group is cleaved to form methacrylic acid units, the carboxylic acid moiety interacts with a neighboring acid and/or ester group to form an anhydride

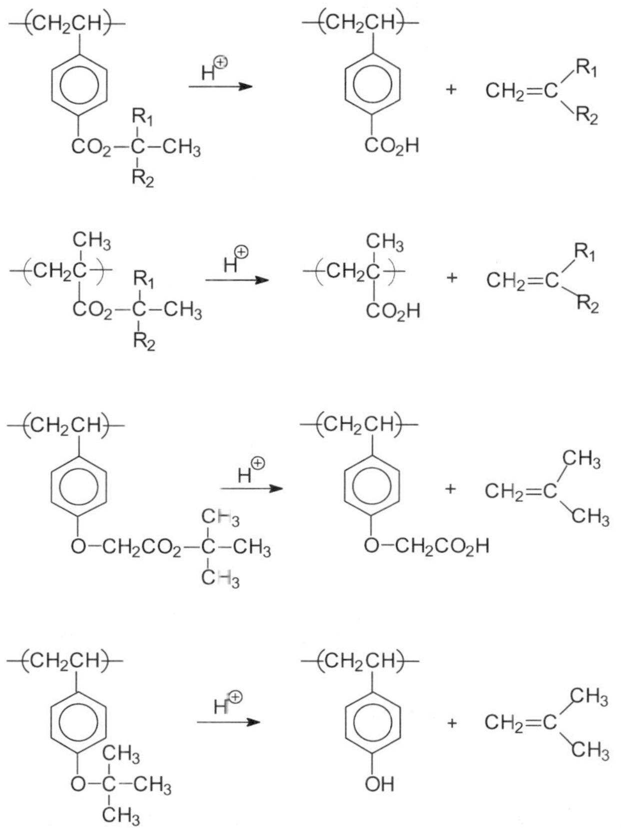

Fig. 22 Acid-cleavable esters and ethers

ring above 160 °C [123]. A TAGA study has indicated that the thermolysis of poly(*tert*-butyl methacrylate) (PTBMA) occurs via two intermediates, *tert*-butyl and *tert*-butoxy, to generate isobutene, acetone, methyl isopropenyl ether, and 2,2-dimethylpropanal in a fashion similar to PBOCST (Fig. 17) [96]. A large number of acid-labile ester groups have been developed especially for use in 193 nm resists in the form of polymethacrylates or polynorbornenes (see later). PHOST has been also protected with a *tert*-butoxycarbonylmethyl [25, 124] and *tert*-butyl [125] groups. In the former protection scheme, a carboxylic acid group is produced as a base soluble functionality upon acidolysis while the latter is similar to *t*BOC and is converted to PHOST. Certain polymethacrylates have been later shown to undergo a small degree of deprotection as well as main chain scission in the absence of acid when irradiated with a high dose of electron beams [126].

Poly(4-vinylbenzoate)s and poly(4-vinylbenzoic acid) are very opaque below 300 nm (OD=1.1/µm and 3.4/µm at 248 nm, respectively) and thus cannot

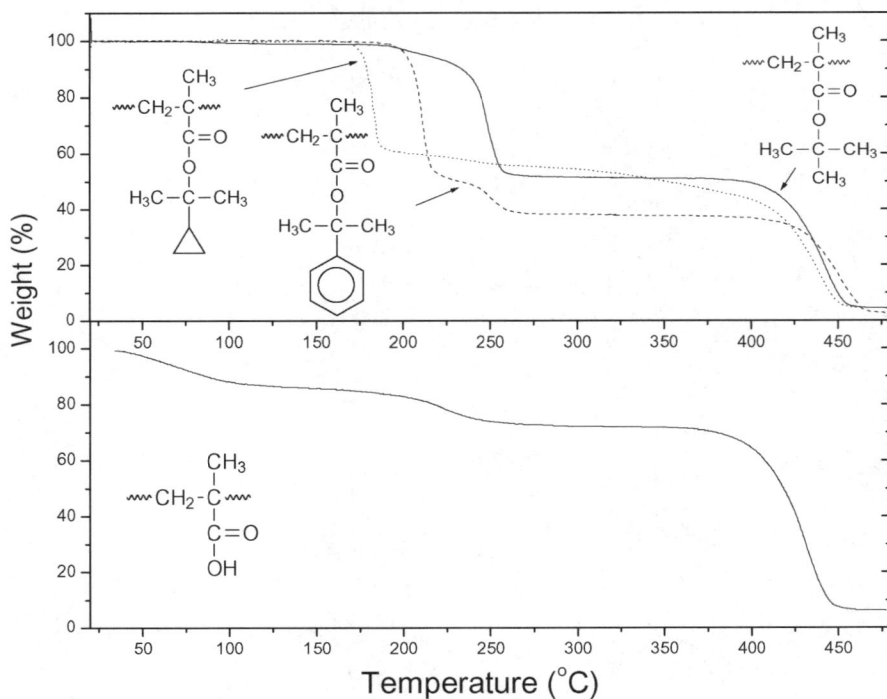

Fig. 23 TGA curves of polymethacrylates and poly(methacrylic acid) [121]

be used as a single layer deep UV resist polymer [119]. The benzoic acid polymer has a high T_g of 250 °C. These polymers can be readily prepared by radical polymerization [127]. Furthermore, methacrylates undergo anionic (sometimes living) polymerization. *tert*-Butyl 4-vinylbenzoate [128] and 4-*tert*-butoxystyrene [125] are compatible with living anionic polymerization.

4.1.3
Hydrolysis

Hydrolysis of trimethylsilyl (TMS) ether [129, 130] and alcoholysis of a tetrahydropyranyl (THP) [131, 132] group have been also employed in acid-catalyzed conversion to PHOST (Fig. 24) (or novolac). Another acetal-protected PHOST, poly[4-(1-phenoxyethoxy)styrene], was prepared by radical polymerization of the corresponding monomer and also by chemical modification of PHOST [133]. This acetal polymer produces a phenolic polymer and phenol upon acidolysis (Fig. 24).

Acetals and ketals have attracted a great deal of attention recently as protecting groups of PHOST due to their lower activation energies of deprotection than *t*BOC and *tert*-butyl esters. While the majority of chemical amplification resists require PEB to accelerate acid-catalyzed reactions, deprotection of ac-

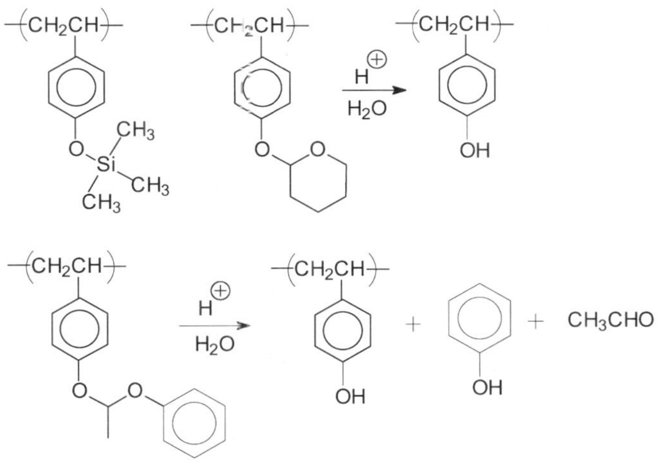

Fig. 24 Hydrolysis of protected poly(4-hydroxystyrene)

etal or ketal systems proceeds at room temperatures as soon as acid is generated by irradiation, thus minimizing some of the ill effects associated with PEB (as described in more detail later). However, such protecting groups tend to lack storage stability. The influence of the acetal structure on the hydrolytic stability (thermal stability in the presence of a phenolic hydroxyl functionality) has been recently investigated in detail [134].

4.1.4
Polyhydroxystyrenes

PHOST is the resin for deep UV lithography as much as novolac resins are for near UV lithography. All the advanced 248 nm chemical amplification resists, both positive and negative, are built on this structure at least in part. PHOST provides aqueous base developability, which is mandatory in today's semiconductor manufacturing, dry etch resistance, and high deep UV transmission. Because of its very important and unique role in chemical amplification resists and also in order to facilitate better understanding of the subsequent sections, this phenolic polymer is separately described.

Figure 25 presents several synthetic approaches to PHOST. The first PHOST commercially available in the early 1980s was prepared by Maruzen Petrochemical in Japan (Lyncur M) by direct radical polymerization of 4-hydroxystyrene without isolating or purifying the monomer after catalytic dehydrogenation of 4-ethylphenol, which is still the major source of PHOST today. The polymerization behavior of hydroxystyrene (vinylphenol) was investigated by Sovish [135], Overberger [136], and Kato [137]. As the phenolic OH group is a radical scavenger, this process yields low molecular weight polymers with M_w of a few thousands and a dark red color. Due to the presence of quinone structures,

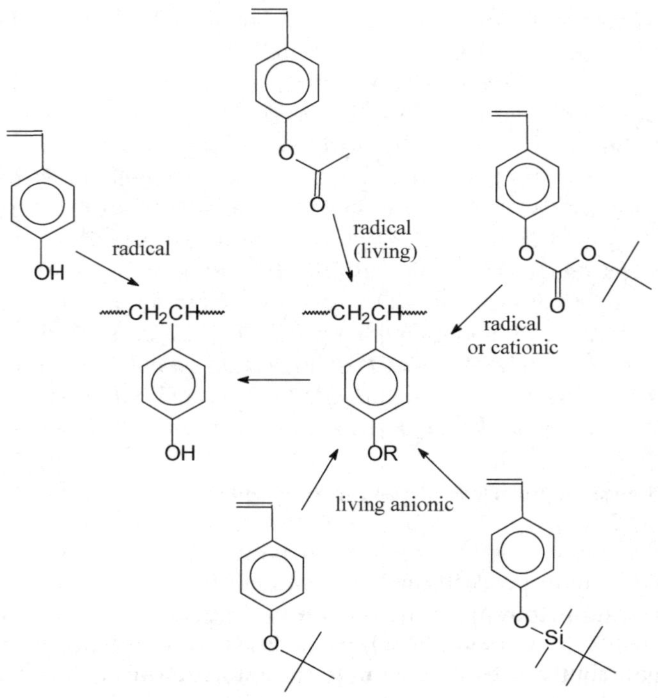

Fig. 25 Synthetic procedures of PHOST

the polymer absorbs significantly in the deep UV region. To improve its deep UV absorption, the polymer has been subjected to catalytic hydrogenation, resulting in high deep UV transmission, accompanied by introduction of a pendant cyclohexanol structure (Lyncur PHM-C). The PHOST most predominantly used today for deep UV resist formulation is this hydrogenated product. Storage stability of 4-hydroxystyrene monomer is poor as it undergoes thermal oligomerization rapidly, which was reported more than 40 years ago [135, 136]. An apparent activation energy of radical polymerization of 4-HOST was reported to be 18.0 kcal/mol [137]. Cationic polymerization of HOSTs has been reported by Kato [138].

The first transparent PHOST was prepared by thermolysis or acidolysis of PBOCST as mentioned earlier [97]. This monomer BOCST can be prepared by the Wittig reaction on a protected 4-hydroxybenzaldehyde with a high yield due to the good stability of the tBOC group toward a base catalyst [97]. PBOCST with M_n of >50,000 or <10,000 can be readily synthesized with AIBN, BPO, or other radical initiators. The protected polymer thus obtained can be readily converted to PHOST simply by heating the polymer powder to ~200 °C (with some C-alkylation, Fig. 16) or by treating the polymer with acid such as acetic acid or HCl in solution. The acid-labile tBOC group can be made compatible with cationic polymerization by using liquid sulfur dioxide as the solvent [97].

Another commercial procedure which has been worked out by Hoechst-Celanese (TriQuest, now DuPont) is radical polymerization of 4-acetoxystyrene followed by base hydrolysis with ammonium hydroxide. Their synthetic approach to the monomer is 1) Fries rearrangement of phenyl acetate to form 4-hydroxyacetophenone, 2) protection of OH with the acetyl group, 3) reduction to carbinol, and 4) dehydration. The Wittig process results in poor yield due to base hydrolysis of the protecting group during the olefination reaction. Radical polymerization of 4-acetoxystyrene has been reported by several groups [139–141] and NMR analysis of its polymer by Trumbo [142]. "Living" radical polymerization of 4-acetoxystyrene with a TEMPO-adduct as the initiator followed by base hydrolysis produced narrow polydispersity PHOST with M_w/M_n of 1.1–1.4 [143]. The slow polymerization process due to an extremely low radical concentration is a drawback. The narrow polydispersity polymers have been found to provide 10–20 °C higher glass transition temperature (T_g) than the conventional polymers with M_w/M_n of 2.0–2.4 [143]. The T_g of conventional PHOST changes from 140 to 180 °C when M_n is below 10,000 and plateaus at ca. 180 °C above M_n of 10,000 (Fig. 26) [143]. The radius of gyration of PHOST (M_w=8000) in a 200-nm-thick film was estimated by small angle neutron scattering (SANS) to be less than 3 nm [144].

Synthesis of monodisperse PHOST was first reported by Nakahama et al. [145], who demonstrated that 4-*tert*-butyl(dimethyl)silyloxystyrene undergoes living anionic polymerization with butyllithium as the initiator in tetrahydrofuran (THF) at cryogenic temperatures under high vacuum and that desilylation of the resulting polymer with HCl cleanly yields monodisperse PHOST. It

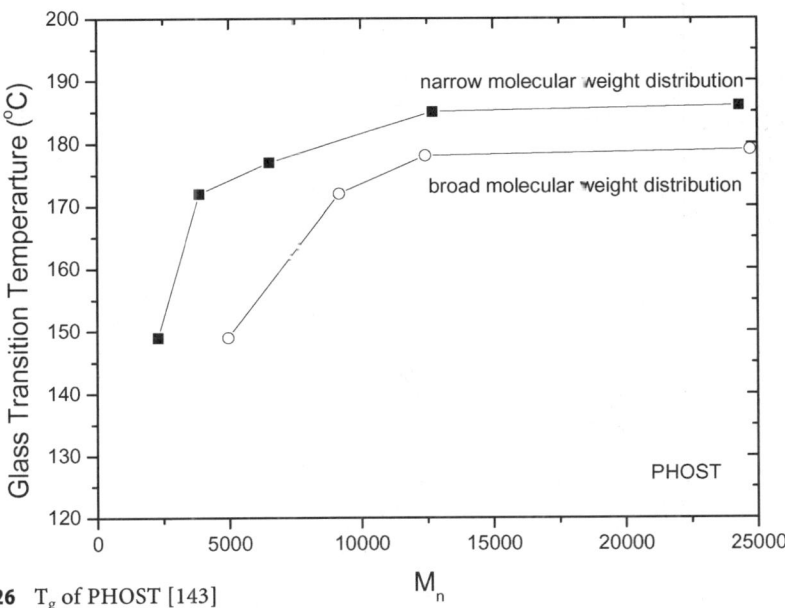

Fig. 26 T_g of PHOST [143]

has been later reported that replacement of THF with cyclohexane allows the anionic polymerization to be carried out in a flask at *room temperatures* even without drying the reagents [99]. Narrow polydispersity of <1.1 was readily achieved with quantitative conversion in a few minutes.

Monodisperse PHOST has become commercially available from Japan. The starting monomer for living anionic polymerization appears to be 4-*tert*-butoxystyrene, which may be synthesized by reacting a styrenic Grignard reagent with di-*tert*-butyl peroxide [125]. Deprotection requires a strong acid such as BBr_3.

PHOST dissolves much more rapidly than the novolac resins in an aqueous base developer and is thus more difficult to inhibit its dissolution by adding a small molecule such as DNQ. Molecular mechanics and dynamics calculations have indicated that the OH groups of PHOST are located on the periphery of the molecule and directed outward [146, 147]. Consequently, these OH groups do not take part in any significant intramolecular (intrachain) hydrogen bonding but have a great deal of intermolecular (interchain) coupling through hydrogen bonds, accounting for faster dissolution and higher T_g. The dissolution rate (Å/s) in a 0.21 N tetramethylammonium hydroxide (TMAH) aqueous solution of the PHOST film cast from ethyl lactate and baked at 150 °C for 60 s is plotted as a function of the number-average molecular weight (M_n) in Fig. 27 [143]. Below M_n of 10,000 the dissolution rate is a strong function of M_n and the dissolution rate of this phenolic polymer is very well correlated with

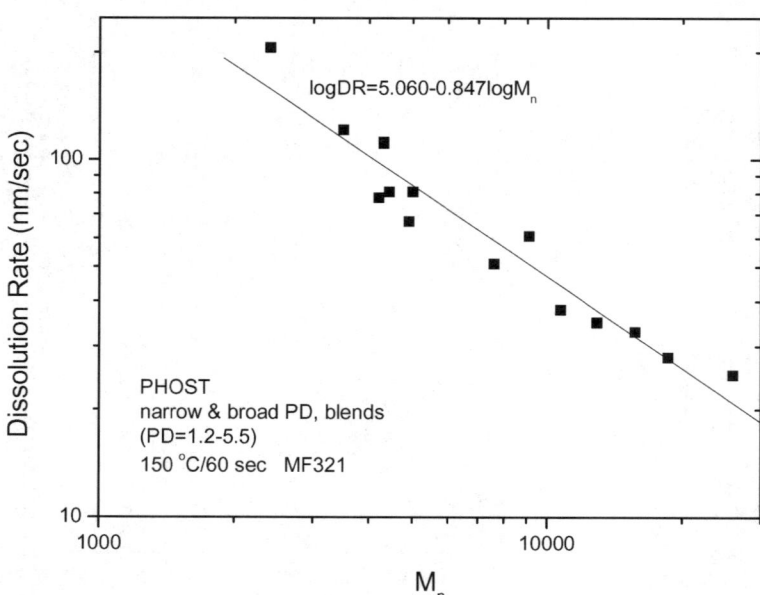

Fig. 27 Dissolution rate of PHOST in 0.21 N TMAH as a function of M_n [143]

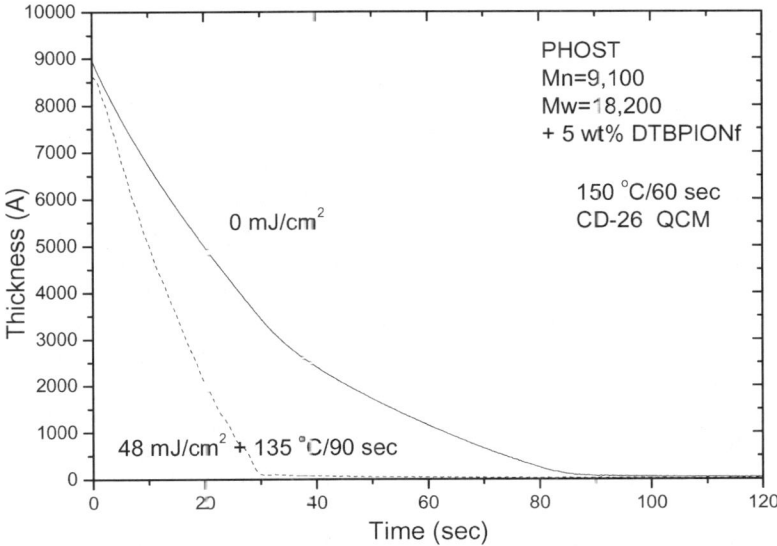

Fig. 28 Dissolution behavior of a PHOST containing 5 wt% of di-(*tert*-butylphenyl)iodonium nonaflate in 0.26 N TMAH [148]

M_n irrespective of the weight-average molecular weight (M_w) or polydispersity ($D=M_w/M_n$). As mentioned earlier, addition of a certain PAG to PHOST results in a reduced dissolution rate in aqueous base. A film of PHOST (M_n=9100 and M_w=18,200) containing 5 wt% of di-(*tert*-butylphenyl)iodonium nonaflate dissolves only at about 100 Å/s in industry standard 0.26 N TMAH solution as the dissolution kinetics curve measured on a quartz crystal microbalance (QCM, see later) indicates (Fig. 28) [148]. Exposure of the binary film to 48 mJ/cm² of deep UV radiation (followed by PEB at 135 °C) results in an increase in the dissolution rate to ca. 300 Å/s. A high exposure dose (a high acid concentration) and PEB at a high temperature (150 °C) can crosslink PHOST, which could show up as a negative behavior in positive imaging, reducing the contrast [148].

A solution presented to this fast dissolution problem is partial protection with an acid-labile lipophilic group, which will be described in detail later.

meta-Isomer, which dissolves more slowly than the *para*-PHOST in aqueous base, has been also prepared and used as a matrix resin [149–151]. *ortho*-Isomer is almost insoluble in an aqueous base solution. The dissolution rate of PHOST can be adjusted by partial protection with an acid-inert group such as methyl or isopropyl carbonate as well as with an acid-labile group as mentioned earlier. Substitution of the 3-position of 4-hydroxystyrene with a methyl group also reduces the dissolution rate significantly and such a phenolic polymer has been evaluated as a matrix resin for chemical amplification resists [149, 152, 153].

4.1.5
Copolymers

A copolymer approach can provide more flexibility to the resist design because all the necessary functions do not have to reside on one component. Today's advanced positive deep UV resists are exclusively based on this concept with 4-hydroxystyrene as one component. However, early copolymer systems and some of the 193-nm resists consisted of lipophilic components only. Incorporation of 4-acetoxystyrene to poly(4-*tert*-butoxycarbonyloxystyrene sulfone) has already been mentioned. This section deals with copolymer resists composed of lipophilic comonomers first and then the currently dominant hydroxystyrene copolymers. Co- and terpolymers for ArF excimer laser lithography will be described in a separate section.

Early examples of copolymer resists were alternating copolymers of styrene with maleimide (Fig. 29). In one case, the acidic NH group of maleimide was protected with *t*BOC [102, 103]. In another case, *N*-phenylmaleimide was copolymerized with 4-*tert*-butoxycarbonyloxystyrene. An alternating copolymer of styrene with *t*BOC-protected *N*-4-hydroxyphenylmaleimide was also synthesized as a resist polymer [101]. These copolymers tend to have high T_g. A number of alternating copolymers of *N*-substituted maleimides and 4-substituted styrenes have been synthesized and evaluated as chemical amplification resist resins [103]. A photochemical acid generator structure has been incorporated into the alternating copolymer through sulfonate-N linkage [52]. One example is a terpolymer of *N*-tosyloxymaleimide, *N*-*tert*-butoxycarbonylmaleimide, and 4-*tert*-butoxycarbonyloxystyrene, which generates tosic acid upon irradiation and thus functions as a one-component chemical amplification resist.

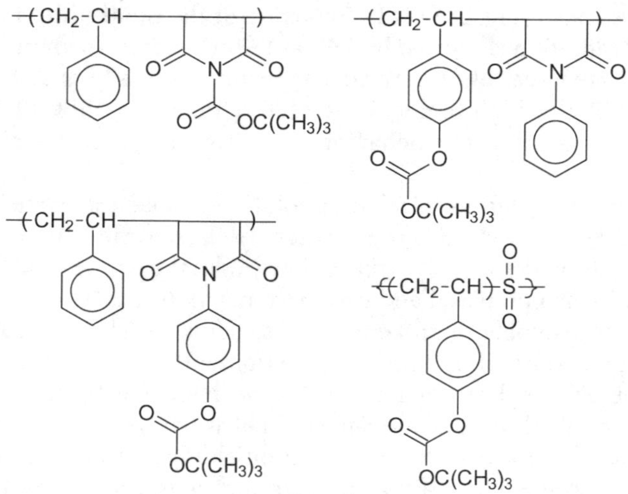

Fig. 29 Copolymers for deep UV lithography

The resist system that placed an emphasis on the importance of the copolymer approach was a copolymer of α,α-dimethylbenzyl methacrylate with α-methylstyrene (Fig. 30) [122], which is almost alternating due to the electron-poor and -rich natures of the comonomers and the lack of homopolymerizability of the latter. Necessary attributes are carried by several components of the resist. Styrene and *tert*-butyl methacrylate were also used as comonomers. What is noteworthy in this resist system is that dual tone imaging is still possible, although the concentration of the polar group, carboxylic acid, generated in the exposed film is always <50%. Generation of <35% of the polar group solubilizes the copolymer in aqueous base, which is a sharp contrast with the *t*BOC resist that requires >90% conversion for base solubility. Conversion of carboxylic ester to carboxylic acid provides a greater polarity or solubility change than that of phenyl carbonate to phenol, allowing positive or negative imaging at a lower degree of deprotection. Another important message of this work is that sufficient dry etch resistance can be achieved as long as the concentration of the aromatic component in the polymer is >50%. This copolymer approach has been extended to a 1:1 copolymer of tetrahydropyranyl methacrylate with styrene (Fig. 30) [154], which was reported to provide positive images by development with 0.08 mol/L TMAH aqueous developer. These two copolymer resists employed triphenylsulfonium hexafluoroantimonate as PAG. A block copolymer of benzyl methacrylate with styrene (Fig. 30) has also been synthesized by group transfer polymerization and evaluated for chemically amplified imaging [155].

The polymers employed to formulate today's deep UV positive resists are predominantly 4-hydroxystyrene copolymers (Fig. 31), which has resulted from realization that only a small amount of a protecting group can make the resin insoluble in aqueous base. For example, as mentioned earlier, more than 90% deprotection is needed to render the *t*BOC resist soluble in aqueous base, meaning that 10% protection sufficiently reduces the dissolution rate of PHOST. Thus, the partial protection approach has overcome the problem associated with the much faster dissolution rate of PHOST than novolac resins in aqueous base and allowed one to make chemical amplification positive resists compatible with the industry standard 0.263 N TMAH developer which had been selected for the

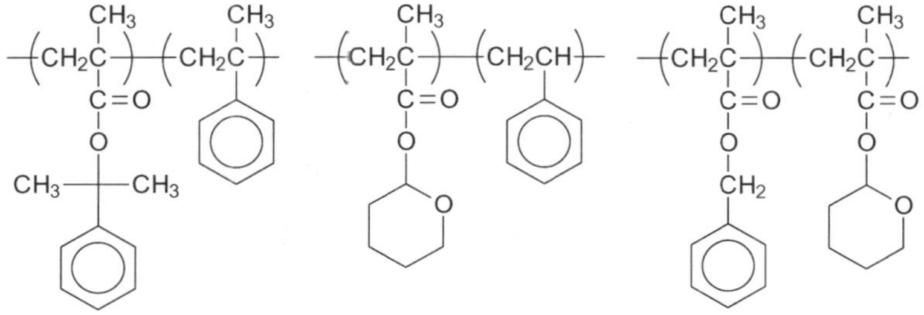

Fig. 30 Methacrylate copolymers for acid-catalyzed deprotection

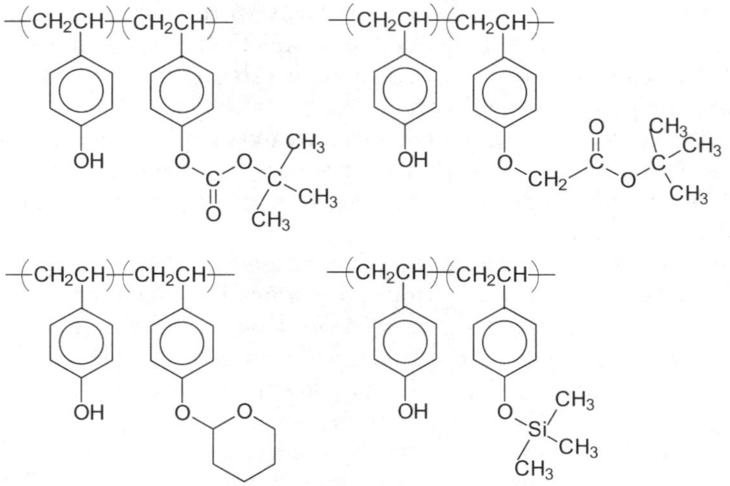

Fig. 31 Hydroxystyrene copolymers for positive deep UV lithography

novolac/DNQ resist. Although these copolymers could be prepared by direct copolymerization of protected monomers with 4-hydroxystyrene or by copolymerization of two protected hydroxystyrene monomers followed by selective deprotection (for example, radical copolymerization of di-*tert*-butylmalonylmethylstyrene with 4-acetoxystyrene followed by base deacetylation [156] and living anionic block copolymerization of hydroxystyrenes protected with two different silyl groups followed by selective desilylation [99] (Fig. 32), the primary synthetic methodology has been partial protection of PHOST [98]. The *t*BOC group has been the most popular protecting group in the copolymer approach [157, 158]. The concentration of the protecting group in the copolymer is typically 20–30% and is adjusted according to the developer strength and other factors. As mentioned above, the PHOST most frequently employed as the starting polymer for deep UV resists contains a significant amount of a pendant cyclohexanol structure. Narrow polydispersity PHOST has been also utilized to prepare copolymers [158]. Although partial protection typically employs a base to abstract the acidic phenolic hydrogen, protection with ketal requires acid. If the polymer reaction for partial protection is heterogeneous, blends could form instead of random copolymers because PHOST and fully protected polymer are phase-incompatible in many cases [159]. One such example is PBOCST and PHOST, which phase-separate even in solution in good solvents. Other protecting groups employed in the partial protection of the hydroxyl group include tetrahydropyranyl [132] and *t*BOC-methyl [124] groups.

The phenolic OH group is hydrogen-bonded with the ester or ether oxygen in these copolymer films. This interaction, however, reduces the thermal stability of the protecting groups significantly; the acid-labile group co-exists with an acidic phenolic OH group. Figure 33 presents TGA curves of PBOCST and

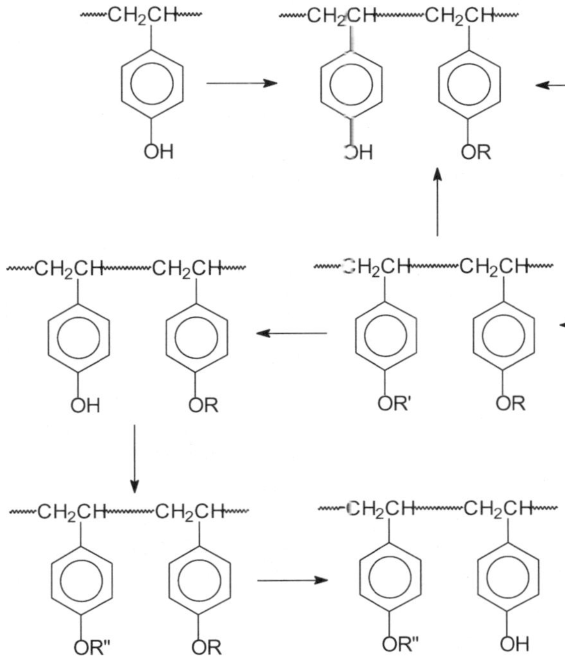

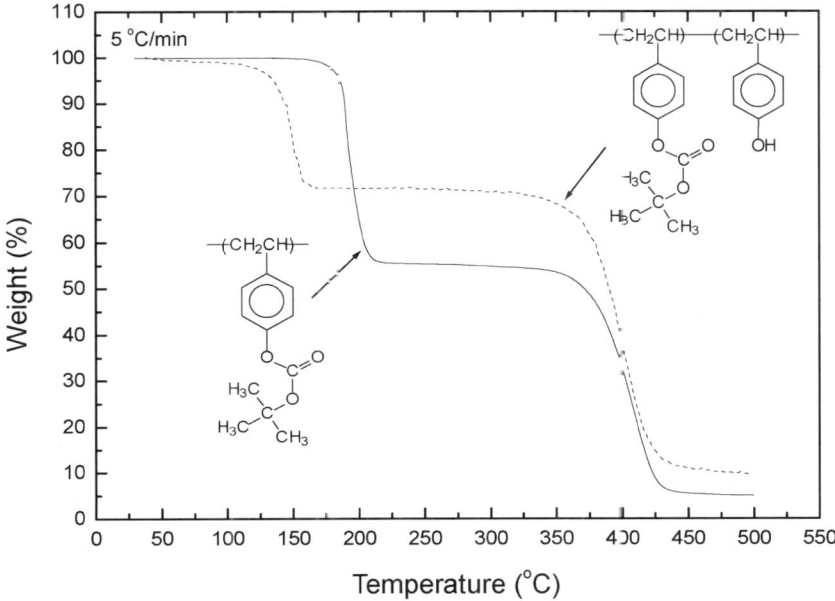

Fig. 32 Synthesis of hydroxystyrene copolymers

Fig. 33 TGA curves of PBOCST and partially protected PHOST [159]

PHOST partially-protected with tBOC as an example [159]. While the tBOC group in the homopolymer is stable to ca. 190 °C, introduction of a phenolic OH group results in as much as 60 °C reduction in the thermal deprotection temperature. Thus, the thermal deprotection is autocatalytic. All the protecting groups shown in Fig. 31 exhibit thermal deprotection at ca. 130 °C in the presence of the phenolic OH functionality. Thus, bake temperatures must be set below the decomposition temperature of 130 °C for these copolymer resists. Care must be exercised in handling these copolymers as their storage stability in the powder form or in resist formulation could be much shorter than the homopolymer systems. Acetals have been also employed as a protecting group of carboxylic acid.

An interesting concept has been proposed, which is an acid-labile crosslinked unit based on acetal [160]. When PHOST is reacted with vinyl ether in the presence of an acid catalyst, the phenolic hydroxyl groups are converted to acid-cleavable acetal groups. The molecular weight and molecular weight distribution remain unchanged, suggesting that no crosslinking has occurred. If PHOST is replaced with PHM-C (hydrogenated and containing cyclohexanol units) in the same reaction scheme, higher molecular weight fractions are produced, increasing the polydispersity from 2 to 3–7. The acetalization is accompanied by partial crosslinking through acetal linkage (Fig. 34) but the polymers thus produced are still soluble. It has been demonstrated by gel permeation chromatography (GPC) analysis of the exposed crosslinked acetal resist that

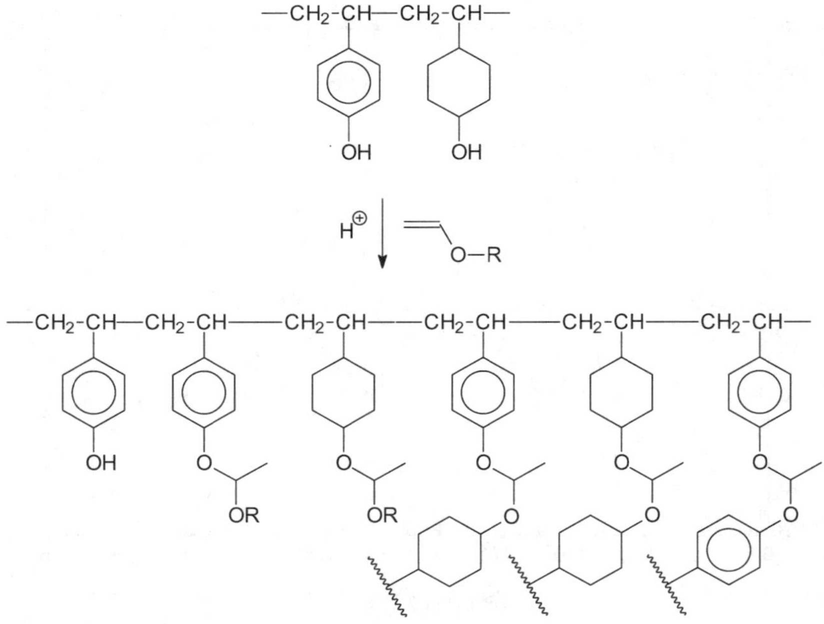

Fig. 34 Protection of PHOST with acetal accompanied by crosslinking

the acetal linkages including the crosslinks are cleaved by the photochemically generated acid to reproduce the original PHM-C. The crosslink increases the thermal flow resistance by more than 10 °C and reduces undesired dissolution in the unexposed regions.

The crosslinking of PHOST with vinyl ethers has been investigated by other workers also [161, 162]. In one application, PHOST was partially reacted with 2-chloroethyl vinyl ether to incorporate vinyl ether as a pendant group (Fig. 35) [161a]. The lithographic behavior of this system is complicated due to the diverse reaction modes of the vinyl ether groups. As illustrated in Fig. 35, when the concentration of photo-generated acids is low, the pendant vinyl ether groups react very quickly with acids to give small amounts of carbocations, which initiate cationic polymerization of remaining vinyl ether groups to form crosslinked networks, resulting in negative-tone imaging. Such negative-tone imaging mechanisms based on cationic polymerization will be discussed later

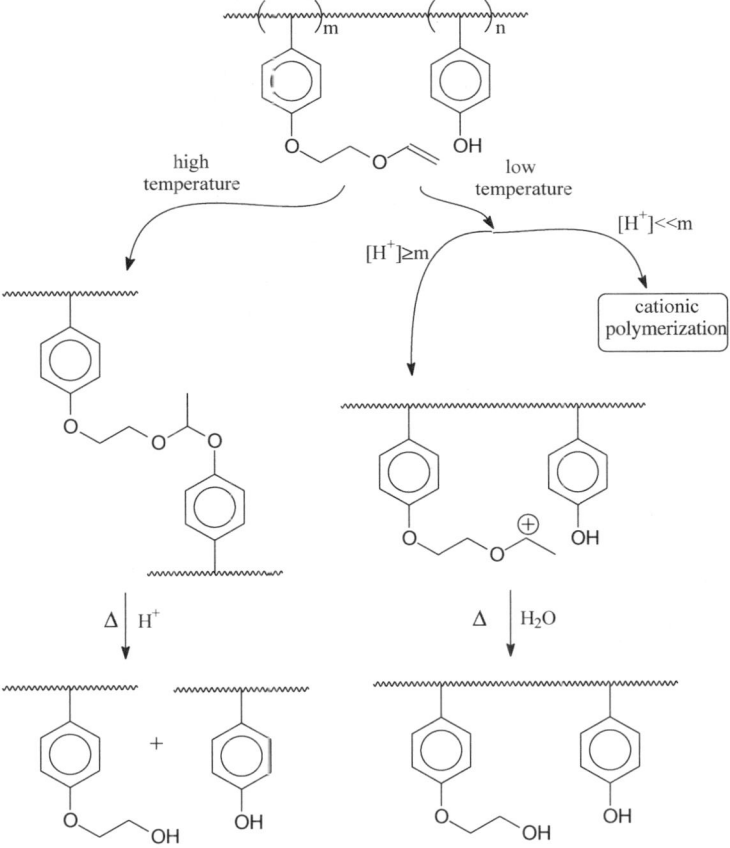

Fig. 35 Reactions involving vinyl ether

in more detail. When the concentration of acid is increased, the formation of carbocation is increased and the concentration of the vinyl ether groups decreased. As a result, the rate of cationic polymerization is reduced and the carbocation undergoes hydrolysis, exhibiting the positive-tone imaging behavior. The hydrolysis and cationic polymerization (crosslinking) are competing reactions. When phenol groups co-exist and the PAB temperature is high, the vinyl ether groups react with the phenolic hydroxyl groups to give crosslinked networks. As mentioned earlier, the acetal crosslinks are cleaved with photo-generated acid to produce PHOST and polystyrene with pendant alcohol. Thus, positive imaging results (Fig. 35). In other applications bulky vinyl ethers (Fig. 36) were blended with phenolic resins such as novolac [161b] and PHOST [161c–e]. Acid-breakable resins that can be converted to polyphenol fragments by acidolysis have been prepared through co-condensation reactions between polyphenol and aromatic multi-functional vinyl ether compounds (Fig. 36) [162]. A condensation reaction with a polyphenol/vinyl ether ratio of 100/50 gave a high molecular weight (M_w>5000) resin, which was insoluble in 0.26 N TMAH aqueous solution. The acid-catalyzed fragmentation

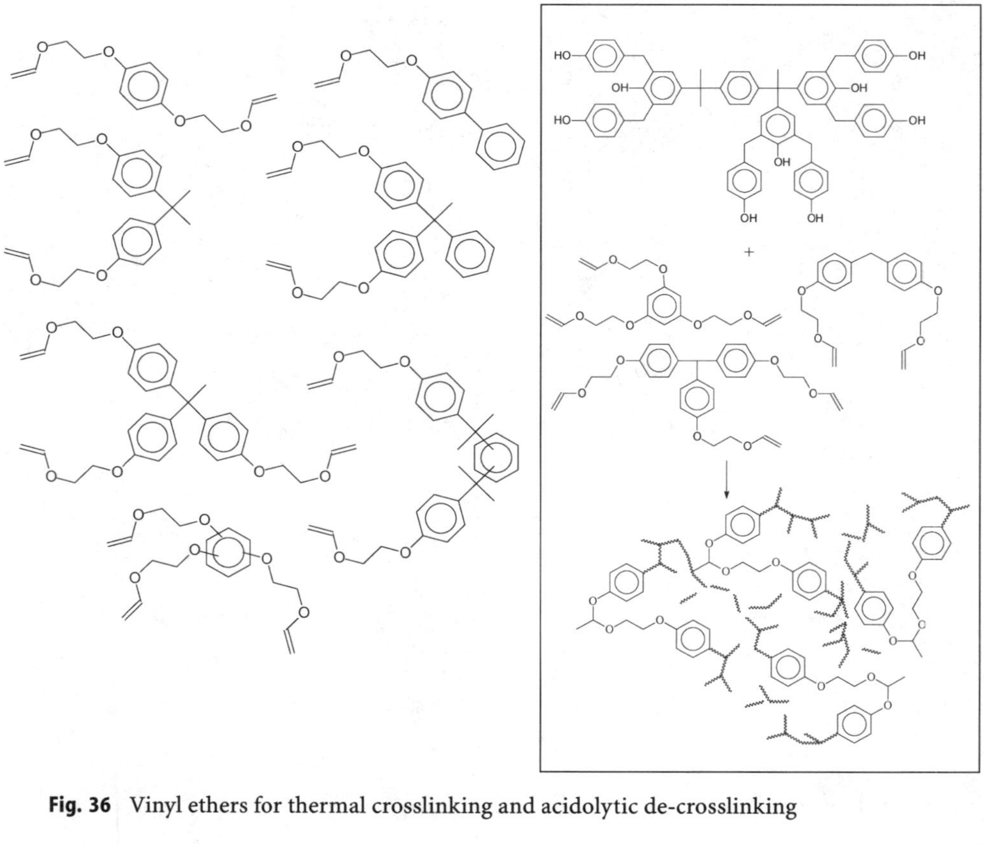

Fig. 36 Vinyl ethers for thermal crosslinking and acidolytic de-crosslinking

of the polymer in resist film was confirmed by GPC analysis and the resist exhibited a high resolution of 80 nm line/space patterns with a high sensitivity of 5.0 µC/cm^2 of 30 kV e-beam radiation.

A new type of copolymer resist named ESCAP (environmentally stable chemical amplification photoresist) has recently been reported from IBM [163], which is based on a random copolymer of 4-hydroxystyrene with *tert*-butyl acrylate (TBA) (Fig. 37), which is converted to a copolymer of the hydroxystyrene with acrylic acid through photochemically-induced acid-catalyzed deprotection. The copolymer can be readily synthesized by direct radical copolymerization of 4-hydroxystyrene with *tert*-butyl acrylate or alternatively by radical copolymerization of 4-acetoxystyrene with the acrylate followed by selective hydrolysis of the acetate group with ammonium hydroxide. The copolymerization behavior as a function of conversion has been simulated for the both systems based on experimentally determined monomer reactivity ratios (Table 1) [164]. In comparison with the above-mentioned partially protected PHOST systems, this copolymer does not undergo thermal deprotection up to 180 °C. Furthermore, as mentioned earlier, the conversion of the *tert*-butyl ester to carboxylic acid provides an extremely fast dissolution rate in the exposed regions and a large

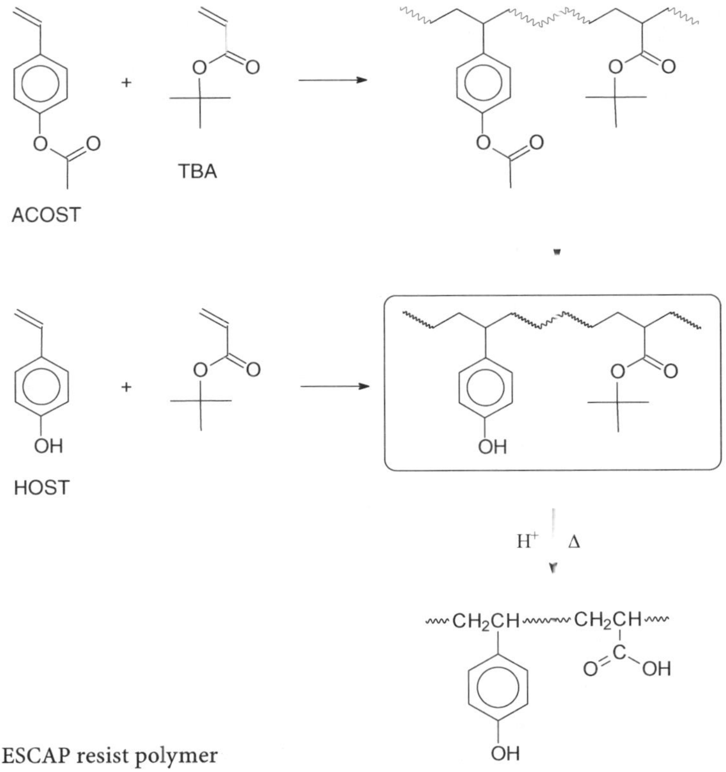

Fig. 37 ESCAP resist polymer

Table 1 Monomer reactivity ratios for HOST derivatives (M_1) and TB(M)A (M_2). Determined by nonlinear regression[a] [164]

	Fig. 1. 4ACOST	3ACOST	BOCST	4HOST	3HOST
Fig. 2 TBA	r_1=1.140 (1.119)		r_1=1.412 (1.208)	r_1=0.179 (0.179)[b]	
	r_2=0.297 (0.294)		r_2=0.248 (0.206)	r_2=0.202 (0.189)	
TBMA	r_1=0.792 (0.766)	r_1=0.903 (0.912)	r_1=1.162 (1.068)	r_1=0.159 (0.152)[c]	r_1=0.474 (0.482)
	r_2=0.603 (0.594)	r_2=0.571 (0.559)	r_2=0.623 (0.545)	r_2=0.410 (0.394)	r_2=0.517 (0.526)

[a] The values in parentheses calculated by the Kelen-Tüdös method.
[b] In isopropanol; r_1=0.342 (0.333) and r_2=0.218 (0.209) in PGMEA.
[c] In isopropanol; r_1=0.236 (0.234) and r_2=0.496 (0.494) in PGMEA.

developer selectivity. Highly branched *tertiary* alkyl leaving groups (Fig. 38) were employed in the ESCAP-type copolymer [165]. The thermal decomposition temperatures and activation energies were lower with more branched ester. The lithographic performance was reported to be higher for the highly branched ester due to its enhanced dissolution switching potential. Deep UV resist systems built on the ESCAP concept are the current workhorse in device

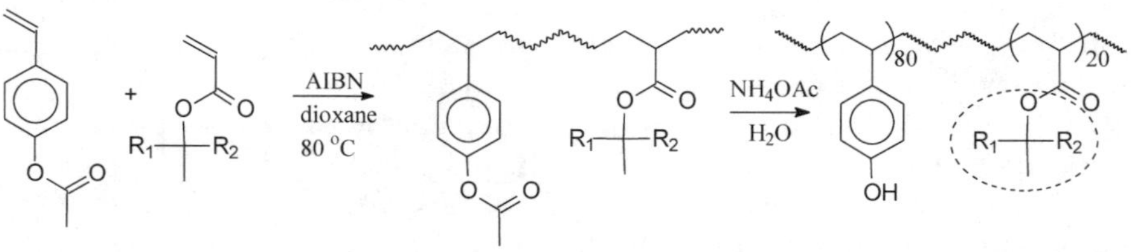

R_1	R_2	Molecular volume (Å^3)	M_w	M_w/M_n	T_g	T_d	OD_{248}	E_a (kcal/mol)
CH_3	CH_3	151	16600	1.7	162	212	0.15	41
CH_3	CH_3CH_2	166	16100	1.7	156	206	0.17	32
CH_3	$(CH_3)_2CH$	182	18100	1.7	151	204	0.15	30
$(CH_3)_2CH$	$(CH_3)_2CH$	215	17500	1.7	137	171	0.16	29

Fig. 38 Highly branched *tertiary* esters employed in ESCAP-based resists [165]

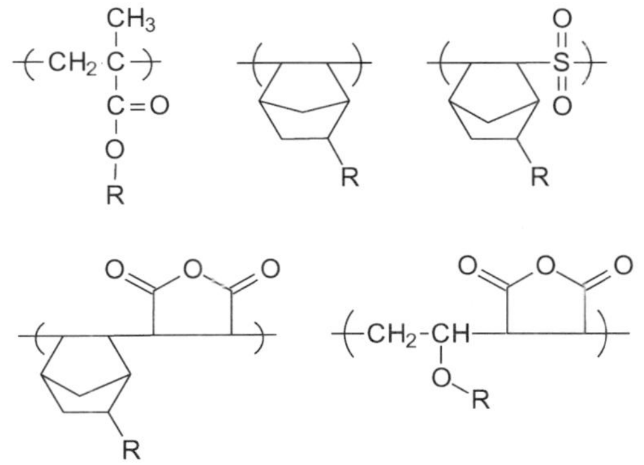

Fig. 39 193 nm positive resist platforms

manufacturing at 248 nm, being evaluated for EUV lithography, and are commercially available from resist companies, which will be described in more detail later.

Polymers for use in 193 nm lithography are co-, ter-, and tetra-polymers of 1) methacrylates, 2) norbornenes, 3) norbornene-maleic anhydride, 4) norbornene-sulfur dioxide, and 5) vinyl ether-maleic anhydride (Fig. 39). While 1), 3), 4), and 5) are prepared by radical polymerization, all-norbornene polymers 2) are synthesized by transition-metal-mediated addition polymerization [166–168]. Norbornenes (Fig. 40) are sluggish to undergo radical [168, 169] and cationic [170] polymerizations. Their ring-opening metathesis polymerization (ROMP, Fig. 40) [171] has never produced worthy resist polymers. The C=C double bonds introduced in the ROMP polymer backbone must be hydrogenated to reduce the 193 nm absorption and the ROMP polymers tend to have low T_g. However, the major problem for the ROMP polymers was their unacceptable swelling in aqueous base development. While polymethacrylate systems contain etch-resistant alicyclic structures in the ester side chain, norbornene-based systems carry the alicyclic unit in the backbone. Essentially all the 193 nm re-

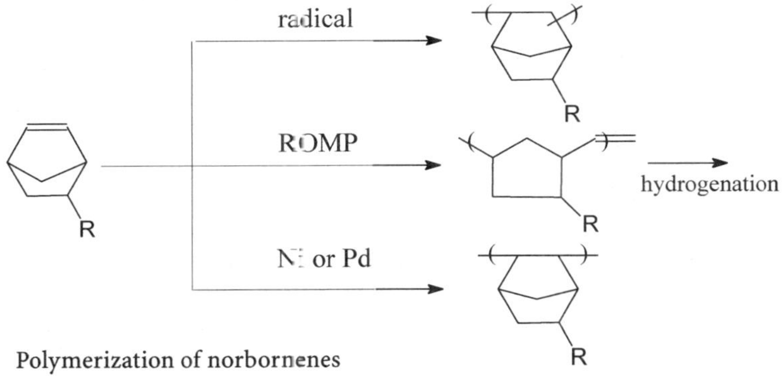

Fig. 40 Polymerization of norbornenes

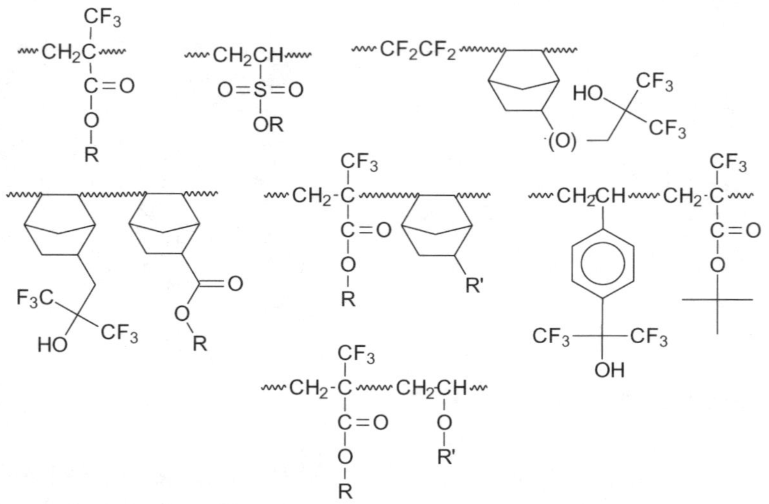

Fig. 41 157 nm positive resist platforms

sist polymers contain a small amount of carboxylic acid in place of opaque phenol groups. A more detailed description of 193 nm resist polymers is provided later.

As described in detail later, the polymers for use in 157 nm lithography universally contain fluorine because only fluoropolymers provide low enough absorption at 157 nm. The acid group of choice is hexafluoroisopropanol, which has a pKa similar to that of phenol, as mentioned earlier. Several platforms are available (Fig. 41): 1) tetrafluoroethylene-norbornene, 2) 2-trifluoromethylacrylate-norbornene, 3) 2-trifluromethylacrylate-styrene, 4) 2-trifluoromethylacrylate-vinyl ether, 5) all-norbornene, 6) methacrylate, and 7) cyclopolymers.

4.1.6
Blends and Dissolution Inhibitors

The novolac/DNQ design has been adopted into chemical amplification resists. Instead of the photoactive dissolution inhibitor, a radiation-inert but acid-sensitive dissolution inhibiting molecule is added to a phenolic polymer together with PAG. The small lipophilic carbonate, ester, or ether molecules inhibit the dissolution of the phenolic polymer in aqueous base through an hydrogen-bonding interaction in the unexposed regions and are converted to base-soluble phenolic or carboxylic acid groups in the exposed regions. Since the acidic molecules generated in the exposed area of the phenolic film are sometimes dissolution promoters or no longer dissolution inhibitors, the exposed regions dissolve rapidly in aqueous base, providing positive tone images. Since, as mentioned earlier, it is much more difficult to inhibit the dissolution of PHOST, novolac resins were extensively used as the matrix resin for small *tert*-butyl

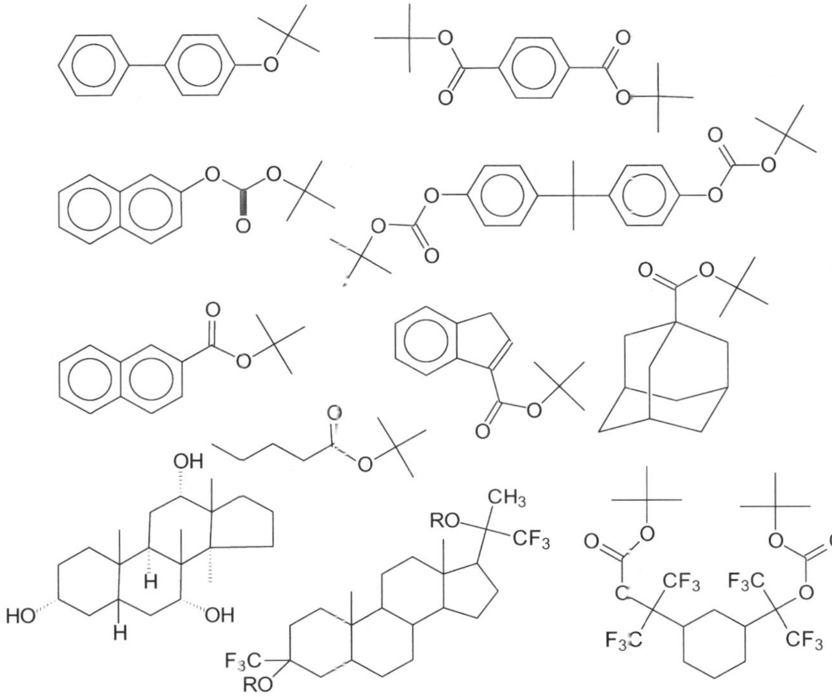

Fig. 42 Dissolution inhibitors

ester, carbonate, and ether dissolution inhibitors (Fig. 42) [172–174]. Thus, the three-component blend systems consisting of a purely phenolic resin, small dissolution inhibitor, and PAG were not suitable for deep UV lithography due to high absorption of the novolac resin but were pursued more seriously in X-ray lithography. In fact, Hoechst developed a three-component RAY-PF resist for X-ray, which employed a small acetal compound as an inhibitor of a novolac resin [174]. The dissolution inhibiting acetal is catalytically hydrolyzed in the phenolic matrix resin to alcohol and aldehyde which are not dissolution inhibitors. Due to its lower activation energy of deprotection, the acetal system undergoes acid-catalyzed decomposition to almost completion at ambient temperatures within 30 min in the presence of water.

The primary function of PAG is to generate acid upon irradiation. However, other properties of PAGs must be taken into consideration in designing new resists. One such property is a dissolution inhibition effect. Some of the PAGs are extremely good dissolution inhibitors of phenolic resins, much better than DNQ, as mentioned in 3 [44]. One good example is sulfonium and iodonium salts [44, 89]. Triphenylsulfonium and diphenyliodonium salts can inhibit dissolution of a novolac resin very efficiently at a loading much lower than that needed for DNQ. Furthermore, a two-component system comprising a novolac resin and few percent of triphenylsulfonium hexafluoroantimonate provides positive images upon development with aqueous base (without PEB) at a low 254 nm dose of ~20 mJ/cm^2. Thus, the sulfonium salt behaves just like DNQ as mentioned earlier. Other dissolution inhibiting PAGs include disulfones [49] sulfonyl-substituted diazomethanes [175], and some sulfonate esters [176].

The deprotection chemistry has been incorporated into the acid generator structure itself [177]. Phenolic hydroxyl groups pendant from triphenylsulfonium salts were protected with *t*BOC (Fig. 43). This dissolution inhibiting PAG mixed with PHOST becomes base soluble through photochemically-induced acid-catalyzed deprotection and thus the exposed area dissolves rapidly in aqueous base, which was named SUCCESS and promoted by BASF. A similar approach has been later reported on *o*-nitrobenzyl sulfonate acid generators, in which a *tert*-butyl ester was attached to the benzene ring for acid-generation and acid-catalyzed deprotection on one molecule (Fig. 43) [178].

Although it is generally difficult to find polymer pairs that mix homogeneously, PHOST partially protected with THP or *tert*-butoxycarbonylmethyl group has been reported to be miscible with a novolac resin or PHOST and to function as a dissolution inhibitor [124, 132]. Another acid-labile polymeric dissolution inhibitor is a terpolymer of *tert*-butyl methacrylate (TBMA), methyl methacrylate (MMA), and methacrylic acid (MAA), which is miscible with a novolac resin [179] while PTBMA is not [159]. This methacrylic terpolymer is only marginally miscible with PHOST but more compatible with C- and

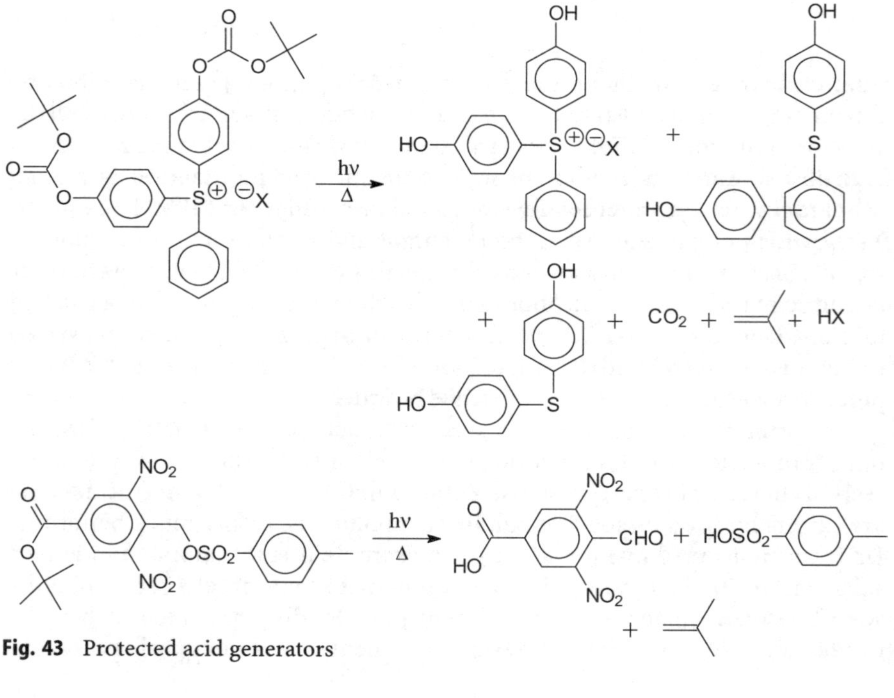

Fig. 43 Protected acid generators

O-methylated PHOST [179]. This terpolymer was originally developed as a chemically amplified laser resist for circuit board application [180] and then as a single layer 193 nm positive resist [181], which will be described in more detail later. Another interesting three-component approach is the use of a *N*-acetal polymer as a dissolution inhibitor of poly(3-methyl-4-hydroxystyrene) [182]. A deep UV resist consisting of poly(3-methyl-4-hydroxystyrene-*co*-4-hydroxystyrene), poly(*N*,*O*-acetal), bis(arylsulfonyl)diazomethane, and a photobase was reported from Hoechst (currently Clariant). The function of the photobase is described later. A copolymer of 4-hydroxystyrene with styrene was also employed as a matrix resin.

Hexakis(4-*tert*-butoxycarbonylphenoxy)cyclotriphosphazene was employed as a dissolution inhibitor of a novolac resin for a design of a three-component positive resist [183].

The copolymer and blend approaches have been combined to design three-component chemical amplification positive resists, which consist of PHOST partially protected with an acid-labile group (see Fig. 31), an acid-labile dissolution inhibitor, and PAG [124, 157, 158]. In this scheme both the matrix polymer and

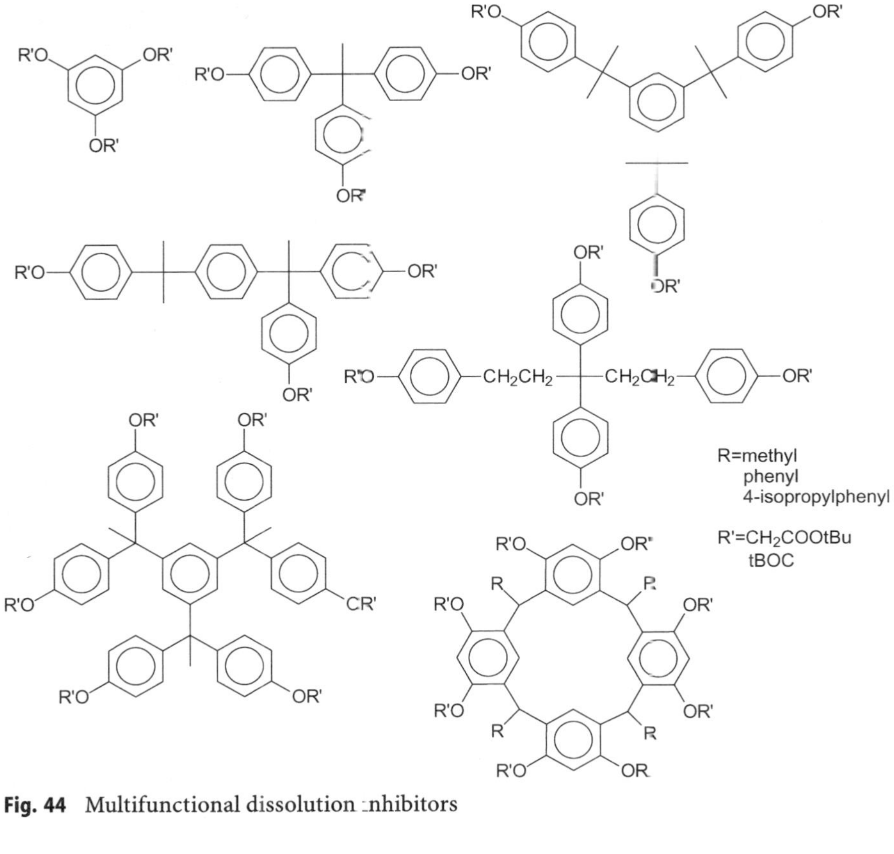

Fig. 44 Multifunctional dissolution inhibitors

the small lipophilic compound are converted to base soluble forms by photochemically-triggered acidolysis.

Multi-functional phenolic compounds have been protected with tBOC or tert-butoxycarbonylmethyl to serve as acid-labile dissolution inhibitors. Calixarenes [184], cyclic oligomers produced by condensation of phenols with aldehydes, have been investigated as lithographic resist materials. For example, calix[6]arene derivatives, prepared by condensation of p-cresol and formaldehyde, have been reported as high resolution negative electron-beam resist (not chemically amplified) [185]. Calix[4]resocinarenes, cyclic tetramers which can be synthesized by condensation of resorcinol with aldehydes (Fig. 44), have eight OH groups pointing outward. Protected calx[4]resorcinarenes exhibited a stronger inhibition effect than a bifunctional dissolution inhibitor such as bisphenol A protected with tBOC (Fig. 40) [186]. An inhibitor molecule capable of efficiently shielding the hydrophilic channels in the polymer matrix is the most effective in dissolution inhibition. A series of multi-functional (1–6) dissolution inhibitors were evaluated in novolac resins and in PHOST partially protected with tBOC and the effects of the number of functionality and degree of protection on the inhibitor molecule have been investigated (Fig. 44) [187]. An interesting new design of acid-labile dissolution inhibitors has been proposed, which involves base hydrolysis of deprotected inhibitor molecules based on bisphenol in the exposed regions during aqueous base development (Fig. 45) [188]. Acid-catalyzed deprotection of the tBOC group takes place upon PEB, which renders the exposed area more soluble in aqueous base. Furthermore, the

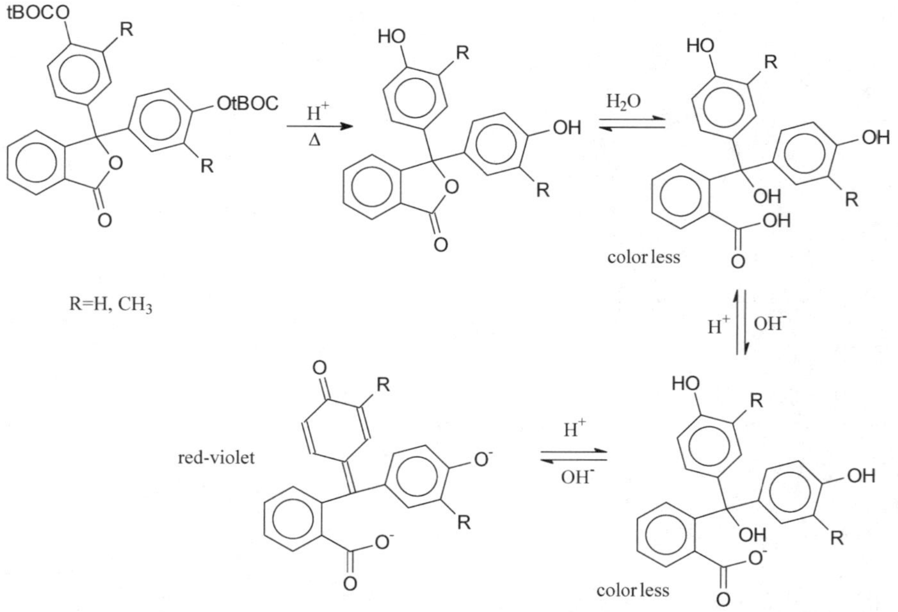

Fig. 45 Dissolution inhibitor that undergoes selective base hydrolysis

lactone ring in the 1-(3H)-isobenzofuranone structure is hydrolyzed during the development process to generate carboxylic acid, which enhances the dissolution rate even more, providing a high development contrast (Fig. 45). The ring-opening during development was confirmed by measuring the UV-visible absorption of a solution containing an exposed resist. A strong absorption appeared at 568.5 nm, corresponding to the colored form of o-cresolphthalein.

This three component approach has become more popular in the design of 193 nm positive resists as described later.

4.1.7
Deprotection Involving Ring-Opening (Mass Persistent Resists)

The acid-catalyzed deprotection mechanism is in general designed to lose volatiles and thickness in the exposed areas. As excessive shrinkage is not desirable, the concentration of the protecting group is kept low to minimize the shrinkage but high enough to maintain a good development contrast. The very first attempt to eliminate or minimize the thickness loss due to liberation of volatile products was to employ *cyclic* esters (lactones) and carbonates [189]. Poly(3-methylenephthalide) is a cyclic analog of poly(α-acetoxystyrene) and poly(*tert*-butyl 4-vinylbenzoate), which has a lactone ring with a *tertiary* carbon located in the polymer backbone (Fig. 46). Similarly, poly(4-methylene-4H-1,3-benzodioxin-2-one) can be considered as a cyclic analog of poly(*t*BOC styrene), a cyclic carbonate bearing a *tertiary* carbon in the backbone. The α, α-cyclization reduces the steric hindrance significantly and these α-substituted styrenes undergo radical homopolymerization very readily to form heterotactic (almost completely atactic) polymers [190]. The cyclic carbonate polymer was confirmed by TGA, IR, UV, and NMR to be converted upon baking at about 200 °C to poly(o-hydroxyphenylacetylene), releasing carbon dioxide [189]. Only one mole of gas is released in this case, while the liberated olefin is now a part of the backbone. However, there was no evidence for rapture of the lactone ring; an unfavored conformation for thermal ring opening and equilibrium between the cyclic benzoate and carboxylic acid in acidolysis. The ring might open upon reaction with acid but might close rapidly.

More lactones were later evaluated for the design of mass persistent photoresists [191], but acid-catalyzed ring opening was not observed as the equilibrium favored the ring-closed form (Fig. 46). A spiro linkage was introduced into the lactone ring to induce strain. Norbornene bearing a lactone ring with two spiro linkages was synthesized and polymerized by ring opening metathesis polymerization. No acid-catalyzed ring-opening was observed. A large ring lactone was attempted but failed. A Fittig bis-lactone containing two lactones and an acetal moiety did not work either (Fig. 46). Four-membered β-lactones were selected next and a model acidolysis of the strained lactone in solution indicated that the ring indeed opened upon treatment with p-toluenesulfonic acid (Fig. 46). Styrene bearing a β-lactone was synthesized and copolymerized with 4-vinylphenylacetic acid and styrene substituted with hexafluoroiso-

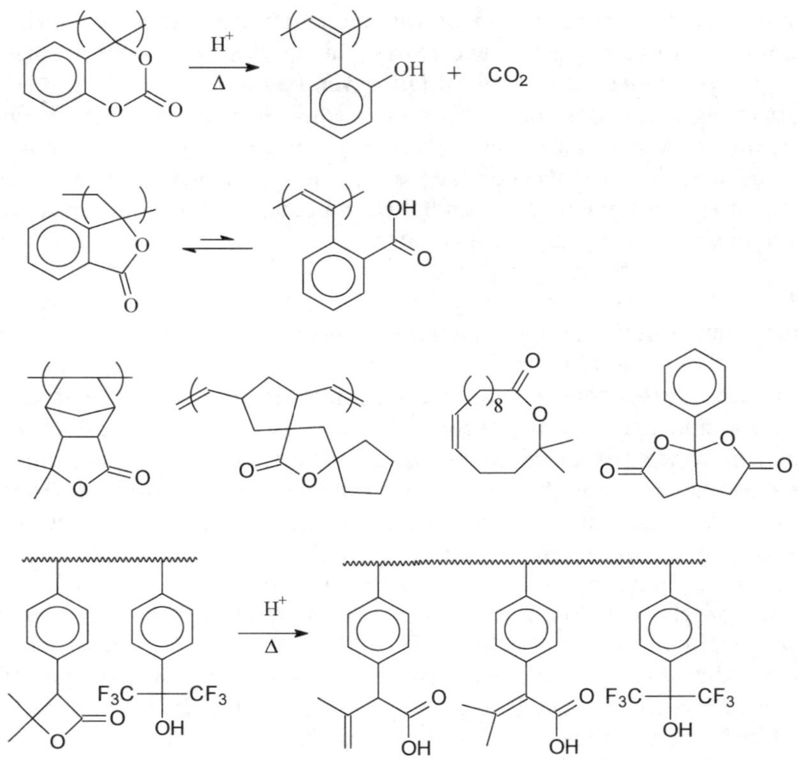

Fig. 46 Mass persistent resist design

propanol. The latter copolymer developed to give a positive image by 248 nm exposure.

4.1.8
Dendritic Polymers and Small Amorphous Materials

In addition to the linear and branched phenolic polymers, hyperbranched dendritic polymers and low molecular weight amorphous materials have been prepared and evaluated for use in chemically amplified resists. Select examples are presented in Figs. 47 and 48, respectively. Dendrimers, first reported by Tomalia [192] in the early 1980s, are nearly completely monodisperse and spherical molecules, with a core of atoms at the center that branch out radially. No chain entanglement is feasible. The small spherical structure with no entanglement has prompted resist chemists to design resist materials based on dendrimers as they have a potential for higher resolution and smaller edge/ surface roughness in comparison with linear polymers [193–195]. However, the dendrimer synthesis is time-consuming and costly, involving exhaustive step growth reactions. Thus, dendritic hyperbranched polymers have been evalu-

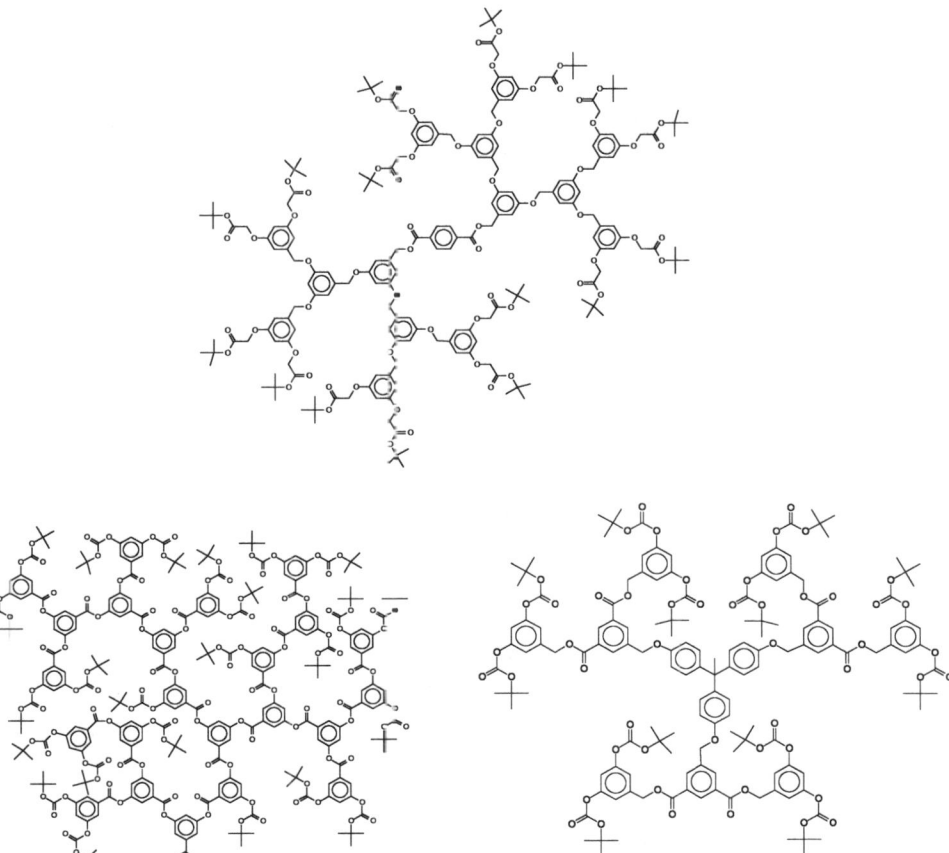

Fig. 47 Dendritic polymers

ated as resist materials instead of true dendrimers as such hyperbranched polymers can be prepared much more easily, sometimes in one pot. A chemically-amplified resist based on a dendritic polymer in roughly 5 nm in diameter has been reported to exhibit much smaller line edge and surface roughness after development than a conventional resist [194]. However, these early dendritic resists were typically developed in a negative mode (see tBOC resist above) and suffered from low T_g, low sensitivity, and poor adhesion.

The high resolution imaging of a calixarene negative e-beam resist (not amplified as mentioned earlier) [185] has spawned a great deal of interest in such low molecular weight amorphous materials because the polymer size would eventually limit the resolution of conventional resists as the feature size shrinks to <50 nm. Low molecular weight organic compounds tend to crystallize and do not form amorphous homogeneous films upon spin-coating from solution, limiting their application methods to vacuum deposition. "Amorphous mole-

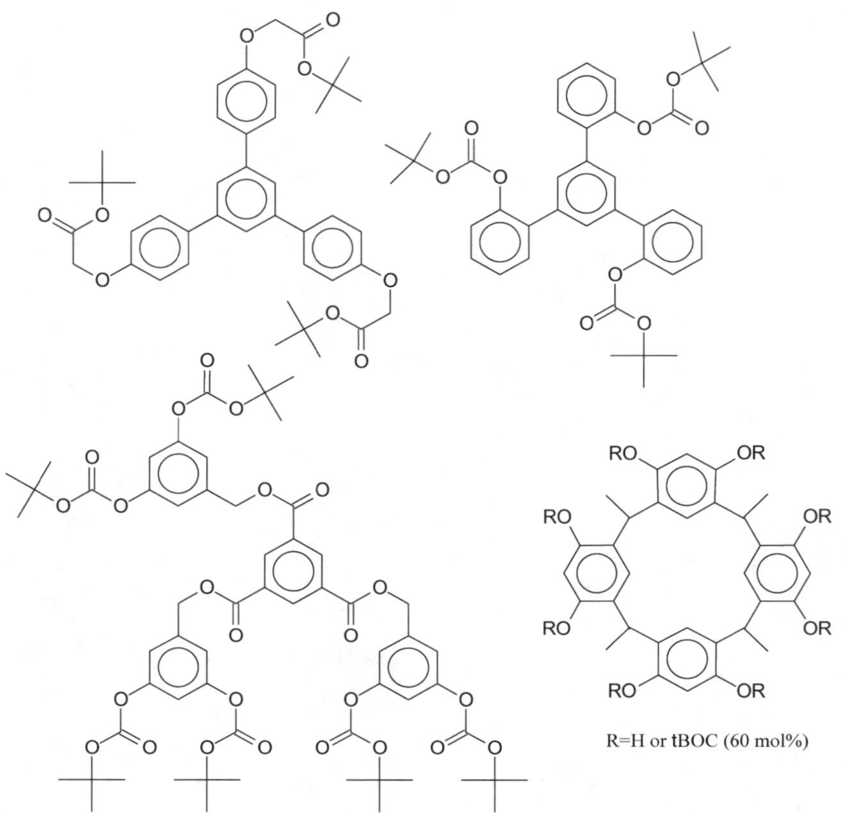

Fig. 48 Small amorphous materials

cules," low molecular weight amorphous materials [196], can be cast into transparent tough films. Such amorphous molecules can be designed by increasing the number of conformers through lowering the symmetry of non-planer molecules. Both positive- and negative-type chemical amplification resists have been reported [197, 198]. However, as the linking unit had a benzoate structure, which absorbs strongly below 300 nm, lithographic imaging was carried out by using near UV or e-beam irradiation.

As a viable approach to achieving resolution below 50 nm with minimal edge roughness, more research efforts are likely to be directed toward dendritic and amorphous molecular resists.

4.1.9
Evolution of 248 nm Positive Resists

The acid-catalyzed deprotection has become the exclusive foundation for all the modern positive resists. The first generation system was built on lipophilic homopolymer or copolymer such as PBOCST (*t*BOC resist), P(BOCST-SO$_2$) (CAMP resist), and P(BOCST-ACOST-SO$_2$) (CAMP6 resist). The very first chemical amplification resist, IBM's *t*BOC resist, was employed in its negative mode, using anisole as a developer, in production of 1 Mbit DRAMs in IBM [43]. Positive imaging with aqueous base as a developer encountered with many problems. It has later become clear that a highly lipophilic resists are difficult to develop in aqueous base due to poor wettability of such hydrophobic films. IBM's APEX resist is an example of the second generation deep UV resists and is based on PHOST partially protected with *t*BOC [199]. This copolymer approach utilizing a high concentration of an acidic group for aqueous base affinity has become the rule of the design of positive resists (Fig. 31). The APEX resist was the first successful chemical amplification positive resist employed in device manufacturing in the entire semiconductor industry and is still commercially available.

However, the shift from the DNQ/novolac resist to chemical amplification was not evolutionary but revolutionary and thus many new problems unique to chemical amplification resists have surfaced. These problems were typically related to the catalytic nature of the imaging mechanisms and were serious enough to retard or even to stop implementation of chemical amplification resists in general manufacturing. Significant progress has been made in solving or minimizing the problems, resulting in industry-wide use of the chemical amplification resists in manufacturing.

From the time of the inception, chemical amplified positive resists suffered from formation of a surface-insoluble layer, which typically showed up as a skin or T-top profile (Fig. 49). The problem has been ascribed to contamination by

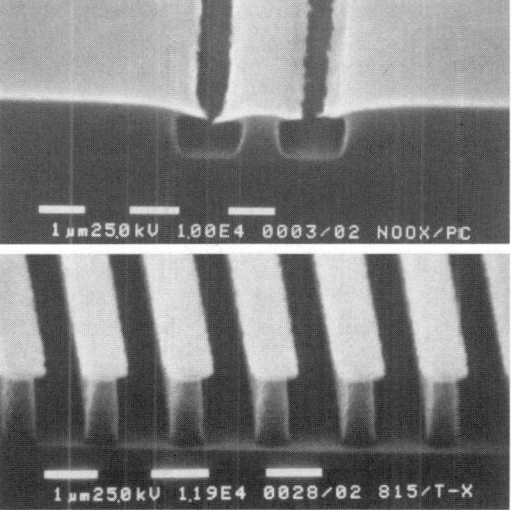

Fig. 49 Skin and T-top profiles in positive-tone images [148]

a trace amount, on the order of 10 ppb, of airborne basic substances such as N-methylpyrrolidone (NMP) and amines [200]. A minute amount of base absorbed by the resist film upon standing after coating (especially after exposure) interferes with the desired acid catalyzed reaction, resulting in formation of a surface insoluble layer in positive resists and in linewidth shift in negative resists. Because the problem is most serious when wafers are not baked immediately after exposure, the skin formation is called a postexposure delay (PED) effect.

The activated carbon filtration system [200] presented in Fig. 50 was employed in the unequivocal identification of the cause of the skin formation [201] and has presented an engineering solution to the formidable problem. While wafers stored in the filtered air printed normally, wafers stored in an airstream containing only a few ppb of NMP exhibited severe skin formation (Fig. 51). Various basic chemicals used in wall paints, cleaning liquids, etc. can contaminate chemical amplification resists. The activated carbon filtration of the air enclosing the coated wafers has permitted IBM to manufacture millions of 1 Mbit DRAMs using the negative tBOC resist by deep UV lithography in the mid-1980s [43] and remained as an important stabilization technique.

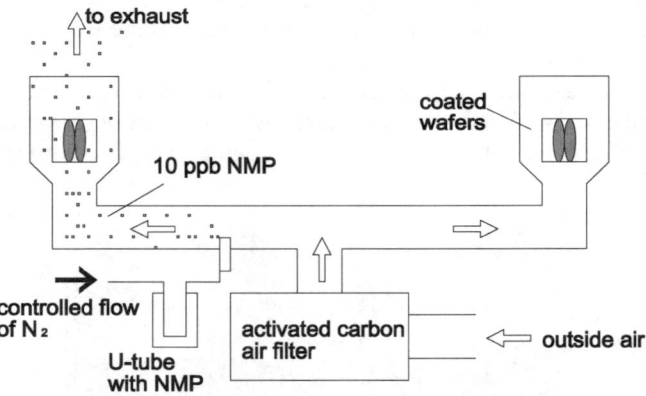

Fig. 50 Activated carbon air filtration

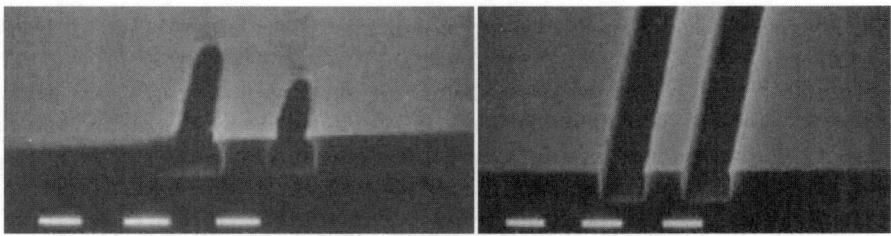

Fig. 51 Effect of activated carbon filtration on image profile: *left* – wafer stored in an airstream containing a minute amount of NMP; *right* – wafer stored in a filtered air [210]

In addition to the air filtration, several techniques to stabilize chemical amplification resists toward airborne contamination have been proposed:

- Activated carbon filtration [200]
- Protective overcoat [202–204]
- Additives [205, 206]
- Delayed acid generation [30, 207]
- Associating PAG [208]
- Low activation energy of deprotection [209]
- Reduction of free volume [163, 210]

Application of a protective overcoat to seal off airborne contaminants was also a popular approach initially. Although many polymers, lipophilic and hydrophilic, have been evaluated as a topcoat, water-soluble poly[(meth)acrylic acid] is most commonly employed, which can be cast from a water solution without interfacial mixing with the resist layer and can be removed during aqueous base development. However, it has been reported that a poly(acrylic acid) overcoat allows diffusion of water, which reportedly contaminates a chemical amplification resist [211]. Poly(α-methylstyrene) has been recommended as a good barrier against both airborne base and water [211].

Some stabilizing additives, mostly basic, have been incorporated into resist formulation, which include NMP, diphenylamine, 4-aminophenol, 2-(4-aminophenyl)-2-(4-hydroxyphenyl)propane, 2-aminobenzoic acid, tri(4-dimethylaminophenyl)sulfonium triflate, dicyclohexylammonium p-toluenesulfonate, etc. The additives must stay in the resist film during PAB with minimal evaporation and be distributed effectively throughout the film. In addition to the PED stabilization, these basic compounds are added to formulation in an attempt to reduce acid diffusion during PEB. In fact all the modern chemical amplification resists contain a small amount of base (called quencher) to enhance the resolution. This resolution enhancement effect of a base additive cannot be over-exaggerated. Such a base additive can also improve the storage stability of highly acid-sensitive resists such as the ones based on low activation energies of deprotection. An attention must be paid to selection of a base for a specific PAG as certain PAGs can be readily decomposed by added base. An interesting concept has been proposed, which is the use of a photobase, a base that can be decomposed to a neutral compound by UV irradiation [212]. A photochemically generated acid diffuses from the exposed to unexposed regions of the film upon standing after exposure, which results in a linewidth shift. A photobase in the unexposed area neutralizes the incoming acid, thus maintaining the intended exposed/unexposed boundary while the base is decomposed in the exposed area (Fig. 52). An example of photobase is triphenylsulfonium hydroxide.

Chemical amplification resists require PEB in general to accelerate acid-catalyzed reactions and thus the acid generated by irradiation could be deactivated (by airborne contamination) upon standing before PEB. Then, generation of acid when it is needed (at the time of PEB) could eliminate/minimize the

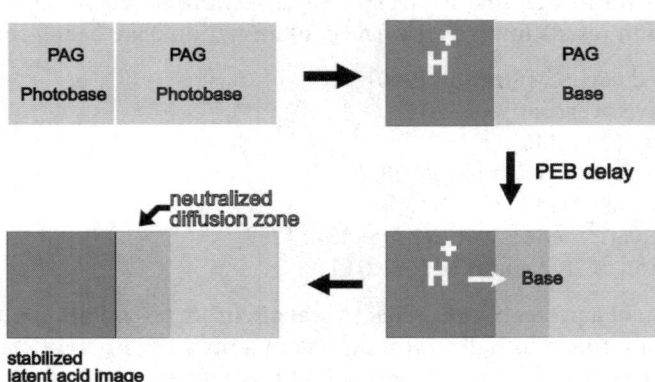

Fig. 52 Photo-decomposable base

PED problem. The idea is to generate an acid by heating a precursor molecule produced by irradiation. A couple of examples which could be used for this purpose have been reported. Electron-absorbed (e-beam exposed) sulfone compounds reportedly have a long life and decompose to sulfinic acid upon heating, contributing to a better PED stability than triphenylsulfonium triflate [207]. α,α-Bis(arylsulfonyl)diazomethanes undergo a photolysis reaction similar to the Wolff rearrangement via a carbene intermediate, which is believed to be partially converted to a highly reactive sulfene to give the expected complex sulfonic acid or to the corresponding thiosulfonic acid esters. The thioesters decompose at elevated temperatures typical for PEB to yield sulfonic acids, which could reduce the PED problem [30].

In 2-nitrobenzyl benzenesulfonate PAGs, 4-nitro, 4-chloro, and 4-methoxy derivatives, which are assumed to be strongly associating with incoming base diffusants, have been reported to offer better PED stability [208].

Some of the stabilization techniques discussed above provide improvement but not a cure and are difficult to implement. More fundamental approaches have been proposed; reduction of an activation energy of deprotection and reduction of a free volume of a resist film.

Chemical amplification resists require PEB to accelerate acid-catalyzed reactions due to fairly high activation energies in general. However, as described later, the polyphthalaldehyde resist provided self-development by depolymerization during exposure at room temperature, which is due to the low activation energy needed to cleave the acetal bond with acid (and the low ceiling temperature) [24, 25]. Thus, the use of protecting groups with low activation energies such as ketals and acetals has produced a new generation of aqueous base developable positive resists based on partially-protected PHOST. Since the acid-catalyzed deprotection proceeds at room temperature as soon as acid is generated by irradiation, PEB is not mandatory and therefore these resist systems are stable toward airborne contamination [209]. However, the ketal-protected materials may be too unstable hydrolytically, making the synthesis difficult and the shelf life (polymer powder, polymer solution, and resist formulation) too short. Thus, more focus has been placed on acetal-protected materials, which could still suffer from the lack of robustness. Another problem with low activation energy systems is linewidth slimming; instead of forming a skin layer or T-top profile, the linewidth becomes smaller when development is delayed due to diffusion of acid into the unexposed regions. Strong standing wave patterns on the resist sidewall can be a problem also in the low activation energy resists. Standing wave patterns are produced by constructive and destructive interactions of the incoming light and the light reflected from the resist/substrate interface and are indicative of high resolution, indicating a limited acid diffusion in the case of chemical amplification resists. However, the rugged sidewalls are not suitable for subsequent substrate etching and therefore PEB is applied to smooth out the standing wave patterns; by slightly diffusing the acid or inducing thermal flow. In the case of the low activation energy resist, a polymer with a higher T_g is produced at room temperature in the exposed regions. An attempt to induce a slight thermal flow would require heating the film to the higher T_g and thus would destroy the thermally unstable unexposed film. An anti-reflection coating (ARC) is effective in reducing standing wave patterns on the side wall.

The environmental stabilization technique involving reduction of a free volume of a resist film by annealing [163, 210] is based on the systematic investigation on the propensity of thin polymer films to absorb airborne NMP using a ^{14}C labeling technique [213]. This study has clearly indicated that low T_g polymer films absorb much less NMP when the bake temperature is set (at 100 °C) due to good annealing and reduced free volume (Fig. 53). A separate investigation has revealed that the NMP uptake is not governed by the concentration of a residual casting solvent (PGMEA in this case) (Table 2) [214]. As the diffusant mobility (DM) is expressed by $DM=e^{-B/f}$, where f is a fraction of the free volume and B is a constant ranging from 0.5 to 1.0 depending on the size, shape, etc. of the diffusant, a small decrease in the free volume can be translated into a profound reduction of the diffusivity of small molecules in polymer films. Annealing can be achieved by reducing T_g of resist polymer or by increasing the PAB temperature [163, 210]. T_g can be reduced by adding a plasticizer, by decreasing molecular weight, or by employing a *meta* isomer [210]. The additive

Table 2 Residual PGMEA in polymer films baked at 100 °C for 5 min [214]

Polymer	Epoxy Novolac	P3tBOC ST	P4tBOC ST	P(TBMA-MMA)	P(TBMA-MMA-MAA)	P3Me4 HOST	P4HOST	Cresol Novolac
Residual PGMEA (wt%)	0.05	0.5	0.6	5.9	8.6	12.8	20.4	21.1

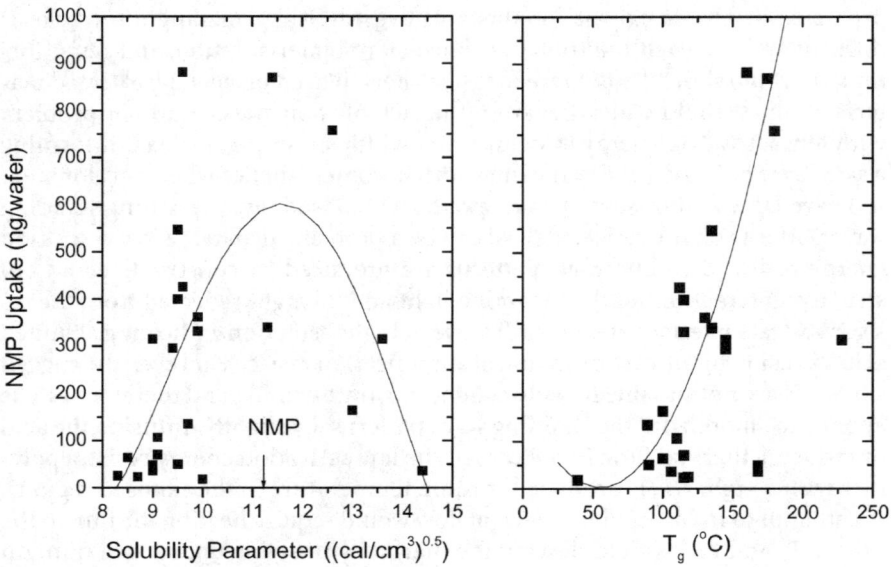

Fig. 53 NMP uptake in polymers [213]

approach mentioned earlier might improve the delay stability due to better annealing resulting from the plasticization effect of the additive. *meta*-PBOCST has been compared with *para*-PBOCST in terms of NMP uptake and imaging stability against airborne NMP [210]. When baked at 100 °C for 5 min, the polymers contain a very small amount (0.55–0.6 wt%) of PGMEA in ~1 μm thickness but absorb a vastly different amount of NMP (931 ng/wafer for *para* and 99 ng/wafer for *meta* from an airstream containing 15 ppb of NMP in 1 h). The *meta* isomer with T_g of 85 °C absorbs about ten times less NMP than the *para* isomer with T_g of 130 °C. The *para* isomer film is slightly anisotropic, indicating poor annealing at 100 °C. Furthermore, the *meta* isomer film has a significantly higher refractive index than the *para* isomer film ($\Delta n_{\parallel}$=~0.008, $\Delta n_{\perp}$=~0.005), indicating that the former is more annealed and densified than the latter. The refractive index difference of the *para* and *meta* isomer films (~0.4%) suggests that the *para* isomer film has ~40% more free volume than the *meta* isomer. Because the mobility of small molecules in polymer films is proportional to $\exp(-Bf^{-1})$, the *meta* isomer allows contaminants to diffuse much more slowly than the *para* isomer, by a factor of 10^3–10^7, depending on the size of the diffusant (B=0.5–1.0). The robustness of the *meta* isomer system toward NMP contamination has been lithographically demonstrated. Poly-(BOCST-*co*-HOST) positive resist systems also demonstrated smaller NMP uptake and greater contamination resistance of imaging for the *meta* isomer than for the *para* isomer.

As low T_g polymers could suffer from thermal flow of relief images during high temperature substrate fabrication processes, an approach involving high temperature PAB to reduce the free volume has been applied to the design of a production-worthy positive deep UV resist [163]. However, while this annealing process requires a resin and PAG that are stable thermally and hydrolytically, the commonly used positive resist resins (Fig. 31) cannot be heated to their T_g. These partially-protected PHOST resins have rather high T_g of ~150 °C due to the high T_g of PHOST (180 °C). Although many of these protecting groups are stable thermally to 200 °C when placed in a homopolymer, they undergo thermal deprotection at ~130 °C in the presence of the acidic phenolic OH group as is in the copolymers. Thus, the thermal deprotection temperature is lower than T_g in these partially-protected PHOSTs and therefore the positive deep UV resist films cannot be annealed and are consequently susceptible to airborne base contamination. Copolymers of tert-butyl (meth)acrylate and HOST have been reported to be exceptionally stable hydrolytically with the thermal deprotection commencing at ~180 °C while T_g is typically in the range of 150 °C in the case of the acrylate copolymers [163]. Thus, the copolymer of TBA with HOST is a rare example of the positive resist resins that can be heated to or above its T_g without premature decomposition of the protecting group and thus can be annealed well for the free volume reduction. Figure 54 compares NMP absorption from an airstream containing 10 ppb NMP by the APEX resist and the new ESCAP resist based on poly(TBA-co-HOST) [163, 210, 213]. The latter absorbs NMP much more slowly than the former (7.8 vs 18 ng/min) even when baked at 100 °C, below its T_g of 150 °C, which is due to incorporation of the acrylate structure (solubility parameter consideration, Fig. 53). Baking the film at 170 °C, above its T_g of 150 °C, significantly reduce the NMP uptake down to 1.8 ng/min,

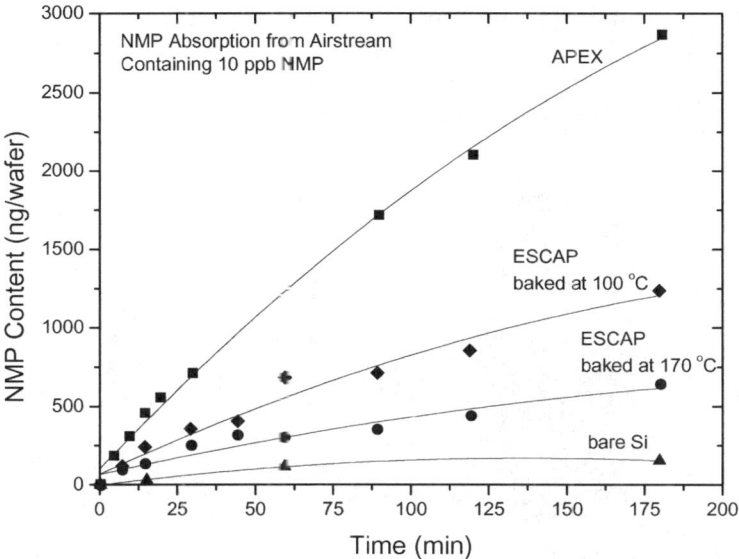

Fig. 54 NMP uptake kinetics of resist polymers

which is close to the value for bare silicon (1.1 ng/min). While the thermal and hydrolytic stability of PAGs varies widely from below 100 to above 300 °C, the high temperature bake process requires a non-volatile bulky PAG which is stable to ~150 °C and generates a bulky non-volatile acid [31]. The poly(TBA-co-HOST) resist (ESCAP) has demonstrated a superb PED stability and resulted in standard commercial products [215].

Another important issue in developing 248 nm resists was compatibility to 0.26 N (1.38 wt%) TMAH developer. This developer strength was selected for the novolac/DNQ resists and has become the industry standard. All the advanced 248 nm resists based on high activation energy (ESCAP-type) and low activation energy (acetal) systems currently on the market are engineered to be compatible with the industry standard 0.26 N TMAH developer. The 248 nm positive resists implemented in manufacturing worldwide for the 180 nm node are capable of resolving 125 nm (half wavelength) equal line/space patterns, and being used in 130 nm device production (Fig. 55). The half wavelength lithography has been achieved by increasing NA of the lens, improving resist materials, and applying resolution enhancement techniques (RET) [216] such as off-axis illumination, use of phase-shift masks [217], etc. The critical resist parameters to control are the dissolution contrast (n) and the minimum dissolution rate of the resist (R_{min}). Higher maximum dissolution rates (R_{max}) can lead to improved underexposure latitude which is very important for printing dark field features such as trenches and contact holes. Conversion of *tertiary* ester to carboxylic acid embodied in the ESCAP resist design provides a high and controllable dissolution contrast and high adjustable R_{max}, allowing preparation of several versions for specific applications by varying the resin composition and formulation. Furthermore, innovative process tricks to enhance the resolution limit has been proposed and implemented in manufacturing, which includes a thermal flow process to shrink contact holes (Fig. 56) [218]. The aerial image contrast degrades the fastest for contact holes. To overcome this problem and to enhance the resolution, a non-lithographic thermal flow

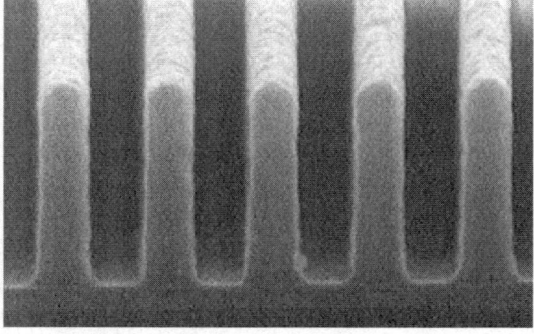

Fig. 55 Half wavelength lithography with ESCAP (125 nm line/space patterns generated by 248 nm irradiation) [22]

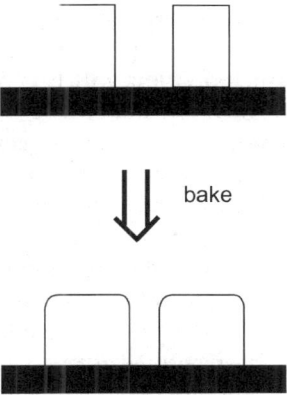

Fig. 56 Thermal flow process

process has been used in production of 64 Mbit and 256 Mbit DRAMs. For example, 200 nm contact holes were flowed down to 80 nm hole size [219]. However, there is a minimum pitch limit of 400 nm and the process is difficult to control. The process control has been improved by inducing light crosslinking.

The 248 nm positive resists based on the ESCAP concept is the workhorse of the industry and has been applied to 157 nm lithography in a thin form as well as to EUV lithography. Although the commercial 248 nm resists diluted to cast thin films to compensate for their high absorption at 157 nm provided valuable information about 157 nm imaging [220], new transparent materials are needed for 157 nm photolithography as described in detail later. Thinned mature 248 nm positive resists continue to be used in EUV lithography [221], but sensitivity enhancement and reduction of line edge roughness (LER) are the major issues. Projection electron beam lithography (PEL) employs 248 nm resists, especially acetal-based systems such as KRS [222], for sensitivity. However, to make PEL economically feasible in manufacturing, its wafer throughput must be increased, most likely through sensitivity enhancement of the 248 nm resist systems.

4.1.10
Polymethacrylates and Norbornene Polymers for 193 nm Lithography

Photolithography is further moving to a shorter wavelength for a higher resolution and ArF (193 nm) excimer laser lithography is maturing as a manufacturing technology. The first major concern from the material point of view was the high absorption expected with almost all organic compounds at this wavelength. Thus, a single layer lithography employing wet development was not considered feasible and an initial attempt was focused very much on photochemical ablation [223] and top-surface imaging (TSI) involving oxygen reactive ion etching (RIE), which takes advantage of the strong UV absorption of organic polymers and resists. The top-surface imaging will be discussed in detail later. Etching of organic polymers with 193 nm ArF excimer laser radiation was demonstrated by Srinivasan and Mayne-Banton in 1982 [223] and the early ArF excimer laser imaging was primarily centered around this photochemical ablation phenomenon. Upon irradiation with the ArF laser, organic materials undergo an ablative photodecomposition. Poly(methyl methacrylate) was again the first resist to be imaged using an ArF excimer laser with the imaging mechanism based on main chain degradation [224]. Furthermore, PMMA and poly(methyl methacrylate-*co*-methacrylic acid) were imaged by using 157 nm F_2 excimer laser [225]. As discussed in detail in this section, the 193 nm imaging systems which have been studied at an accelerating rate still very much consist of polymethacrylates, though the imaging mechanism has shifted from inefficient main chain cleavage to chemically-amplified deprotection involving pendant groups.

A breakthrough came from the observation that polymethacrylates are highly transparent at the ArF excimer laser wavelength (OD<0.1/µm, Fig. 57) [181]. As mentioned earlier, however, polymethacrylates are not resistant to dry etching conditions in general. Another breakthrough was a discovery that incorporation of an alicyclic structure into polymethacrylates improves dry etch resistance significantly without sacrificing transparency [225]. Thus, 193 nm resist activities have been very much centered around polymethacrylates bearing pendant alicyclic groups. Incorporation of an alicyclic structure into a polymer

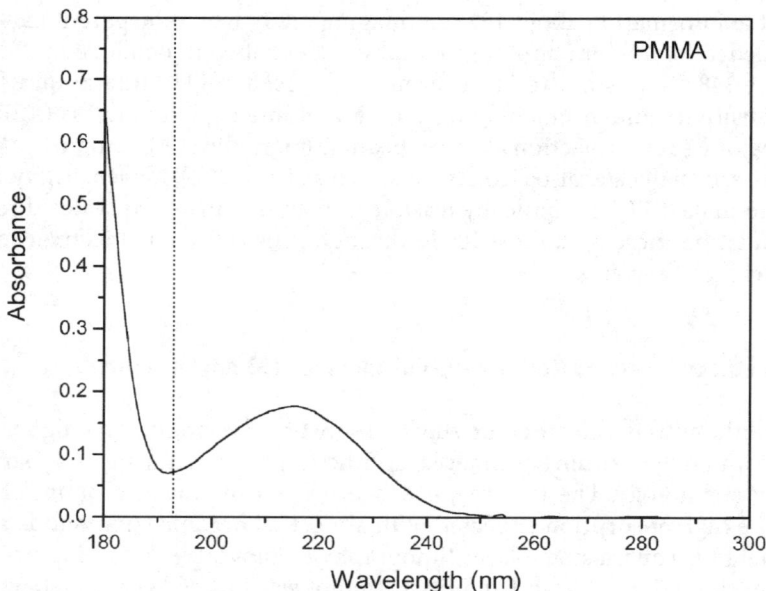

Fig. 57 193 nm absorption of PMMA

backbone has been also attempted. Reflecting the importance of incorporating dry etch resistance in 193 nm resists, serious attempts were made to correlate the etch rate with the polymer composition and structure. The first model employed to predict etch resistance was the one that had been reported by Gokan, Esho, and Ohnishi 20 years ago [227]. They showed that the etch rate (V) of organic polymers under ion bombardment has a linear dependence on $N/(N_C-N_O)$, where N, N_C, and N_O are the total number of atoms, number of carbon atoms, and number of oxygen atoms, respectively, per repeat unit, indicating that the dry etch resistance is determined by the effective carbon content in polymer:

$$V \propto N(N_C-N_O)$$

Thus, aromatic polymers are more resistant to dry etching than aliphatic polymers in general and in aliphatic polymers cyclization can reduce the number of hydrogen atoms and therefore N/N_C. As this model showed a poor relationship to chemical structure for etch rates in largely chemical etching, an additional parameter based on polymer structure, called the ring parameter (r) was introduced by Kunz et al. [228], which is defined as

$$r = M_{CR}/M_{TOT}$$

where M_{CR} and M_{TOT} are the mass of polymer existing as carbon atoms in a ring structure and the total polymer mass, respectively. The normalized etch rate in chlorine-based high density plasmas has been expressed by a third-order polynomial of r. Additionally, an incremental structural parameter model for predicting RIE rates of 193 nm resist polymers has been proposed [229].

4.1.10.1
Polymethacrylates

A terpolymer of *tert*-butyl methacrylate (TBMA), methyl methacrylate (MMA), and methacrylic acid (MAA) was initially developed as a chemically amplified thick laser resist for circuit board fabrication [180]. The terpolymer containing 1 wt% of di(*tert*-butylphenyl)iodonium trifluoromethanesulfonate (triflate) has demonstrated excellent imaging at 193 nm when developed with a dilute (0.01 N) TMAH aqueous solution [181]. This terpolymer single layer resist was employed for exposure tool testing.

Several polymethacrylates, primarily alkyl esters, were compared with a novolac resist in terms of etch rates in CF_4 and Ar plasmas. The alkyl groups examined included methyl, *tert*-butyl, cyclohexyl, norbonyl, adamantyl, and benzyl [226]. The polymerization of alicyclic methacrylates was pioneered by Otsu, who has demonstrated facile polymerization of bulky methacrylates and reported high T_g of this class of polymethacrylates [230]. It has been found that alicyclic polymers exhibit better dry etch resistance than acyclic esters and that the dry etch durability is increased by an increase in the number of rings. Thus, poly(adamantyl methacrylate) is as stable as a novolac resist under dry etch conditions. A 30/70 copolymer of adamantyl methacrylate with *tert*-butyl methacrylate has 80% transmission/µm at 193 nm and etches only 20–30% faster than a novolac resist in Ar or CF_4 plasma. The copolymer containing 15 wt% of triphenylsulfonium hexafluoroantimonate was evaluated using a KrF excimer laser stepper and proposed as an ArF excimer laser resist [226]. However, incorporation of the highly lipophilic adamantyl group results in poor aqueous base development including poor adhesion and crack formation. Thus, an increase in hydrophilicity has been attempted in the design of 193 nm resist systems in addition to the use of isopropanol/TMAH as the developer [231]. The mixed developer was also employed in development of a ArF resist based on a copolymer of 3-oxocyclohexyl methacrylate and adamantly methacrylate (Fig. 58) [232]. A 1-µm-thick film of a 49/51 copolymer has a transmission of ca. 53% at 193 nm. A 0.4-µm-thick resist film containing 2 wt% of triphenylsulfonium hexafluoroantimonate with 60% transmission was evaluated as an ArF excimer laser resist. The oxocyclohexyl ester is cleaved with a photochemically-generated acid to produce carboxylic acid {poly(methacrylic acid)} and cyclohexenone. The adamantyl ester has been rendered acid-cleavable and the oxocyclohexyl group replaced with a mevalonic lactone ester which is also a quaternary ester (Fig. 58) [233]. With 2 wt% triphenylsulfonium hexafluoroantimonate, a 0.4-µm-thick resist film demonstrated 0.15 µm reso-

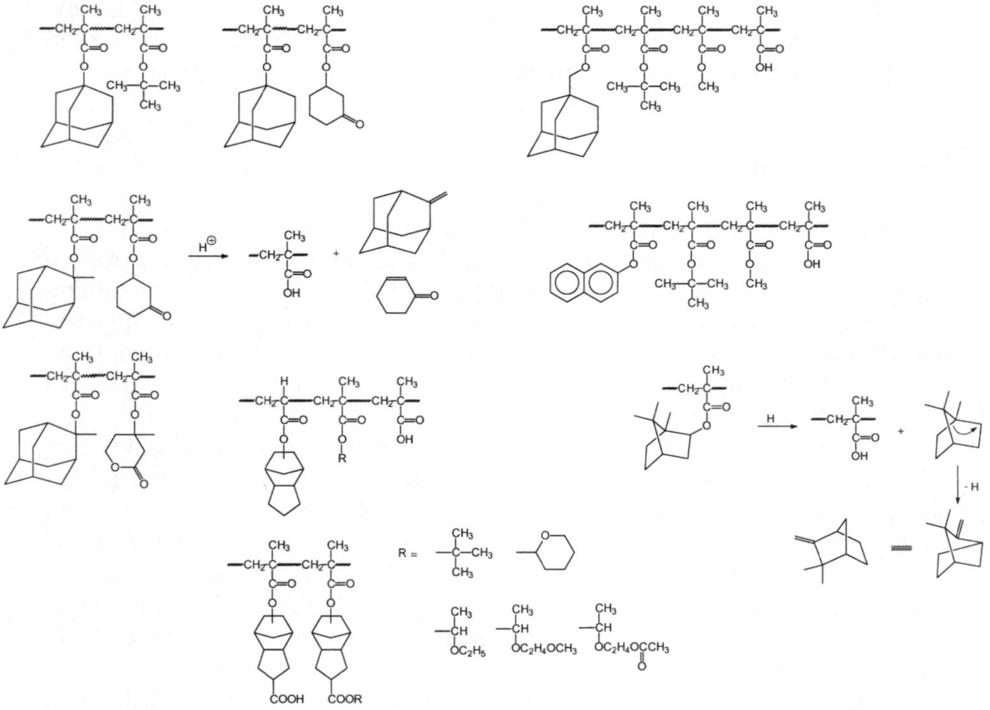

Fig. 58 ArF resists based on methacrylates

lution at 4.8 mJ/cm² on a 0.55 NA ArF excimer laser stepper, using the standard 2.38 wt% TMAH solution. The resist has successfully delineated 0.12 µm line/space patterns using an alternating phase shift mask at 193 nm. The etch rate of the resist relative to a novolac resist was 1.1–1.2 in CF_4 and Ar plasmas.

The methacrylate terpolymer resist has been rendered dry etch resistant by incorporating adamantylmethyl methacrylate [234]. While the acceptable etch resistance required >50 mol% adamantylmethyl methacrylate, it was difficult to control the hydrophobic/hydrophilic balance and T_g (the bulky ester groups increase T_g to >200 °C), resulting in poor resolution and adhesion. The same problems were encountered with isobornyl (meth)acrylate systems, which are acid-sensitive (Fig. 58) [234]. The high hydrophobicity of the homopolymer was reduced by incorporating methacrylic acid as a comonomer, resulting in excessively high T_g of >200 °C and therefore severe stress cracking of the film upon contact with any solvents. The T_g of the polymer was adjusted to an acceptable range of ~170 °C by copolymerization of isobornyl methacrylate with acrylic acid to from a 80/20 copolymer, which could dissolve in a 0.26 N TMAH solution upon exposure/PEB in the presence of di(4-*tert*-butylphenyl)iodonium triflate. However, a high PEB temperature of >120 °C was required for the acid-catalyzed cleavage and rearrangement (Fig. 58), that could undergo reverse reactions with the carboxylic acid and may be the reason for its low contrast [234].

An alternative strategy has been developed, which involves three components; a methacrylate polymer, an alicyclic dissolution inhibitor, and a PAG (see above) [234b, 235]. Bile acid esters (5*B*-steroids) were converted to methyl (passive, acid-inert) and *tert*-butyl (active, acid-sensitive) esters for use as dissolution inhibitors which could provide dry etch resistance. Removing the hydroxyl group improves the dissolution inhibition effect; lithocholate is more inhibiting than ursocholate that inhibits more than cholate. The active inhibitors provide much greater dissolution rate ratios as expected and therefore are more preferred. The TBMA-MMA-MAA terpolymer containing 33% of an active alicyclic dissolution inhibitor and the iodonium PAG showed improved dry etch resistance and good imaging quality (0.05 N TMAH) but suffered from low T_g due to plasticization by the steroid. The plasticization effect of the steroid has been utilized in lowering the excessively high T_g of the isobornyl methacrylate tetrapolymer mentioned above, yielding a dry etch resistant (1.2 times novolac in Cl_2 plasma) material with good imaging quality. The androstane structure has been attached to the methacrylate monomer for radical polymerization (Fig. 59) [236].

Another interesting acrylate polymer for 193 nm lithography is a terpolymer of tricyclo[5.2.1.0^{2,6}]decanyl acrylate, tetrahydropyranyl methacrylate, and methacrylic acid (Fig. 58) [237]. In conjunction, new alkylsulfonium salts for use in ArF excimer laser lithography have been synthesized [238]. Methyl(cyclohexyl)(2-oxocyclohexyl)sulfonium and methyl(2-norbornyl)(2-oxocyclohexyl)sulfonium triflates are highly transparent with their absorption coefficients of 1125 and 1650 L/mol*cm at 193 nm in sharp contrast with

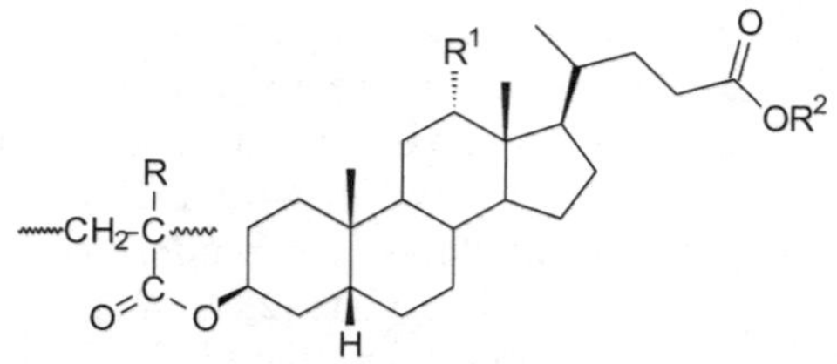

Fig. 59 Androstane methacrylates

R=CH$_3$, H
R^1=OH, H
R^2=H, CH$_2$OC$_2$H$_5$, C(CH$_3$)$_3$

triphenylsulfonium triflate (54,230 L/mol*cm). The former decomposes at a lower temperature than the latter (142 vs 151 °C). The terpolymer with a 50/30/20 composition (selected for imaging) containing 1 wt% of the more stable PAG has a transmission of 64.4% for 1-μm-thickness at 193 nm and a 1.42 times faster etch rate than a novolac resist in CF$_4$ plasma. The developer employed was 0.0476 wt% TMAH solution. The balance of the imaging quality and dry etch resistance is challenging.

To overcome the limitation imposed by the multi-component polymer approach, a carboxylic functionality has been incorporated into the alicyclic structure and partially protected with the *tert*-butyl or tetrahydropyranyl group (Fig. 58) [239]. The etch rate of the polymers was improved to 1.23–1.25 from 1.42 times novolac in CF$_4$ plasma. As the copolymers chosen for lithographic evaluation contained 50–60 mol% carboxylic acid, the developer employed was a weak (0.0476 or 0.119 wt%) TMAH solution. Other acid-labile protecting groups (Fig. 58) have been examined in terms of their effect on the dissolution behavior of the above terpolymers and copolymers containing the tricyclodecanyl group [240]. The polarity (hydrophilicity) of the protecting group can significantly affect the dissolution behavior and rate of the copolymers and terpolymers as well as the resists based on these polymers. A similar observation was made with 50/50 copolymers of MMA and methacrylates bearing *tert*-butyl, isobornyl, tetrahydropyranyl, and tetrahydrofuranyl protecting groups [241].

In the development of 193 nm resists, providing hydrophilicity to the lipophilic alicyclic structure was the most important issue to improve adhesion and imaging performance. Solubility parameters of polyacrylates bearing substituted adamantane were calculated using CAChe in an attempt to predict the hydrophilicity of the polymers [242]. Table 3 summarizes calculated solubility parameters. The starting alcohols for the synthesis of the acrylate monomers were prepared by aerobic autooxidation employing *N*-hydroxyphthalimide according to the procedure developed by Ishii et al. [243] and commercialized by Daicel Chemical Ind., Hyogo, Japan (Fig. 60). The calculated solubility parameter of polyacrylates exhibited a good relationship with the retention time of corresponding acrylate monomers in aqueous chromatography (Fig. 61) except for lactones. The retention time suggests that the lactones are more

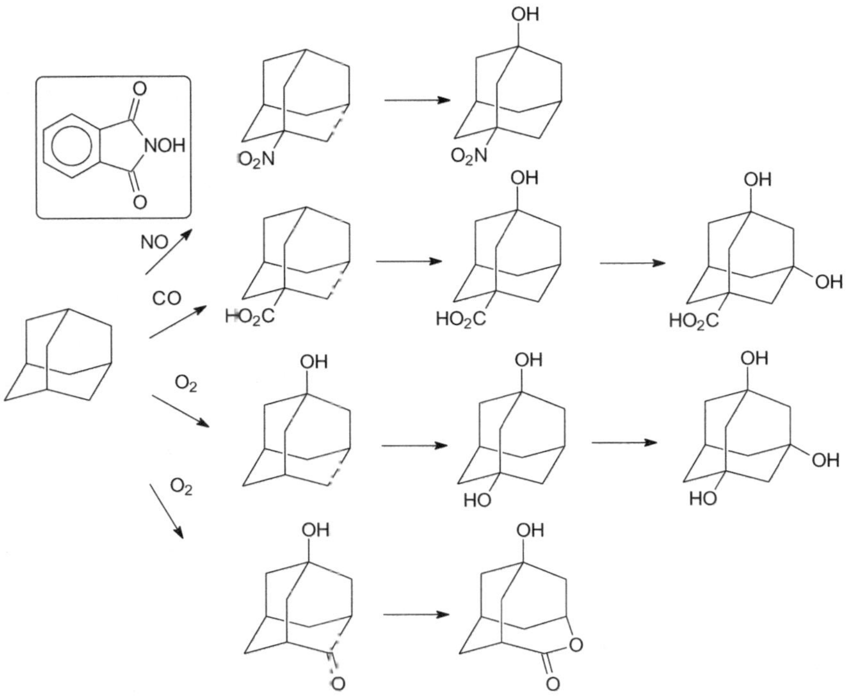

Fig. 60 Auto-oxidation reactions to produce polar alicylic compounds

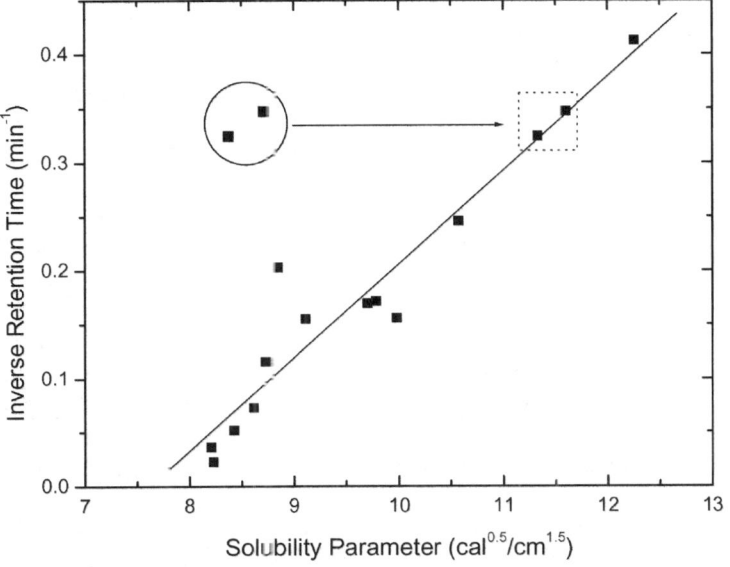

Fig. 61 Retention time vs solubility parameter [242]

Table 3 Calculated solubility parameter (SP) and measured T_g of polyacrylates [242]

monomer	SP of polymer (cal$^{0.5}$/cm$^{1.5}$)	polymer T_g (°C)	monomer	SP of polymer (cal$^{0.5}$/cm$^{1.5}$)	polymer T_g (°C)
CH$_2$=CHCO$_2$-adamantyl	8.45	126.3	CH$_2$=CHCO$_2$-adamantyl-O	8.63	128.5
CH$_2$=CHCO$_2$-adamantyl-OH	10.58	155.8	CH$_2$=CHCO$_2$-adamantyl lactone	8.70 (11.60)[a]	135.1
CH$_2$=CHCO$_2$-adamantyl-CO$_2$H	9.71	144.0	CH$_2$=CHCO$_2$-adamantyl	8.35	120.5
CH$_2$=CHCO$_2$-adamantyl-ketone	9.12	137.0	CH$_2$=CHCO$_2$-adamantyl-OCOtBu	8.24	124.5
CH$_2$=CHCO$_2$-adamantyl-OH, CO$_2$H	11.28	165.8	CH$_2$=CHCO$_2$-tetrahydropyranyl	8.87	99.2
CH$_2$=CHCO$_2$-adamantyl-OH, OH	12.26	179.7	CH$_2$=CHCO$_2$-lactone	8.39 (11.33)[a]	115.0

[a] Values in parentheses are solubility parameters of hydrolyzed structures (Fig. 62)

hydrophilic than the calculated solubility parameter predicts, which is likely to be due to hydrolysis of the lactone ring in aqueous solution (Fig. 62). Adopting the hydrolyzed structure for the lactone, the calculated solubility parameter showed a good linear relationship with the inverse retention time (Fig. 61).

An alternative approach has been proposed. The very strong absorption of aromatic compounds at ~190 nm is due to the π-π* electronic transition of the aromatic ring. Although PHOST exhibits an extremely high OD of 35/μm at 193 nm, however, the absorption falls quickly below 180 nm. Molecular orbital calculations have indicated that the absorption window can be red-shifted to

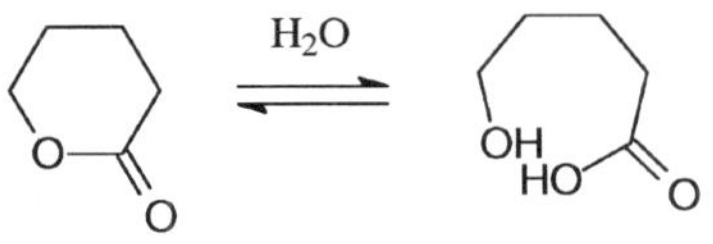

Fig. 62 Hydrolysis of lactone ring

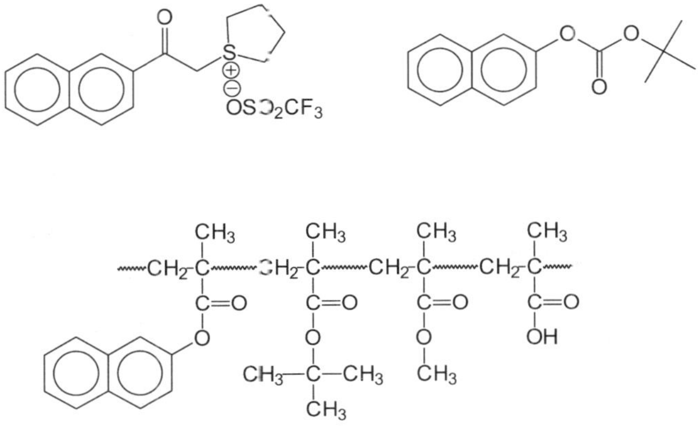

Fig. 63 Naphthalene-based 193 nm resist materials

193 nm by extending conjugation [244]. Thus, an acid generator, dissolution inhibitor, and polymethacrylate incorporating a naphthalene ring have been prepared for use in 193 nm lithography (Fig. 63) [244]. The PAG and dissolution inhibitor based on naphthalene show significantly lower absorptions at 193 nm than the benzene counterparts. However, a copolymer of 2-naphthyl methacrylate and TBMA containing 50 mol% of the naphthalene group for acceptable etch resistance has a high OD of 4.7/µm at 193 nm.

4.1.10.2
Norbornene-Maleic Anhydride Copolymers

Although polymers containing an alicyclic structure in the backbone were investigated along with polymethacrylates with a pendant alicyclic group in the measurements of the etch resistance, such polymers have been later utilized as imaging materials (Fig. 64). Norbornene (NB) undergoes radical copolymerization with electron-deficient maleic anhydride (MA) to yield an alternating copolymer with high T_g while its radical homopolymerization (or copolymerization of two norbornene monomers) is extremely sluggish. The NB-MA copolymer film exhibits only 10% faster etch rates than a novolac resist in Ar or CF_4 plasma [226] but is too lipophilic to be used in aqueous base development [245]. Acrylic acid (AA) was introduced at 5–20% (in feed) into the alternating copolymer as a "defect" to provide base solubility (Fig. 64) [245].

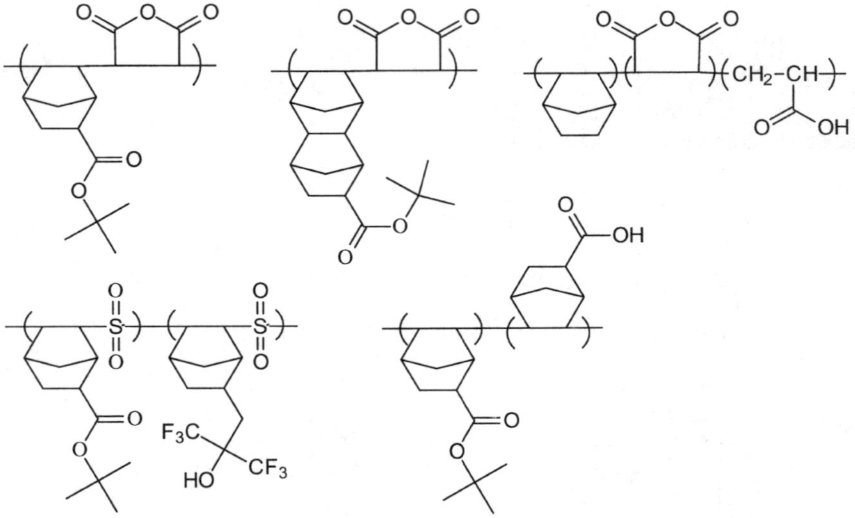

Fig. 64 Norbornene-based 193 nm resist polymers

Compositions resulting from 15 and 17.5% acrylic acid in feed displayed the most useful development behavior (900–1600 Å/min in 0.131 N TMAH) and therefore were evaluated lithographically using steroid dissolution inhibitors (20–35% loading) (Figs. 42 and 65) and triphenylsulfonium hexafluoroarsenate or triflate as PAG. The *tert*-butyl cholate dissolution inhibitor was initially used in conjunction with a novolac resin [173]. Adhesion failure was a major problem. A later version incorporated TBA also. Oligomeric dissolution inhibitors based on the steroid have been evaluated (Fig. 65) [246] and their interaction with the matrix polymer investigated [247]. Solid-state ^{13}C NMR with cross polarization and magic-angle spinning has been employed to study the chain dynamics and length scale of mixing in the tetrapolymer resist formulation [248]. Two-dimensional wide line separation NMR has been used to measure the chain dynamics via the indirectly detected proton line shapes, showing that the polymer does not experience large amplitude atomic fluctuations even at 155 °C. Proton spin diffusion experiments have demonstrated that the polymer and dissolution inhibitors are mixed on a molecular length scale. A new acid generator called "sweet PAG" was proposed for use in the NB-MA-AA-TBA tetrapolymer resist [249]. Bis(4-*tert*-butylphenyl)iodonium cyclamate generates a zwitterionic sulfamic acid upon exposure to 248 and 193 nm radiation (Fig. 66), while it is expected to behave as a photodecomposable base. Because of the zwitterionic nature, sulfamic acids have larger pK_a values (6.5 in dimethyl sulfoxide) than those of sulfonic acids (1.7 for methanesulfonic acid in dimethyl sulfoxide). The expected high pK_a of sulfamic acids would render these acids ineffective in deprotecting even low activation energy protecting groups such as acetal as well as high activation energy protecting groups. However, conju-

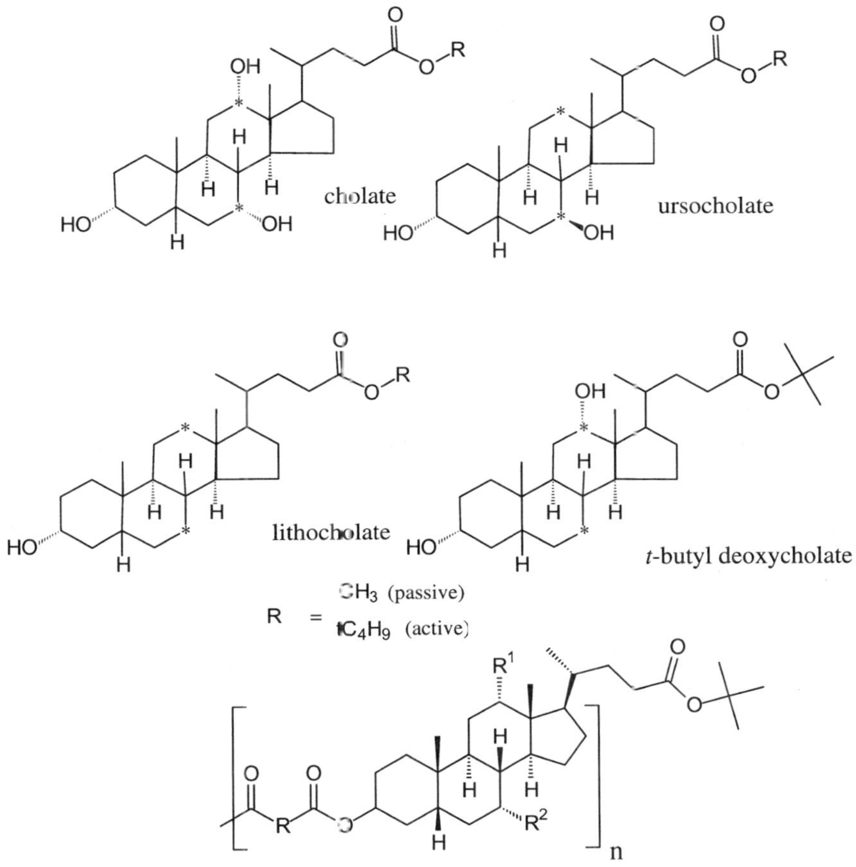

Fig. 65 Steroid dissolution inhibitors for 193 nm lithography

gate acids are known to show a substantial decrease in pK_a with a temperature increase. Thus, although compounds capable of generating sulfamic acids cannot serve as the sole PAG component in resist systems employing high activation energy protecting groups, they can be used as non-volatile, low diffusive photodecomposable bases when combined with a PAG capable of generating a strong acid such as perfluoroalkanesulfonic acid. However, a zwitterionic conjugate sulfamic acid is expected to be sufficiently active at typical PEB temperatures in cleavage of low activation energy protecting groups, where sulfamic acid generators fulfill the dual roles of a PAG and that of photodecomposable base.

As was the case with the polymethacrylate-based resists, the CycloOlefin-Maleic Anhydride (COMA) system attracted a great deal of attention primarily due to ease of polymer synthesis (radical copolymerization) and good lithographic performance. Various norbornene derivatives including tetracyclo-

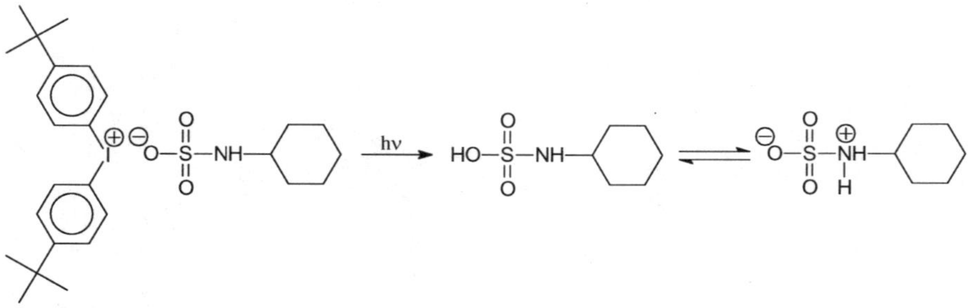

Fig. 66 Sweet PAG

dodecene have been co-, ter-, and tetra-polymerized with maleic anhydride (Fig. 64) [168, 250–252]. M_w is typically <10,000. In almost all the cases, the NB-MA copolymerization was assumed to proceed via charge transfer mechanism and to produce strictly alternating copolymers even when a homopolymerizable third monomer was introduced. To elucidate the copolymerization mechanism, detailed in situ ^{1}H NMR kinetics studies were combined with the analysis of the monomer reactivities by the "mercury" method [253]. Substitution on the 5-position reduces the copolymerization rate and polymer yield; copolymer yields are typically <50% for the substituted norbornenes while unsubstituted NB gives a high conversion of 70–80%. Inverse gated ^{1}H-decoupled ^{13}C NMR analysis of a number of copolymers has indicated that the MA concentration in copolymers is slightly higher than 50 mol% (55–57%). The in situ analysis of terpolymerization of NB, NB bearing *tert*-butyl ester, and MA has clearly in-

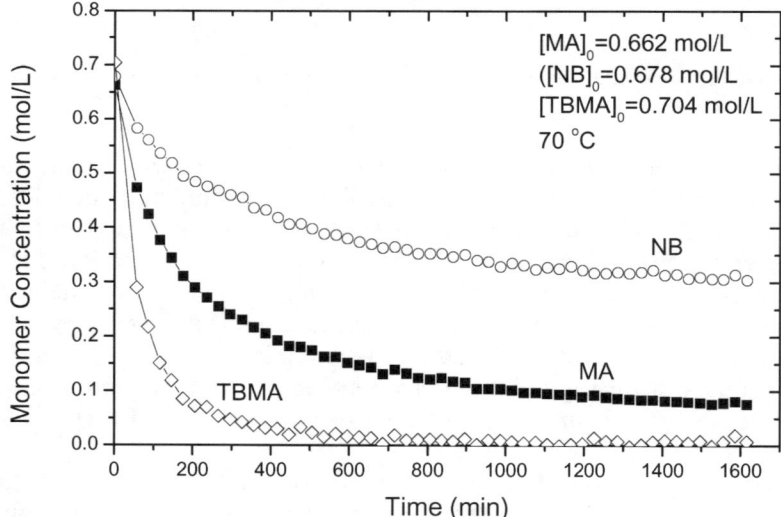

Fig. 67 Kinetics of radical terpolymerization of NB, MA, and TBMA with AIBN in dioxane-d_8 at 70 °C as studied by ^{1}H NMR [253]

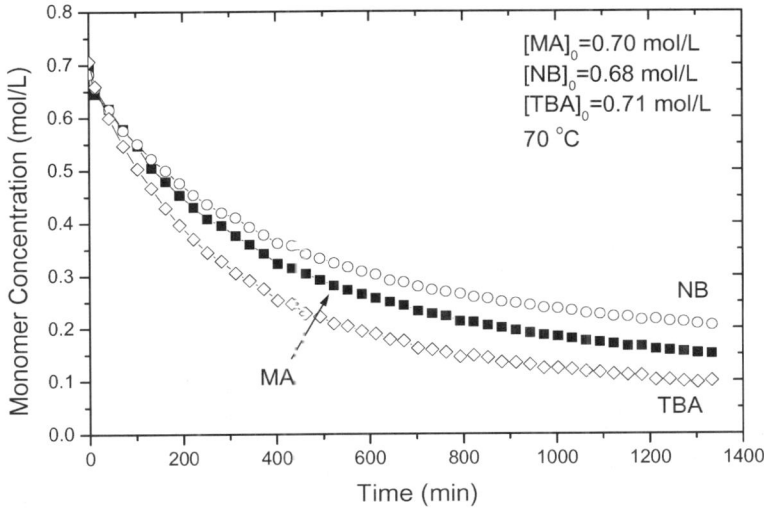

Fig. 68 Kinetics of radical terpolymerization of NB, MA, and TBA with AIBN in dioxane-d_8 at 70 °C as studied by ^{1}H NMR [253]

dicated that the unsubstituted NB reacts faster than the substituted one. The terpolymerization kinetics was later studied by in situ IR analysis also [254]. Furthermore, TBMA (or methacrylic acid) has been demonstrated to be consumed and exhausted rapidly in an early stage of terpolymerization with NB and MA (Fig. 67) while TBA (or acrylic acid) is incorporated into a terpolymer with NB and MA uniformly throughout the entire period of polymerization (Fig. 68) [253]. Acrylates can copolymerize with NB and MA. Thus, acrylic monomers tend to be more uniformly incorporated into the NB-MA polymer than methacrylates. The incorporation of a third homopolymerizable monomer could increase the polymer yield but introduce significant heterogeneity into the polymerization and polymer structure. The stoichiometry of MA and NB could be completely destroyed by the third homopolymerizable monomer and the polymer could contain a much higher concentration of MA than NB. This study suggests that the charge transfer mechanism does not play a dominant role in the NB-MA copolymerization.

The charge transfer mechanism for the NB-MA copolymerization has been discarded through investigation on the basis of the mercury method (Fig. 69) [253]. A cyclohexy radical generated by treatment of cyclohexylmercuric chloride with sodium borohydride in dichloromethane was reacted with MA and NB according to the Giese's procedure [255]. In the mass balance experiments [256] on the system consisting of MA and *tert*-butyl 5-norbornene-2-carboxylate (NBTBE), the yield of the cyclohexyl adduct of MA was analyzed by gas chromatography to be 131–85% whereas NBTBE remained mostly unreacted (81–93%). Thus, MA exclusively reacts with the cyclohexyl radical, leaving NBTBE

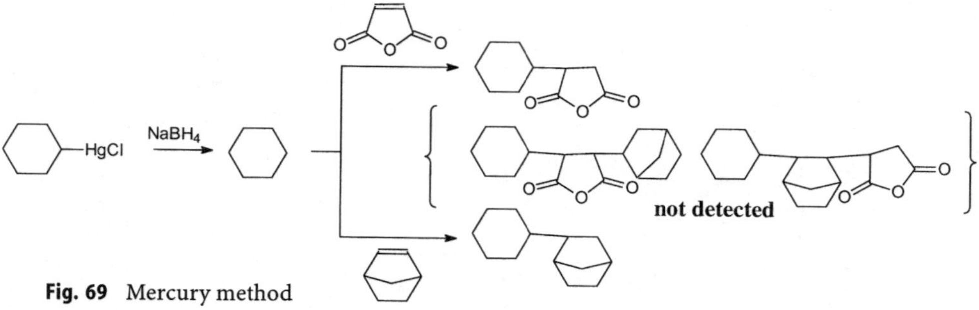

Fig. 69 Mercury method

largely unreacted and the addition of the cyclohexyl radical to a 1:1 charge transfer complex does not constitute a major pathway for the consumption of the monomers.

The NB-MA copolymers tend to absorb more than polymethacrylates at 193 nm, which is less desirable, perhaps due to a high concentration of absorbing end groups (low molecular weights) while the anhydride structure may be only slightly more absorbing than the ester group [257].

Another interesting feature of the COMA polymers and resists is hydrolysis of the anhydride ring, which can be detrimental and beneficial. The anhydride ring can be hydrolyzed during work-up after polymerization, resulting in irreproducible and significant dissolution in aqueous base, and also during storage in a casting solvent (as a polymer solution or as a formulated resist solution), resulting in irreproducible lithographic performance and a poor shelf life [251b,d, 257–260]. The increase in the carboxylic acid concentration in a film of a copolymer of NBTBE and MA cast from PGMEA solution can be clearly detected by IR spectroscopy [257]. Furthermore, accompanying faster dissolution rates in aqueous base indicate hydrolysis during storage of the polymer solution (for example, 524 Å/s in 0.26 N TMAH after one year from 8 Å/s for one-day-old solution) [257]. Thus, the casting solvent for COMA must be dried.

However, this hydrolysis is a reason for good imaging performance of the COMA resist. While a film of a copolymer of NBTBE with MA (free of carboxylic acid) is essentially insoluble in aqueous base, base hydrolysis in the exposed area during development accelerates the dissolution, resulting in a higher development contrast [257]. The exposed area becomes polar due to generation of carboxylic acid through photochemically-induced acidolysis of the *tert*-butyl ester, allowing penetration of the aqueous base developer into the film. The anhydride is hydrolyzed by the base developer, increasing the carboxylic acid concentration and boosting the dissolution rate. The aqueous base cannot diffuse into the nonpolar unexposed area. The base hydrolysis of the COMA polymer film during development has been demonstrated and confirmed by IR spectroscopy (Fig. 70) [257, 261]. IR bands of the anhydride decrease in intensity on the same timescale as the overall film dissolution. The carboxylate ion absorption band increases in intensity synchronously with a decrease in an-

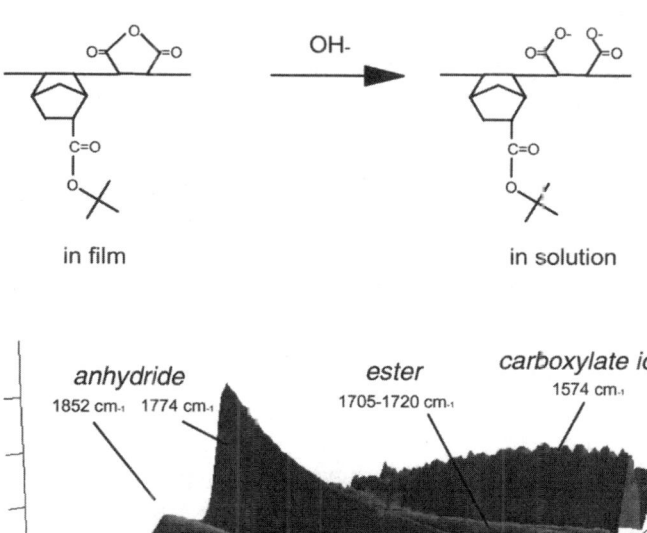

Fig. 70 Hydrolysis of anhydride ring in a NBTBE-MA copolymer during aqueous base development [261]

hydride in a static experiment. Aqueous base penetration into the film is the most important step in base hydrolysis of anhydride during development, which is controlled by polarity of the matrix. A COMA system incorporating a polar group such as acrylic acid could suffer from storage instability and from thinning in the unexposed areas during development. Therefore, incorporation of a hydrophobic additive is required as a barrier for aqueous base penetration and as an anhydride protector rather than as a dissolution inhibitor.

Norbornenes bearing pendant steroid [262] and piperidyl [263] groups have been copolymerized with MA. The latter was an approach to attach a base to polymer for PED stabilization and a control of acid diffusion. They applied the same strategy to the ESCAP 248 nm and methacrylate 193 nm resists [264]. Hybrid systems based on COMA and methacrylates have also been developed [265].

4.1.10.3
Polynorbornenes and Other Polymers

Poly(2,3-norbornene) has an etch rate that is 15% slower than that of a novolac resin in an aggressive chlorine plasma and is another primary 193 nm resist platform [241]. Vinyl addition polymerization of norbornenes resulting in poly(2,3-norbornene) was first mentioned in the early 1960s using classical TiCl$_4$-based Ziegler systems, affording materials with only very low molecular weight (<1000) in low yields [266]. Zirconocene/methaluminoxane systems have been reported to give high polymer, although they exhibit low activity and require a large excess of methaluminoxane [267]. Electrophilic palladium (II) complexes effect "living" polymerization of norbornene [268, 269]. A new family of single component and multi-component catalysts for vinyl addition polymerization of norbornenes has been developed by BF Goodrich Company (currently, Promerus) based on nickel and palladium (Fig. 71) [270]. The molecular weight can be controlled by adding a chain transfer agent, producing olefin-terminated polymers. These new catalysts, especially Pd, have such a high tolerance toward functional groups (such as carboxylic acids and esters) that homo- and copolymerization of a wide variety of norbornene monomers bearing functional groups is possible. A large number of all-norbornene 193 nm resist polymers (COBRA) have been prepared by this technology (Fig. 72).

Polynorbornene bearing a pendant *tert*-butyl ester was imaged in a *negative* mode using an onium salt PAG [241]. This highly hydrophobic polymer failed to develop in a positive tone in aqueous base (see the *t*BOC resist). Norbornene bearing carboxylic acid (NBCA) was copolymerized with NBTBE as a polar group [271]. Unlike the phenolic 248 nm system, however, the dissolution rate of the NBCA copolymers in aqueous base is not a smooth function of the copolymer composition and the concentration of the acidic unit cannot exceed ca. 20 mol%. Typically, the NBCA copolymer does not dissolve in aqueous base when the NBCA concentration is below 20 mol% and dissolves extremely rapidly (>10,000 Å/s in 0.26 N TMAH) when the acid concentration is above

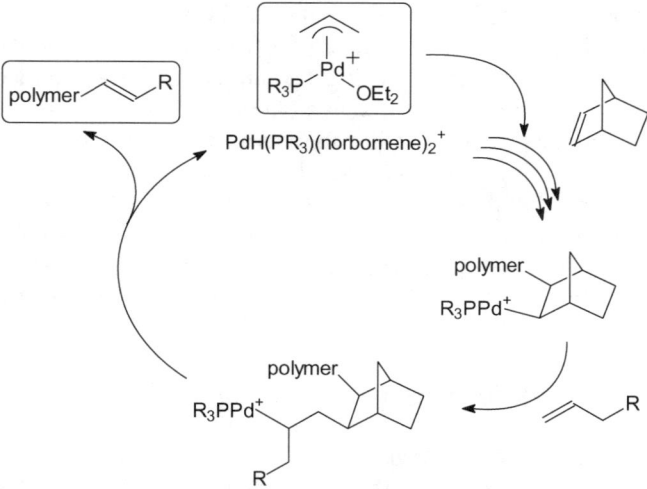

Fig. 71 Vinyl addition polymerization of norbornenes

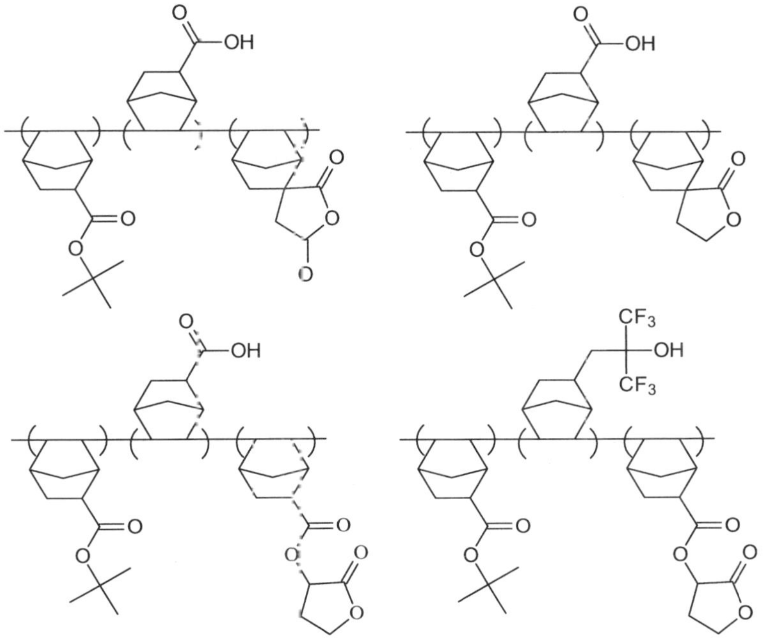

Fig. 72 All-norbornene 193 nm positive resist polymers (COBRA)

50 mol% [258]. What is unique about the NBCA-based COBRA system is that the copolymer film swells a lot in aqueous base in the composition range of 20 mol%<NBCA<50 mol% (Fig. 73) [258]. The frequency decrease in the QCM measurement has been confirmed to indeed correspond to gel layer formation by combining QCM with impedance measurements and interferometry [261]. Important discoveries have been made: a) a four layer (liquid, interfacial layer, polymer film, quartz) model accurately replicates experimental reflectance, b) fitting of experiment to model provides gel and dry layer thicknesses and gel composition (gel contains ca. 55 volume % developer in this case), c) for gel thickness <500 nm, QCM provides accurate measure of mass of a gel layer, and d) the growth rate of the gel layer increases linearly with the hydroxide concentration in developer [261]. This technique has detected previously unnoticed swelling in novolac and 193 nm methacrylate polymers as well. The COBRA resist does not dissolve or swell in the unexposed area because the NBCA concentration is below 20 mol% and dissolves fast in the exposed area because the combined carboxylic acid concentration is above 50 mol%. However, a major problem is its swelling in partially exposed areas. This problem is significantly reduced by adding a dissolution modifying agent (DMA) such as *tert*-butyl cholate (Figs. 42 and 65). This COBRA formulation does not require any added dissolution inhibition and DMA is added to improve lithographic performance through suppression of swelling. The swelling rate is reduced

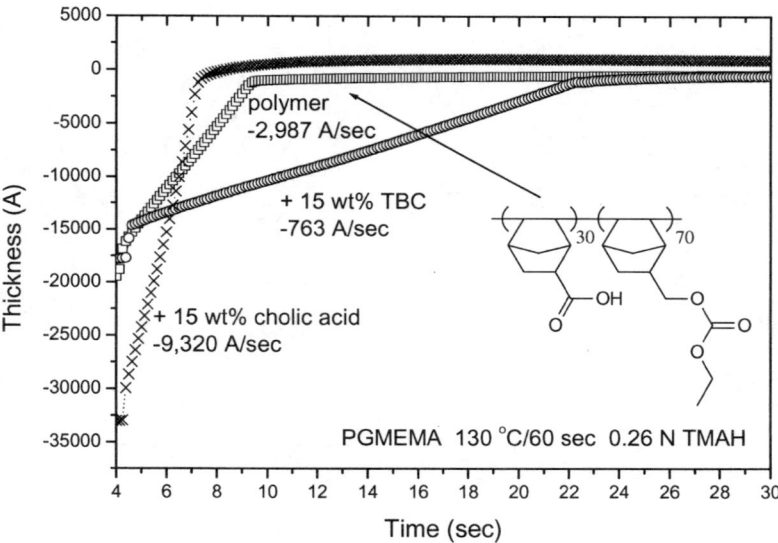

Fig. 73 Suppression and promotion of swelling of NBCA-based polymer by addition of *tert*-butyl cholate and choric acid, respectively [258]

significantly by adding *tert*-butyl cholate while addition of cholic acid results in accelerated swelling as the QCM curves in Fig. 73 indicate [258].

In an attempt to decrease the concentration of the protecting group such as *tert*-butyl and methylcyclopentyl ester (MCP), polar norbornene monomers have been terpolymerized with NBMCP and NBCA (Fig. 74) [258, 272]. The polynorbornenes containing anhydride such as an itaconic anhydride adduct suffered from the problems described for the COMA system [258]. Sulfonamides pendant from norbornene did not provide base solubility. NBTBE copolymers containing up to 80 mol% of methanesulfonamide did not dissolve at all in 0.26 N TMAH. The sulfonamide functionality is not acidic enough in the polycycloolefin to induce dissolution but polar enough to allow penetration of aqueous base developer, inducing hydrolysis of anhydride and promoting swelling [258]. Lactone groups have been found to provide a better stability than anhydride, high polarity, and selective hydrolysis in the exposed area during development [258]. Thus, the COBRA 3000/4000 resist is based on a terpolymer of NBMCP, NBCA, and NB bearing a spirolactone or ester lactone (Fig. 74) [271–273].

Cationic and radical polymerizations of norbornene yield only low molecular weight oligomers [169, 170]. Ring-opening metathesis polymerization (ROMP) of strained cyclic olefins was first reported in the 1950s [171] and is of industrial importance. Attempts were made to prepare 193 nm resist polymer by ROMP of functionalized norbornenes and tetracyclododecenes (Fig. 75) [168, 251]. Hydrogenation of the unsaturated backbone was necessary to afford thermooxidative stability and to reduce OD at 193 nm. Use of tetracyclododecene

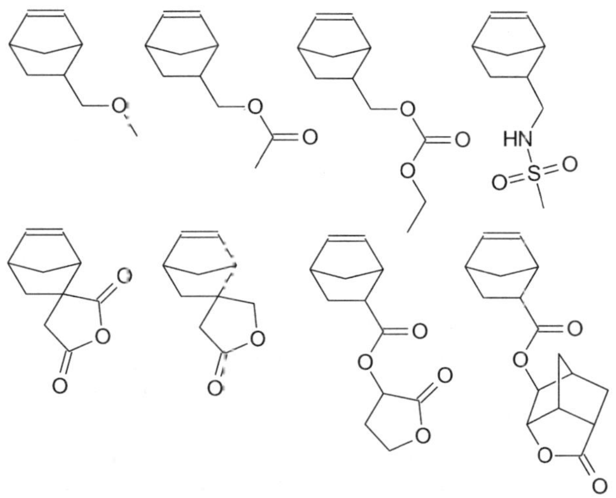

Fig. 74 Polar norbornene monomers employed in terpolymerization

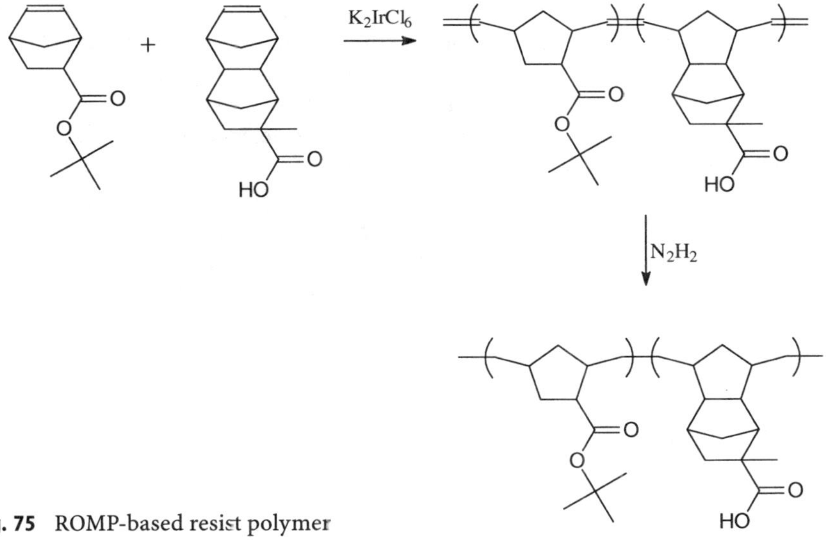

Fig. 75 ROMP-based resist polymer

helped increase T_g. However, the 193 nm resist platform based on ROMP was quickly abandoned primarily due to its poor dissolution characteristics, somewhat similar to the all-norbornene system described above.

Another electron-rich monomer, vinyl ether, has been employed for radical copolymerization with electron-deficient maleic anhydride to yield a VEMA system (Fig. 76) [274]. Linear or cyclic alkyl vinyl ethers have been used in conjunction with deprotectable alicyclic (meth)acrylate. The replacement of NB

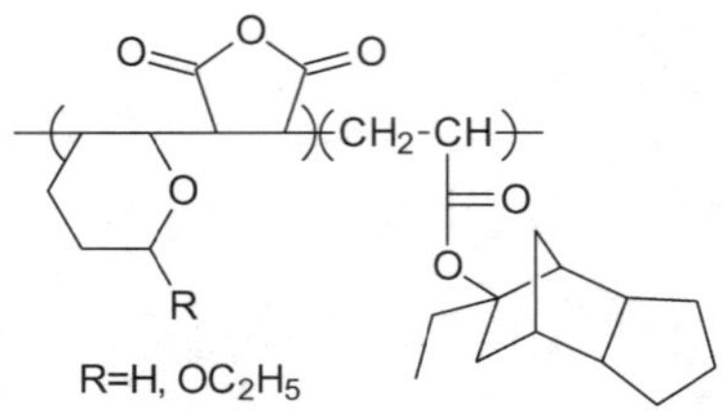

Fig. 76 VEMA 193 nm resist

with vinyl ethers improves copolymer yields, increases the chain flexibility, reduces the very high T_g, and improves adhesion.

In addition to the four major platforms described above, new interesting polymers have been designed and prepared for use in 193 nm lithography. Because the phenolic functionality is too absorbing at 193 nm, ArF resist polymers are built on the use of carboxylic acid. As mentioned earlier in this section, polymers bearing carboxylic acids fail to provide good dissolution properties in many cases. The COMA system is attractive because the polymers can be prepared readily by radical copolymerization. However, the yield is rather low. In an attempt to overcome these two problems, new polymers prepared by radical co- and terpolymerizations of norbornene bearing hexafluoroisopropanol (NBHFA) and sulfur dioxide have been reported (Fig. 77) [115]. As mentioned earlier, hexafluoroisopropanol exhibits pK_a similar to that of phenol and is highly transparent at 193 nm. NBHFA was copolymerized with SO_2 using *tert*-butyl hydrogen peroxide at –45 °C for a few hours to afford a 1:1 alternating copolymer in a quantitative yield. The copolymer exhibited an OD of 0.25/μm at 193 nm and a high dissolution rate of 30,000 Å/s in 0.21 N TMAH. tBOC-protected NBHFA was also copolymerized with SO_2. Terpolymers of NBTBE, NBHFA, and SO_2 were prepared by changing the NBTBE/NBHFA ratio and

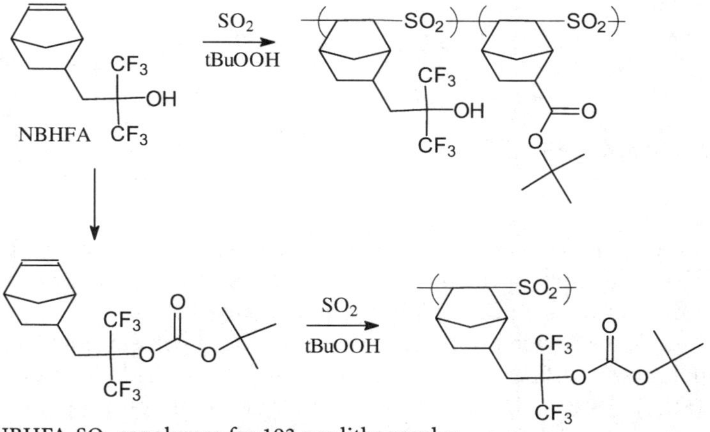

Fig. 77 NBHFA-SO_2 copolymer for 193 nm lithography

Chemical Amplification Resists for Microlithography 117

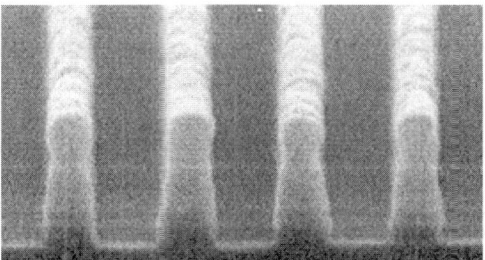

Fig. 78 Dense 105 nm line/space patterns printed by 193 nm lithography [273]

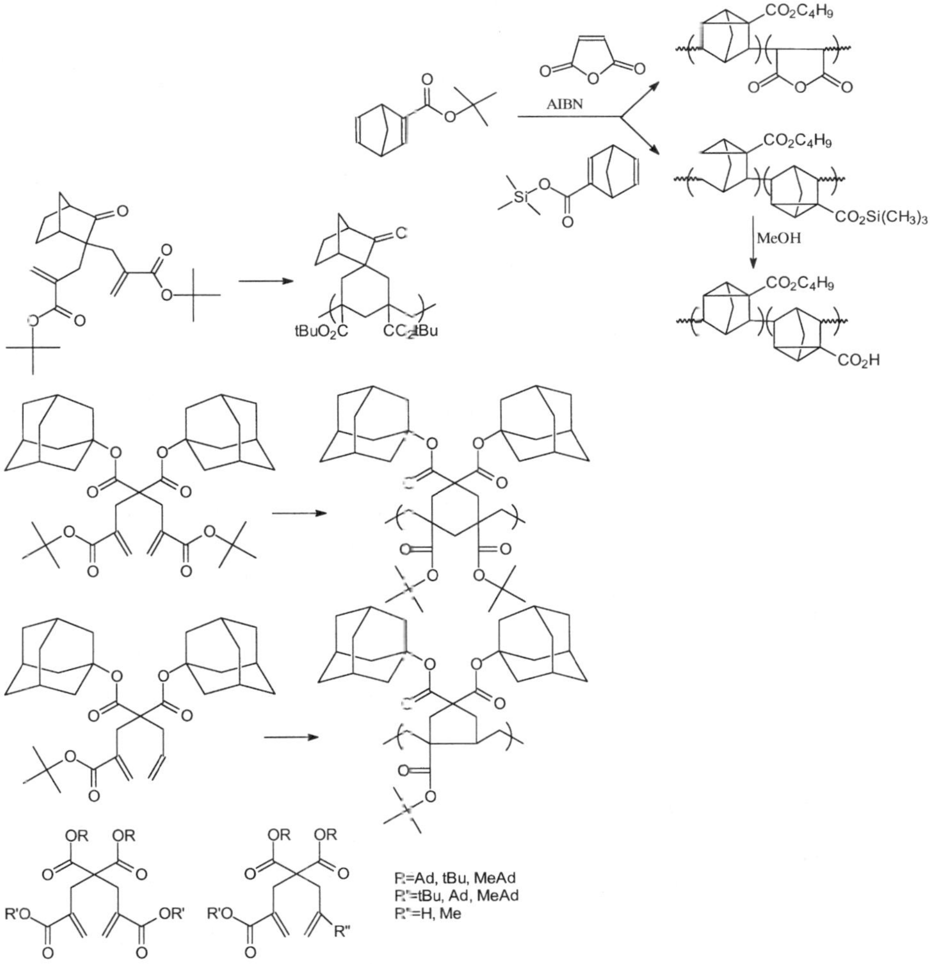

Fig. 79 Radical cyclopolymerization for preparation of 193 nm resist polymers

their dissolution rates investigated. One drawback of the poly(norbornene sulfone) system was its unexpectedly poor dry etch resistance in comparison with its aromatic counterpart, poly(tBOC-styrene sulfone) [104]. However, the use of NBHFA and hexafluoroisopropanol has attracted a great deal of attention recently for the design of 157 nm resist materials and the hexafluoroalcohol has become the acid group of choice in 157 nm lithography, which will be described in more detail later. This surge of interest in the use of NBHFA in 157 nm lithography and its more attractive dissolution behavior than carboxylic acids have revitalized the 193 nm resist design and allowed replacement of NBCA with NBHFA in the COBRA series at last [273]. Dense 105 nm line/space patterns printed in a 193 nm resist are presented in Fig. 78.

While ring-opening polymerization of camphorsultam was attempted futilely to prepare a new polymer containing a bicyclic structure and a new acidic sulfonamide group in the backbone [115b], radical cyclopolymerization was exploited in the synthesis of 193 nm alicyclic polymers (Fig. 79). Transannular polymerization to form polynortricyclene bearing *tert*-butyl ester was utilized in radical copolymerization with MA (Fig. 79) [275]. Radical cyclopolymer-

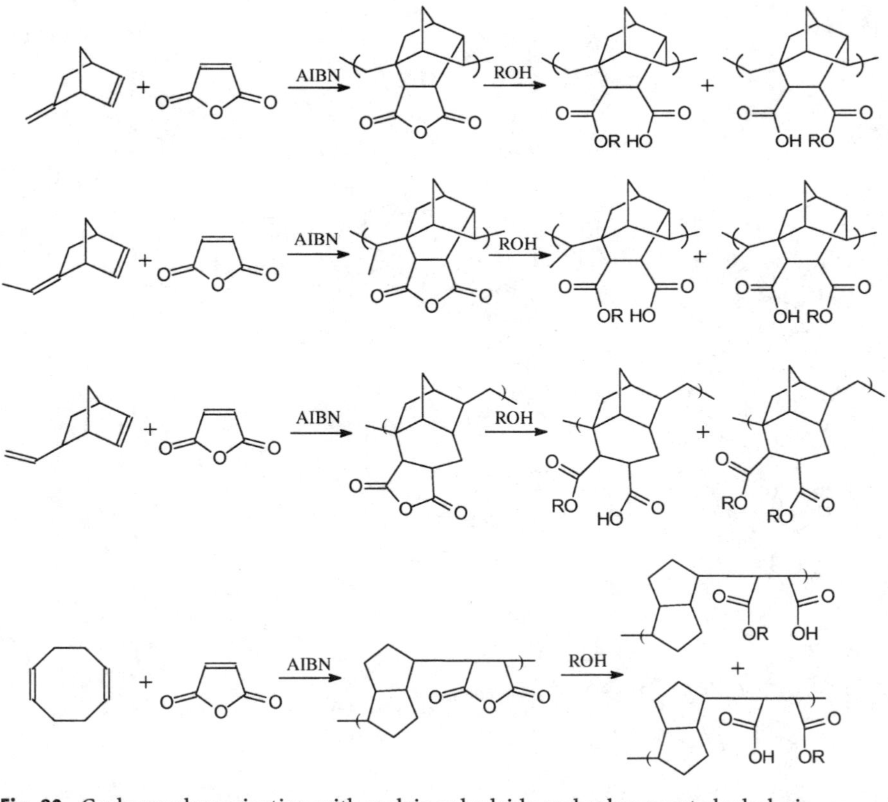

Fig. 80 Cyclocopolymerization with maleic anhydride and subsequent alcoholysis

ization of spirobornanes and acyclic analogs has been exploited (Fig. 79) [275]. Radical copolymerization with maleic anhydride involving cyclopolymerization has been employed in the synthesis of new 193 nm alicyclic polymers (Fig. 80) [277]. In this approach the anhydride was hydrolyzed with alcohol to form a half-ester and the remaining carboxylic acid was protected with acid-labile groups such as acetal for chemical amplification imaging. Molecular weights were typically low (1000–3700) and OD was rather high (>0.6/µm).

4.1.10.4
New Processes

Block copolymers of TBMA with 3-methacryloxypropylpentamethyldisiloxane prepared by group transfer polymerization exhibited better development behavior in aqueous base than that of the corresponding random copolymer (Fig. 81) [278]. The block copolymers with higher silicon concentrations are developable as negative resists using supercritical CO_2. In a similar fashion, block copolymers of THP-methacrylate with fluoroalkyl methacrylates were evaluated as CO_2-developable 193 nm resists [279]. Use of supercritical CO_2 in resist processing has started to gain a ground as it is environmentally friendly. The concept of using supercritical CO_2 in development was first introduced by Phasex and IBM in 1995 [280]. However, very few polymers are soluble in supercritical CO_2, and those that are soluble are usually either fluorinated or siloxane-based. Therefore, supercritical CO_2 development may be more suited for 157 nm lithography as described in more detail later.

As mentioned earlier, the COMA positive resists tend to have higher optical absorption at 193 nm than polymethacrylate and COBRA systems, which would produce a tapered image profile. To overcome this potential problem, the T-top formation by absorption of base into the top layer (see above) has been intentionally incorporated in the lithographic process (amine gradient process) [281]. Poly(acrylic acid-co-methyl acrylate) and L-proline were dissolved in water and spin-cast on a COMA resist. During PEB the amine in the overcoat diffuses into the COMA resist layer and compensates for the acid gradient caused by illumination, providing a vertical profile.

Furthermore, the thermal flow process (REFLOW, Fig. 56) employed in the 248 nm lithography has been applied to the COMA 193 nm resists [282]. However, shrinking hole sizes by thermal flow is more difficult with 193 nm resists due to their high T_g. Similarly, a hole shrink process named RELACS (resolution enhancement lithography assisted by chemical shrink) (Fig. 82), initially developed for KrF positive resists [283] has been utilized in the COMA 193 nm resists [284]. This shrink technique utilizes a crosslinking reaction catalyzed by an acid component existing in a predefined resist film. In the SAFIER (shrink assist film for enhanced resolution) process proposed by Tokyo Ohka Kogyo, an imaged wafer is coated with a SAFIER chemical and then baked to cause the resist to flow. The SAFIER material is rinsed away with water and residual water is removed by baking at 100 °C. The SAFIER material does not chemically

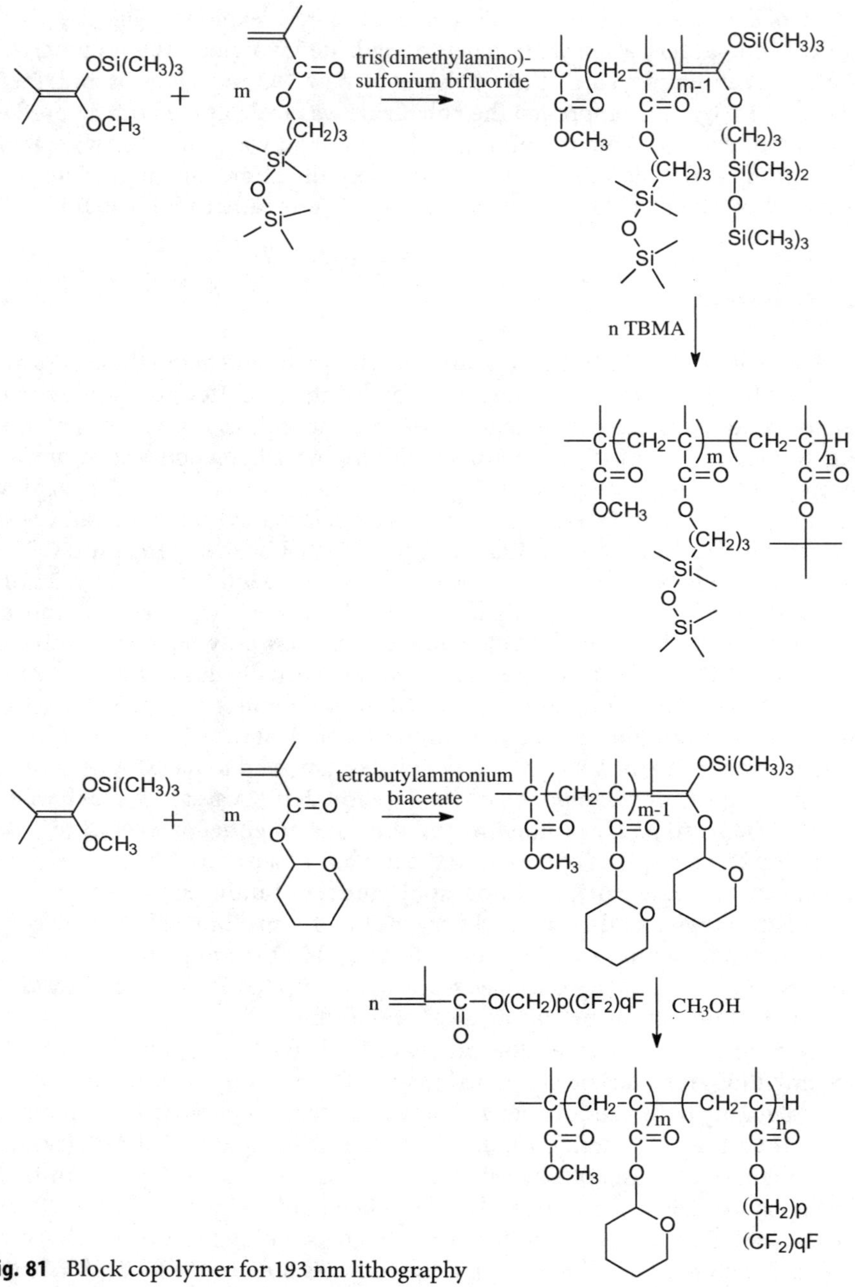

Fig. 81 Block copolymer for 193 nm lithography

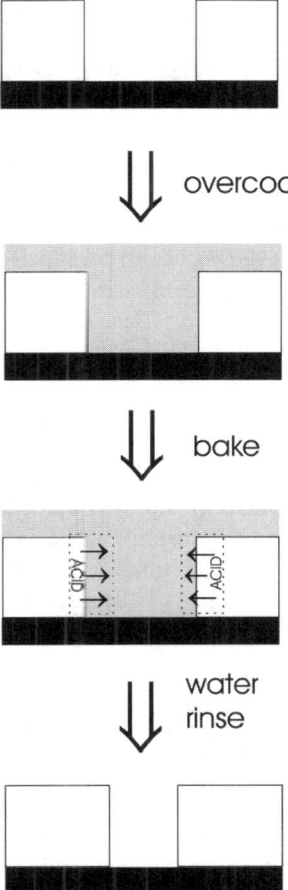

Fig. 82 RELACS process

react with the resist but provides mechanical support to the contact hole walls during the thermal flow process, minimizing pattern profile degradation [285].

For further enhancement of the resolution at 193 nm, immersion lithography employing water as an optical element between the last lens and resist has emerged recently as a new technology in competition with 157 nm lithography. In immersion lithography the effective exposure wavelength is reduced by the refractive index of the immersion fluid (193/1.44=134 nm for 193 nm water immersion).

4.1.11
Fluoropolymers for 157 nm Lithography

The microelectronics industry is interested in maintaining photolithography as the primary manufacturing technology and attempting to migrate to an even shorter wavelength of F_2 excimer lasers at 157 nm for a higher resolution (≤70 nm). However, existing 248 and 193 nm resists are absorbing too much and therefore a search for more transparent materials has been required once again. At this short wavelength, only fluoropolymers and polysilsesquioxanes may provide low enough absorption [286], while mature ESCAP-based 248 nm resists were employed in early imaging experiments in a thin film (<100 nm) [220, 287]. Thus, single layer 157 nm resists must be built on fluoropolymers. The early screening effort was initiated at MIT Lincoln Laboratories and Table 4 summarizes optical densities (ODs) of representative polymers at 157 nm. In addition to the good transparency, compatibility with aqueous base development is a very important consideration. How can hydrophobic fluoropolymers be made soluble in aqueous base? The answer was provided by the earlier 193 nm resist design based on norbornene hexafluoroalcohol (Fig. 21) [115]. While carboxylic acid and phenol absorb strongly at 157 nm, the acidic hexafluoroisopropanol group incorporated in a poly(norbornene sulfone) structure provided good transparency of 3.2/µm [288]. Thus, the hexafluoroisopropanol functionality has become the acid group of choice for 157 nm lithography.

Fluoropolymers for 157 nm lithography can be categorized into two groups: 1) polymers containing F in the backbone, typically prepared by copolymerization involving tetrafluoroethylene (TFE) and 2) polymers containing F in the side chain (Fig. 83). NBHFA has been copolymerized with NBTBE, NBHFA

Table 4 Absorbance at 157 nm of 1-µm-thick Polymer Films [286]

Polymer	Absorption/µm	Polymer	Absorption/µm
Polyhydrosilsesquioxane	0.06	Polynorbornene	6.10
Polydimethylsiloxane	1.61	Polystyrene	6.20
Polyphenylsiloxane	2.68	Polyvinylphenol	6.25
Fluorocarbon 100% fluorinated	0.70	Poly(norbornyl methacrylate)	6.67
Hydrofluorocarbon 30% fluorinated	1.34	Poly(adamantyl methacrylate)	6.73
Partially esterified hydrofluorocarbon 28% fluorinated	2.60	Poly(β-pinene)	7.15
Fully esterified hydrofluorocarbon 31% fluorinated	4.56	Acrylic terpolymer (PTBMA-MMA-MAA)	8.20
Poly(vinyl alcohol) 99.7%	4.16	Polychlorostyrene	10.15
Ethyl cellulose	5.03	Polyvinylnaphthalene	10.60
Poly(methyl methacrylate)	5.69	Poly(acrylic acid)	11.00

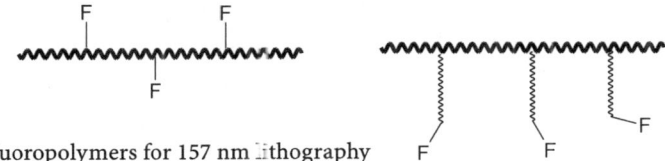

Fig. 83 Fluoropolymers for 157 nm lithography

protected with tBOC, and other norbornene monomers by vinyl addition polymerization using Pd or Ni (Fig. 84) [288, 289]. NBHFA readily undergoes addition polymerization with a Pd catalyst without protecting the acidic OH group to yield a polymer that is highly transparent at 157 nm (1.7/µm) and dissolves at a few hundred nm/s in a 0.26 N TMAH solution [288]. In a fashion similar to the 248 nm APEX resist, partial protection of PNBHFA produces an imageable copolymer. However, in such all-norbornene addition copolymers of NBHFA, the concentration of the ester or carbonate must be below 20 mol% for an acceptable OD [288, 289]. PNBHFA protected with 20 mol% tBOC has a low OD of 1.9/µm at 157 nm owing to the high concentration of fluorine [289]. However, this copolymer generates hexafluoroalcohol upon acidolysis and tends to lack the high contrast provided by conversion of *tertiary* ester to carboxylic acid. Therefore, a *tert*-butyl ester oligomer was added to the copolymer as a dissolution inhibitor to improve the development contrast, which has become the first commercial 157 nm resist marketed by Clariant (Fig. 85) [290]. In an attempt to reduce the 157 nm absorption further, the carbonate was replaced with an acetal protecting group [290]. The TFE-based systems have been promoted by DuPont [291]. A higher concentration of TFE in polymer provides lower 157 nm absorption, lower hydrophilicity, and lower

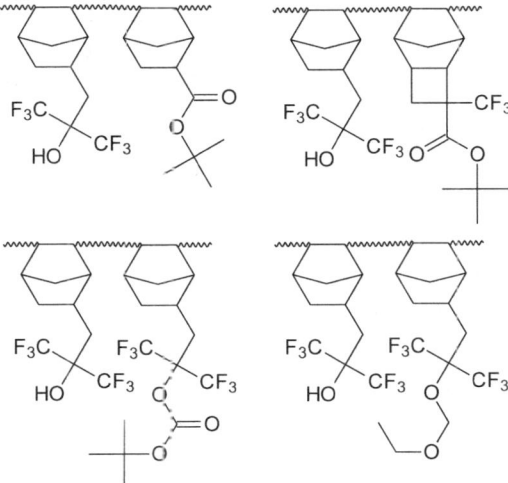

Fig. 84 NBHFA copolymers prepared by addition polymerization

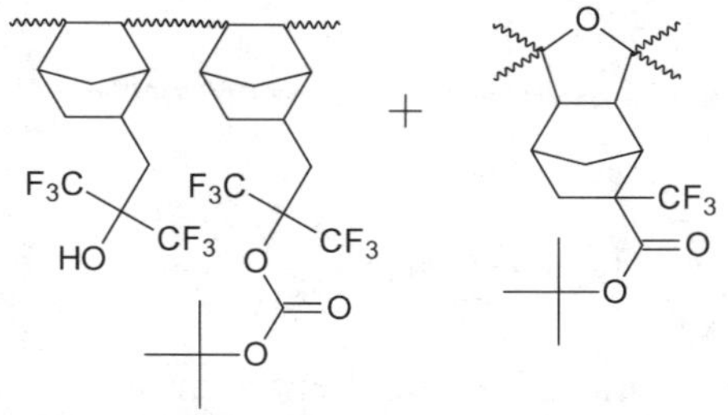

Fig. 85 157 nm resist based on NBHFA addition polymer

dry etch resistance. The polymerization involving TFE requires a special reactor and can be carried out only in a properly-equipped facility. This approach has been also pursued by Daikin Kogyo, Japan [292, 293]. TFE is copolymerized with NBHFA protected with an acid-labile group or terpolymerized with NB or NBHFA and acrylate bearing acid-labile ester (Fig. 86) [291–293].

2-Trifluoromethylvinyl acetate has been copolymerized using a radical initiator with vinyl acetate and the resulting copolymer was hydrolyzed with propylamine to yield poly(vinyl alcohol) containing about 50 mol% CF_3 (Fig. 87) [294]. The CF_3-bearing vinyl alcohol copolymer had an OD of 3/µm (4.2/µm for poly(vinyl alcohol)) and was soluble in 0.26 N TMAH solution. The vinyl alcohol copolymer was protected with a tetrahydropyranyl group (~70% modification) to produce a resist polymer with OD_{157} of 3.2/µm, which was employed in imaging in conjunction with a steroid-based (cholestanol, pregnanediol, etc.) active dissolution inhibitor which generates trifluoromethylcarbinol upon deprotection (Fig. 87) [295]. The distribution of iodonium PAG and the steroid dissolution inhibitor in the THP-protected poly(vinyl alcohol) has been investigated by RBS and segregation was detected in some of the combinations [295].

Interesting hexafluorocarbinol-functionalized block copolymers were prepared by a three-step polymer modification process (Fig. 88) [296]. The fluorocarbinol functional group was incorporated into the polymer backbone

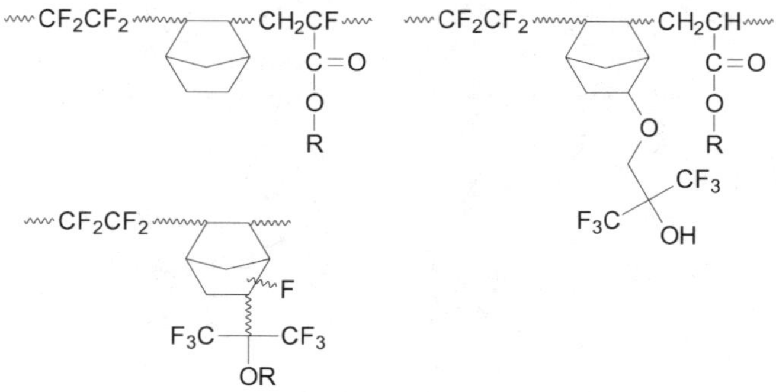

Fig. 86 TFE-based 157 nm resist polymers

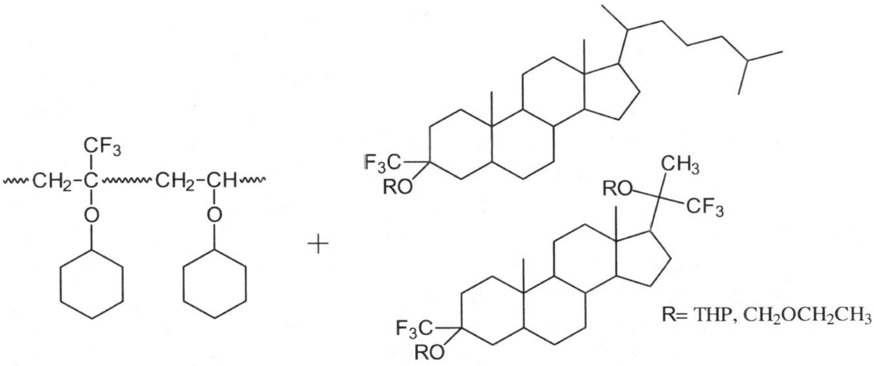

Fig. 87 Trifluoromethylvinyl alcohol polymer and steroid dissolution inhibitors

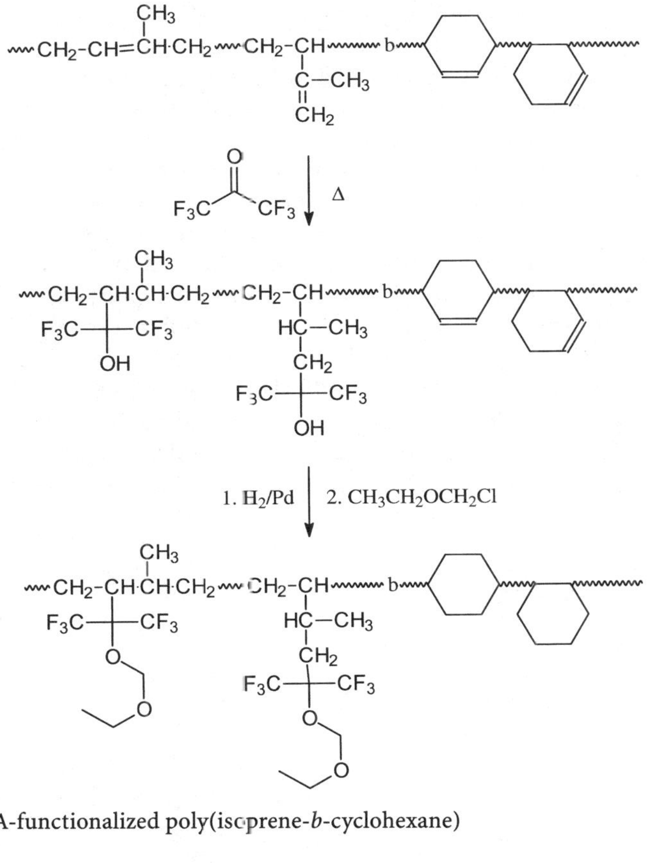

Fig. 88 HFA-functionalized poly(isoprene-b-cyclohexane)

through an ene reaction [297] of hexafluoroacetone with C=C double bonds in polymer. A block copolymer of isoprene and 1,3-cyclohexadienene was prepared by living anionic polymerization with a *sec*-butyllithium/*N,N,N',N'*-tetramethylenediamine system. The cyclic structure was incorporated to raise T_g and dry etch resistance. The polyolefin was treated with hexafluoroacetone at 180 °C for 36 h in a pressure reactor and the resulting polymer was hydrogenated using a supported Pd catalyst to destroy the absorbing C=C double bonds, resulting in only partial hydrogenation of 40%. The partially hydrogenated block copolymer was reacted with ethoxymethyl chloride to produce a resist polymer with OD_{157} of 3.3/µm (Fig. 88).

The hexafluoroisopropanol group has been incorporated in a (meth)acrylate structure (Fig. 89) [298]. 1,4-Bis(2-hydroxyhexafluoroisopropyl)benzene was reduced to the corresponding cyclohexane, which was then reacted with (meth)acryloyl chloride after treatment with *n*-butyllithium to yield an acrylic monomer containing twelve fluorine atoms and a hexafluoroisopropanol group. The acrylate homopolymer prepared by radical polymerization exhibited an unusually low OD_{157} of 1.9/µm, perhaps suggesting the electronic effect of the CF_3 group on the carboxyl group and the volume effect of the CF_3 group (van der Waals volumes for CF_3 and CH_3 are 42.6 and 16.8 $Å^3$, respectively). A methoxymethyl-protected acrylate was synthesized and copolymerized with the unprotected monomer, yielding a copolymer exhibiting low developer selectivity. Therefore, the unprotected monomer was copolymerized with TBMA, tetrahydropyranyl methacrylate, or 2-methyladamantyl methacrylate to generate carboxylic acid upon photochemically induced deprotection.

Another interesting building block of 157 nm resist polymers is a 2-trifluoromethylacrylic group. Poly(methyl 2-trifluoromethylacrylate) (PMTFMA), which was initially prepared as an electron beam resist [299], has been found experimentally [288, 289] and by theoretical calculation [300] to be quite transparent at 157 nm (3.1/µm). Replacement of CH_3 of PMMA with CF_3 reduces the 157 nm OD by a 3–4 order of magnitude. MTFMA is reluctant to undergo radical homopolymerization [301], although it is 54 times more reactive than

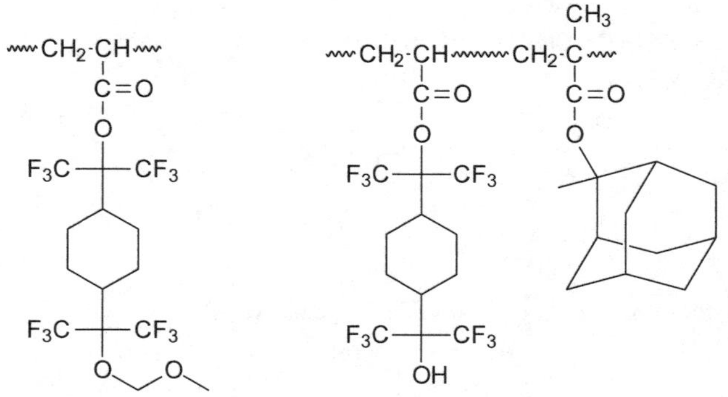

Fig. 89 HFA-bearing (meth)acrylates

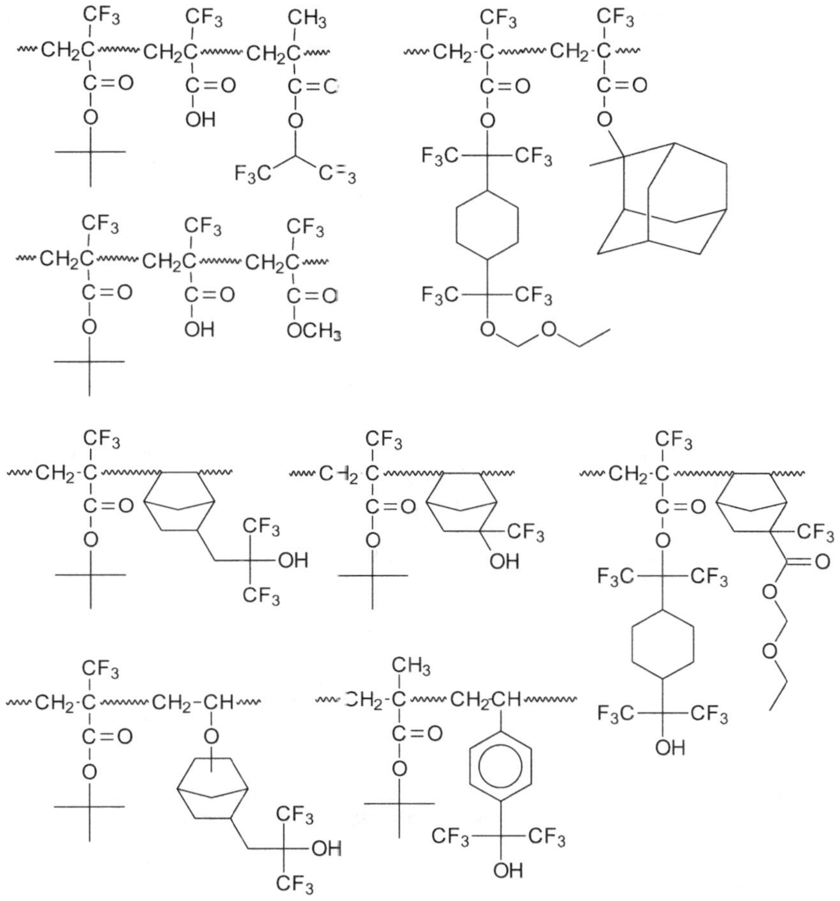

Fig. 90 2-Trifluoromethylacrylate polymers

MMA toward an alkyl radical [302]. Because of its low electron density (e=2.7) [302], this fluoromonomer readily undergoes anionic polymerization with amines and organic and inorganic salts in the presence of 18-crown-6 [301, 303]. However, anionic initiators commonly used for polymerization of MMA such as butyllithium and Grignard reagents fail to polymerize this monomer due to Sn2′ addition-elimination [303, 304]. All-acrylate polymers for 157 nm imaging have been prepared by radical terpolymerization of 2-trifluoromethylacrylate and methacrylate or acrylate [288] (Fig. 90).

The 2-trifluoromethylacrylic monomers have been found to copolymerize with norbornenes such as NBHFA under normal radical polymerization conditions [288]. The copolymerization behavior has been investigated by analyzing the kinetics in situ by ^{1}H NMR [288]. The monomer consumption curves for NB and 2-trifluoromethylacrylic acid (TFMAA) are presented in Fig. 91 and

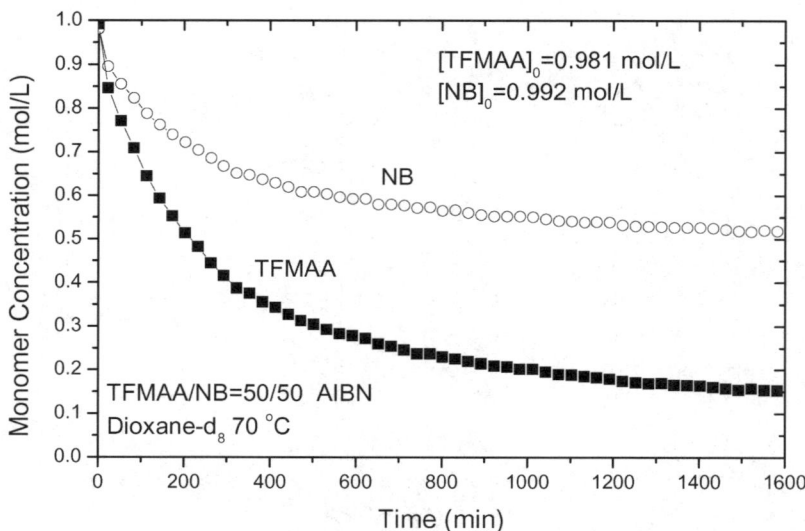

Fig. 91 Monomer consumption kinetics curves for radical copolymerization of NB with 2-trifluoromethylacrylic acid at 60 °C in dioxane-d_8 [288)]

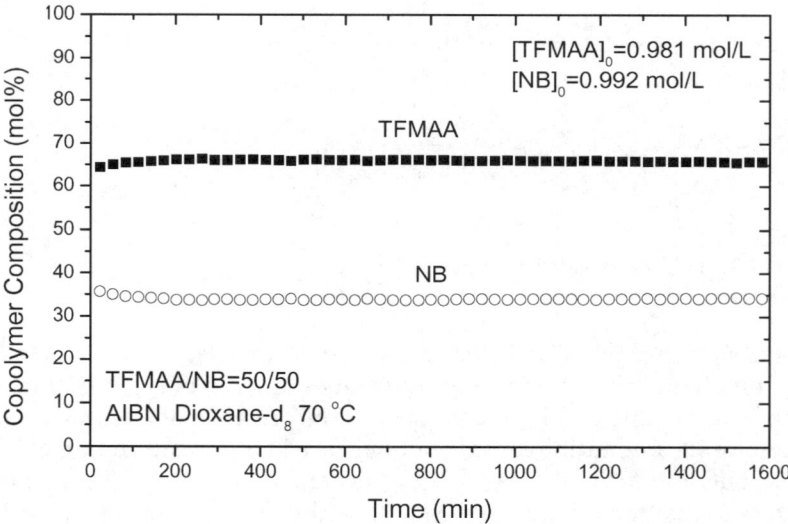

Fig. 92 Time dependence of feed and copolymer compositions in radical copolymerization of NB with TFMAA [288]

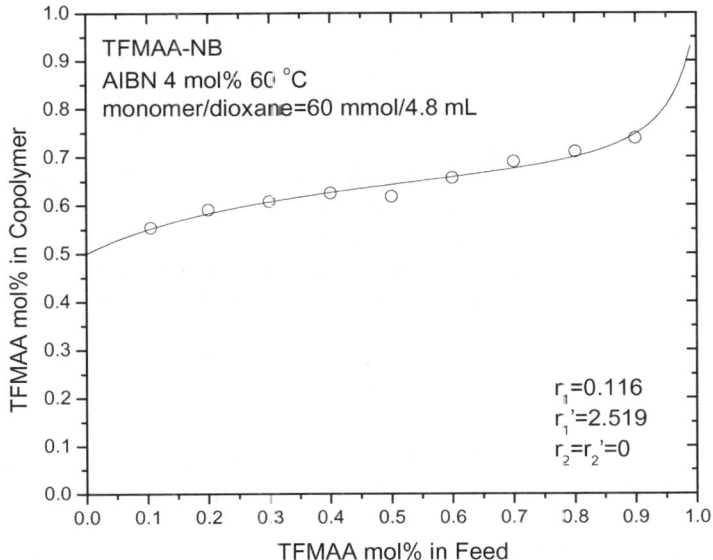

Fig. 93 Copolymer composition curve for radical copolymerization of TFMAA with NB (penultimate model) [305]

the TFMAA concentrations in feed and copolymer in Fig. 92. The copolymer composition is not 1:1 but TFMAA:NB=~2:1. It has been shown that the commonly employed terminal model cannot explain the copolymerization behavior and that copolymerization proceeds according to the penultimate model (Fig. 93) [305]. *tert*-Butyl 2-trifluoromethylacrylate (TBTFMA) has been copolymerized with NBHFA [288, 305] or 5-trifluoromethyl-5-hydroxy-2-norbornene [290] to prepare 157 nm resist polymers with an OD of 2.6–2.7/μm (Fig. 90). The copolymer is quite lipophilic and insoluble in aqueous base (0.26 N TMAH) due to the low concentration of NBHFA (<40 mol%). Because the NBHFA concentration in copolymer cannot be increased by changing the feed ratio, the copolymer was blended with PNBHFA [305]. The blending increases the hydrophilicity and reduces OD to 2.0/μm. The successful radical copolymerization of 2-trifluoromethylacrylic monomers with norbornene derivatives prompted others to seek for electron deficient monomers such as cyanoacrylates for radical copolymerization with norbornenes [306, 307]. However, as cyanoacrylates undergo anionic polymerization very rapidly, the copolymers were contaminated with the homopolymer [306]. 1,2,2-Trifluoroacrylic ester of 2-methyladamantanol has been synthesized and copolymerized by radical initiation with NBHFA to form an alternating copolymer with T_g of 155 °C [307].

Vinyl ethers have been shown to undergo radical copolymerization with 2-trifluoromethylacrylic monomers readily, providing copolymers with high molecular weights in high yields [305]. The incorporation of vinyl ethers is typically 30 mol% like NB derivatives and the copolymerization follows the penultimate model [305]. However, this copolymerization is much more facile than that of norbornenes, producing high molecular weight polymers in high yields even in solution while solution copolymerization of norbornenes with 2-trifluoromethylacrylic monomers is rather sluggish. Dihydrofuran and vinylene carbonate copolymers with TBTFMA were used as resist polymers by blending them with PNBHFA while ethyl vinyl ether and *tert*-butyl vinyl ether copolymers were immiscible with the NBHFA homopolymer [305].

An attempt was made to incorporate the good transparency of the 2-trifluoromethylacrylic unit in the norbornene structure through the Diels-Alder reaction of *tert*-butyl 2-trifluoromethylacrylate [289, 290]. However, norbornene substituted with two groups at the 5-position was reluctant to undergo addition polymerization. Therefore, this monomer was copolymerized with carbon monoxide and the low molecular weight copolymer with OD_{157} of 3.5/μm (Fig. 94) was employed as a dissolution inhibitor of PNBHFA partially protected with *t*BOC and a NBTBE-NBHFA copolymer as mentioned earlier (Fig. 84) [289, 290]. The incorporation of the carbon monoxide copolymer reduces OD in the former case and increases a development contrast through generation of carboxylic acid. Another attempt to polymerize a monomer geminally substituted with CF_3 and *tert*-butoxycarbonyl was to remove the substituents further away from the C=C bond by synthesizing the corresponding tricyclononene analogues (Fig. 94) [289, 290]. This tricyclononene monomer was copolymerized with NBHFA using a Pd(II) catalyst to produce a copolymer with OD of 2/μm at 157 nm. The carbon monoxide copolymer was blended with the tricyclononene copolymer as a dissolution inhibitor.

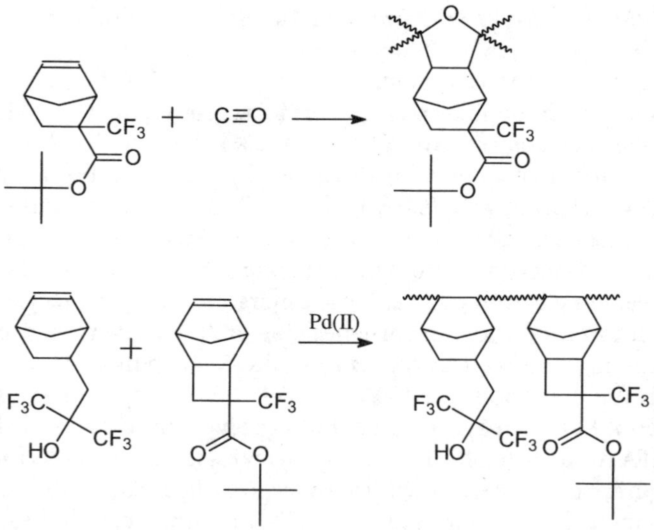

Fig. 94 Polymerization of norbornene derivatives geminally substituted with CF_3 and CO_2tBu

The aforementioned 1,4-bis(2-hydroxyhexafluoroisopropyl)cyclohexane has been combined with the 2-trifluorometylacrylic structure [290, 305, 307]. The fluorodiol was half-protected with an ethoxymethyl group and reacted with 2-trifluoromethylacryloyl chloride in the presence of triethylamine to afford 2-[4-(2,2,2-trifluoro-1-ethoxymethoxy-1-trifluoromethylethyl)cyclohexane]hexafluoroisopropyl 2-trifluoromethylacrylate. This heavily fluorinated acrylate was copolymerized with 2-methyladamanty 2-trifluoromethylacrylate by anionic initiation with potassium acetate/18-crown-6 [307] as described in the literature [303] (Fig. 90). The copolymer (made from a 1:1 feed) was unexpectedly transparent with OD_{157} of 1.6/μm. However, imaging of the copolymer resists was sluggish perhaps due to their low T_g. Radical copolymerization was also performed with norbornene derivatives [290, 305].

Aromatic polymers were believed to be too absorbing at 157 nm. However, polystyrene bearing hexafluoroisopropanol (PSTHFA, Fig. 20) has been found to be unexpectedly transparent at 157 nm with OD of 3.6/μm [288b,c, 308, 309]. STHFA was copolymerized using AIBN with tBOC-protected STHFA to produce PF-APEX. Copying the most dominant 248 nm resist design, STHFA has been copolymerized with TBMA (TBA, isobornyl methacrylate) and TBTFMA to afford by radical polymerization PF-ESCAP (or F-ESCAP) and PF^2-ESCAP resist polymers, respectively (Fig. 95) [288b,c, 308, 310]. In a similar fashion, styrene bearing trifluoroisopropanol was synthesized and polymerized but such polymer did not provide good transparency (OD=4.5/μm) or dissolution behavior (completely insoluble in 0.26 N TMAH while PSTHFA dissolves at 6000 Å/s), which was also the case with norbornene with pendant trifluoroisopropanol [288b,c]. While TBMA copolymerizes with STHFA according to the terminal model, affording copolymers with any compositions, the radical

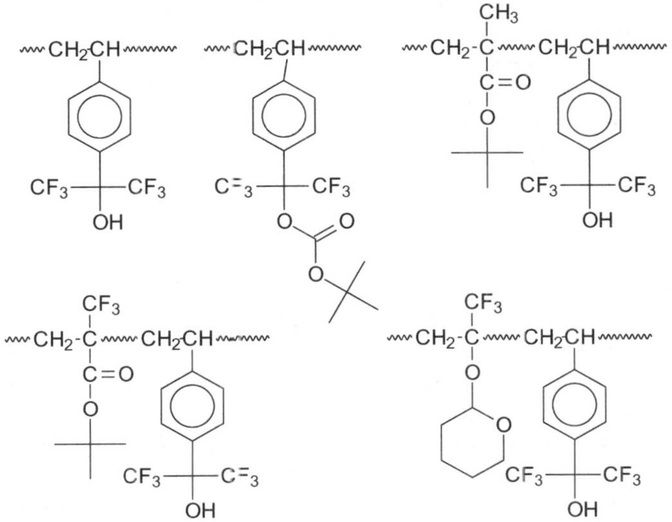

Fig. 95 STHFA polymers

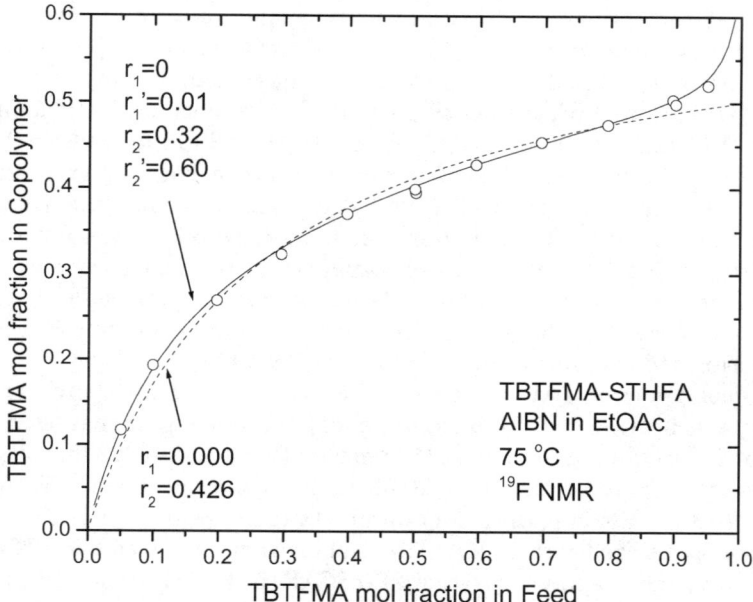

Fig. 96 Copolymer composition curve for radical copolymerization of TBTFMA and STHFA [305]

copolymerization TBTFMA with STHFA can be explained by the penultimate model better than the terminal model (Fig. 96), producing copolymers containing <50 mol% TBTFMA [305]. PF^2-ESCAP is more transparent than PF-ESCAP at 157 nm (3.2–3.6 vs 4.0–4.2/μm) [288b,c, 308]. A copolymer prepared by radical copolymerization of ethoxymethyl-protected STHFA and 2-[4-(2-hydroxyhexafluoroisopropyl)cyclohexane]hexafluoroisopropyl acrylate (1:1 feed) exhibited good transparency of 2.38/μm [298]. STHFA has also been copolymerized with 2-trifluoromethylvinyl acetate at a 1:1 feed ratio and the resulting copolymer (70% STHFA) was hydrolyzed to alcohol, followed by protection with THP. The dissolution contrast was too poor to form a nice pattern [307].

4-Hydroxy-2,3,5,6-tetrafluorostyrene was another aromatic monomer investigated for 157 nm application, which can be easily synthesized from pentafluorostyrene (Fig. 97) [307]. Radical polymerization of this unprotected phenolic monomer was slow. Its *tert*-butyl ether was synthesized in a slightly modified way and polymerized with ease by radical initiation to afford a polymer with OD$_{157}$ of 3.8/μm [307]. The *tert*-butyl ether monomer was copolymerized radically with 2-[4-(2-hydroxyhexafluoroisopropyl)cyclohexane]hexafluoroisopropyl acrylate (Fig. 97), affording a 80:20 copolymer with OD$_{157}$ of 2.1/μm. However, a resist based on the 80:20 copolymer could not be developed even with a 0.52 N TMAH solution for 5 min after exposure to 248 nm irradiation. Imaging was more successful with a 60:40 copolymer but a stronger developer and long development time were required. The copolymers had a low T_g of

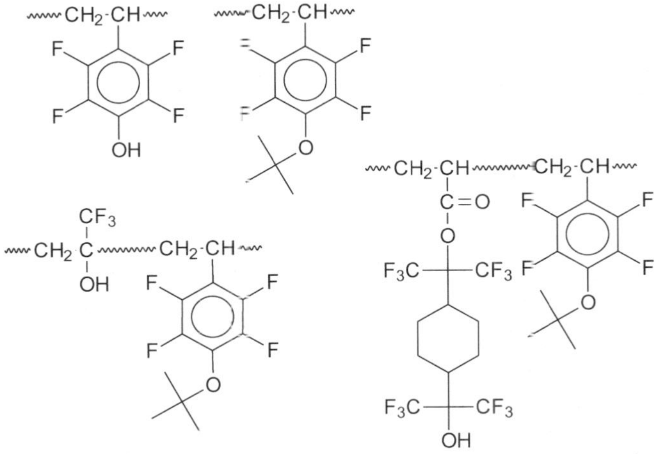

Fig. 97 4-Hydroxytetrafluorostyrene polymers

about 90 °C. The *tert*-butyl ether monomer was copolymerized also with 2-trifluoromethylvinyl acetate [307]. Other fluorinated styrenes have been also evaluated as ESCAP-type 157 nm resists in copolymer with TBA, TBMA, and TBTFMA [311].

In addition to the oligomeric and polymeric dissolution inhibitors discussed earlier, small molecules bearing acid labile groups have been employed in 157 nm resist formulations [295, 312]. Representative examples are shown in Fig. 98. Some are better than others in dissolution inhibition of a copolymer of NBHFA and NBTBE (92:8). What is interesting is that a diazonaphthoquinone PAC developed for mid UV application (Fig. 99) [313] is surprisingly transparent and can inhibit the dissolution of PNBHFA even better than the small acid-labile dissolution inhibitors in Fig. 98 [312]. In contrast, the dissolution of PSTHFA cannot be efficiently inhibited either with diazonaphthoquinone, the small acid-labile lipophilic compounds in Fig. 98, or the carbon monoxide copolymer (Fig. 94) [312].

A new fluoropolymer has been synthesized by cyclopolymerization and demonstrated to have a very low OD_{157} of $<1/\mu m$ for the F concentration of >50 wt% (Fig. 100) [314]. A thick film of 200 nm has been successfully imaged at 157 nm using a resin with OD_{157} of $0.8/\mu m$, resolving 70–65 nm dense patterns (alternating phase shifting). This is an outstanding achievement in transparency and imaging. However, the fluoropolymer exhibits a high dry etch rate of as much as 1.5 times 248 nm resists.

Completely new polymer architecture for 157 nm lithography has been proposed on the basis of quantum chemical calculation of VUV absorption of a number of model compounds [315]. Time-dependent density functional theory (TD-DFT) calculations suggested that sulfonic acid esters are transparent in the 157 nm region. In fact, poly(vinylsulfonyl fluoride) and poly(methyl vinylsulfonate) have been found to show low OD_{157} of 2.1 and $2.2/\mu m$ [316]. Various

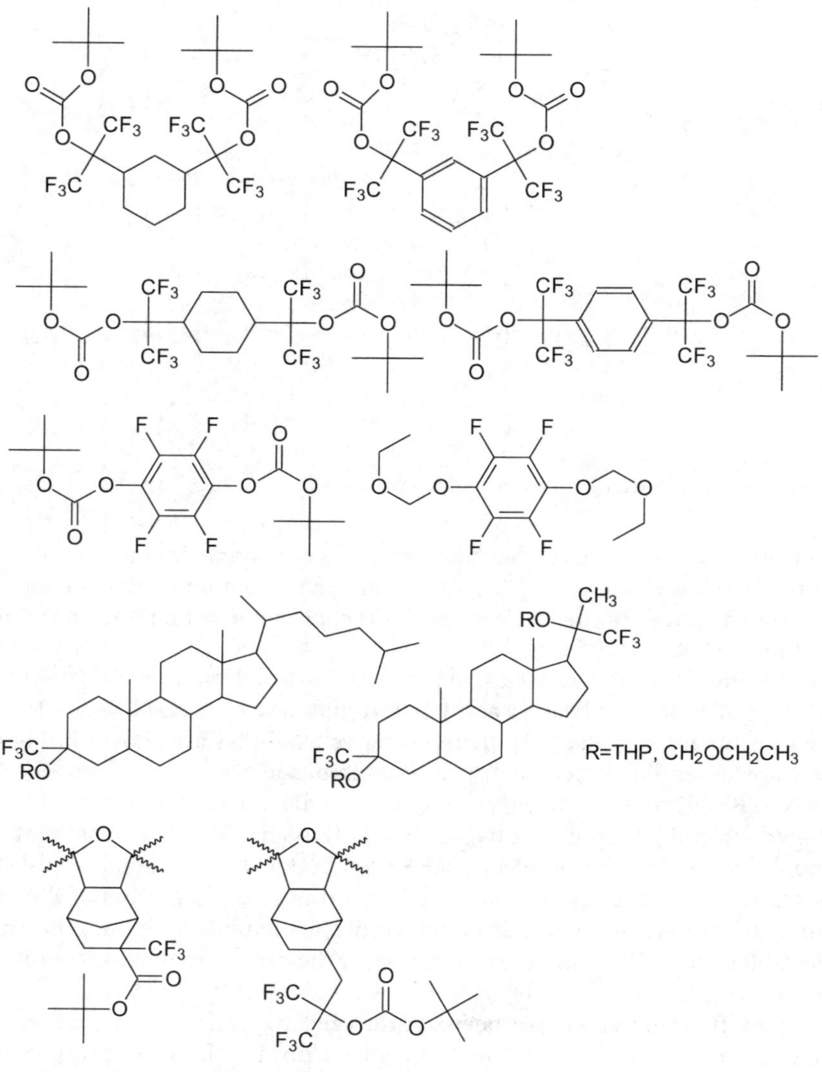

Fig. 98 Fluorinated dissolution inhibitors

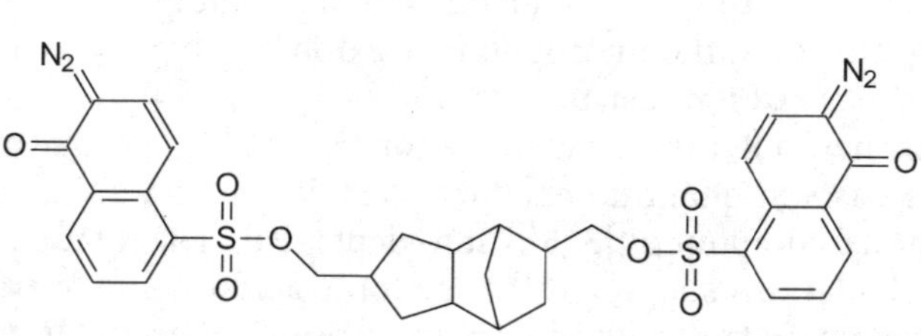

Fig. 99 Diazonaphthoquinone dissolution inhibitor

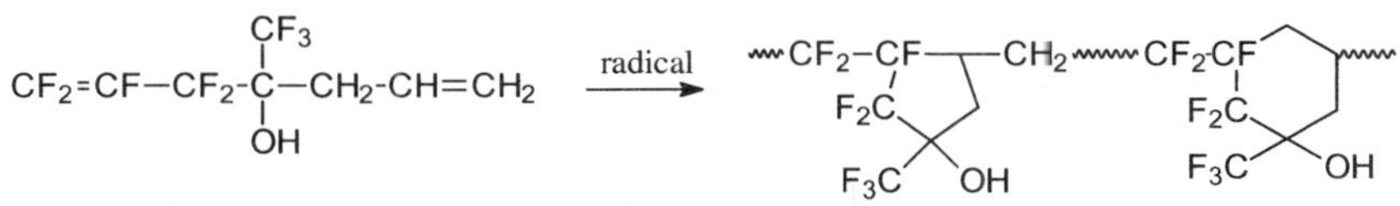

Fig. 100 Fluoropolymer formed by cyclopolymerization

alkyl vinylsulfonates have been synthesized from 2-chloroethanesulfonyl chloride and alcohols in the presence of pyridine and subjected to radical homo- and copolymerizations with STHFA using AIBN as the initiator in bulk (Fig. 101) [316]. Efforts to prepare *tertiary*-alkyl vinylsulfonates resulted in decomposition of products during purification by column chromatography. Small *primary* sulfonates homopolymerized easily, producing high molecular weight polymers in good yields, while large *primary* sulfonates were reluctant to undergo homopolymerization. Small *secondary* sulfonates polymerized below 50 °C but homopolymerization of large *secondary* sulfonates did not occur, producing only vinylsulfonic acid. 1:1 Copolymerizations of *primary* and *secondary* sulfonates with STHFA produced high molecular weight copolymers containing 20–30 mol% sulfonate. The *secondary* sulfonate copolymers were readily hydrolyzed and became soluble in water. The ODs of the copolymers at 157 nm ranged from 2.4 to 3.5/μm. A copolymer of hexafluoroisopropyl vinylsulfonate with STHFA (40:60) with OD_{157} of 2.4/μm was partially protected (32 mol% out of 60%) with *t*BOC to produce an imageable terpolymer with OD_{157} of 2.6/μm.

Copolymers of methacrylonitrile with 4-trimethylsilyloxy-α-methylstyrene, HFA-bearing α-methylstyrene, and trimethylsilyl methacrylate have been proposed as 157 nm positive resists [317].

Candidates and design rules for 157 nm resist polymers were sought by measuring VUV absorption of existing polymers including fluoropolymers or

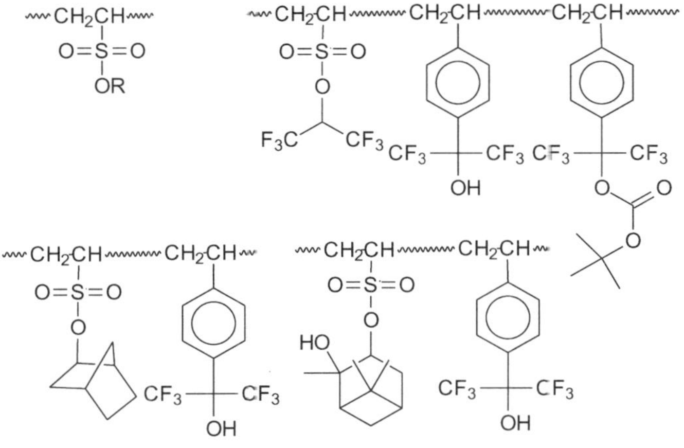

Fig. 101 Vinylsulfonate polymers

of small model compounds in gas phase and by calculating VUV spectra of model compounds. An empirical increment scheme to estimate absorption of polymers at 157 nm have been proposed (STUPID model) [318], which cannot describe all aspects of absorptivity. It is generally difficult to predict how substituents affect the absorbance of molecules because the introduction of such substituents significantly affects energy levels and spatial distribution of occupied and unoccupied molecular orbitals. Mtasuzawa et al. have demonstrated that TD-DFT calculation is useful in aiding the design of transparent materials [300, 319]. In their calculations, the transition energies of the molecules were adjusted using empirical equations because the employed methods had exhibited systematic errors over a wide range of wavelengths (80–310 nm). Ando et al. have recently shown that the TD-DFT calculations employing the B3LYP hybrid functional by combining geometry optimization using the 6-311G(d) basis set and subsequent calculations of transition energies and oscillator strengths using the 6-311++G(d,p) basis set can reproduce observed spectra of model compounds without incorporating empirical equations (Fig. 102) [315]. The calculated absorbance was represented by the oscillator strength divided by the van der Waals volume (nm^3) of molecules. Incorporation of hexafluoroisopropanol effectively reduces absorption owing to its large van der Waals volume (dilution effect). For estimating absorbance in the solid state (polymer film), calculated oscillator strengths should be divided by molecular volume. Fluorinated compounds have larger molecular volumes than those expected from their van der Waals volumes, which further reduces the absorbance. This study has correctly predicted that poly(alkyl vinylsulfonates)

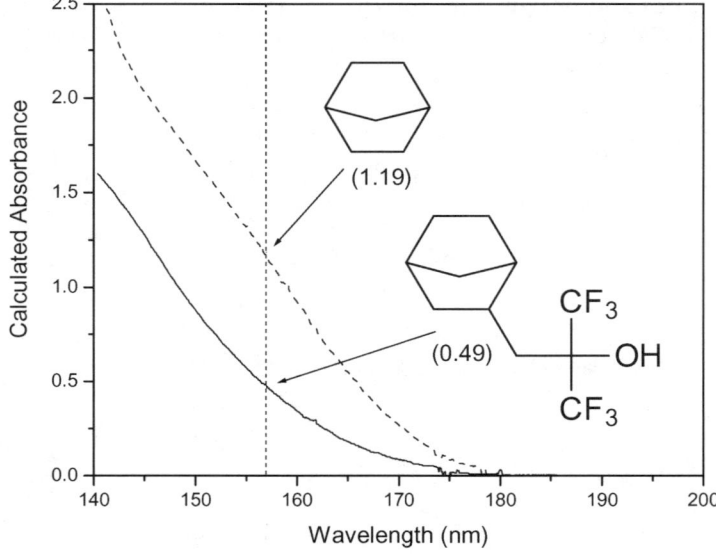

Fig. 102 Calculated VUV absorption for NB and NBHFA [315]

can be a viable platform for 157 nm lithography as mentioned earlier [316]. Highly precise theoretical calculation of VUV spectra has been performed by using a symmetry adapted cluster configuration interaction (SAC-CI) method but was limited to relatively simple molecules [320].

4.1.12
Resists for Next Generation Lithography

While immersion lithography at 157 nm [321] (the gap between the wafer and the objective lens filled with a liquid medium with index of refraction, n, that reduces the effective wavelength, and the resolution is now expressed by R= kλ/nNA) has emerged as a means to extend photolithography, extreme ultraviolet (EUV) lithography at 13.4 nm and electron beam projection lithography (EPL) are considered to be the next generation lithography (NGL) technologies after (or competing with) 193 nm immersion and 157 nm lithography for resolution below 50 nm. X-ray lithography has dropped out of the roadmap recently in US, although this is the only NGL that has demonstrated its feasibility by fabricating a working chip. In the case of the soft X-ray at 13.4 nm, atomic absorption dominates and there is little leverage in polymer structural modifications. Ultra-thin films (80–125 nm) of the mature 248 nm resists are likely to be employed in conjunction with a hard mask. Modification of the existing ESCAP resist for thin film imaging is under way as mentioned earlier. The major issues are a control of LER below 3 nm, sensitivity enhancement for throughput, shot noise limits in extremely sensitive resists (2 mJ/cm^2=1 photon/nm^2), and mask fabrication.

EPL (SCALPEL, scattering angular-limited projection electron-beam lithography developed by Bell Laboratories, and PREVAIL, projection reduction exposure with variable-axis immersion lenses developed by IBM) has been built around the mature 248 nm resists and have the same issues as EUV and some requirements unique to e-beam. Even in the projection mode, high sensitivity is critical for throughput (5–15 μC/cm^2 at 100 keV). The sensitivity enhancement has been achieved by increasing the PAG concentration and changing the resin composition, considering a strong dissolution inhibition effect of PAG [222]. For SCALPEL, the ESCAP-type 248 nm resist has been modified to increase the sensitivity [322]. For PREVAIL a ketal-based system has been primarily employed and optimized for the development of the technology [222]. The major concern in the use of the low activation energy protecting group in e-beam exposure under high vacuum (~10^{-9} torr) was potential outgassing leading to e-beam tool contamination due to deprotection during exposure. However, because the ketal deprotection reaction requires a stoichiometric amount of water (Fig. 103), the polymer essentially remains intact during exposure in low humidity vacuum. Only after the exposed resist film has been removed from the e-beam chamber into contact with moist ambient air, the deprotection reaction proceeds. The effect of humidity on the degree of deprotection of a ketal resist system (KRS) is presented in Fig. 104 as investigated by

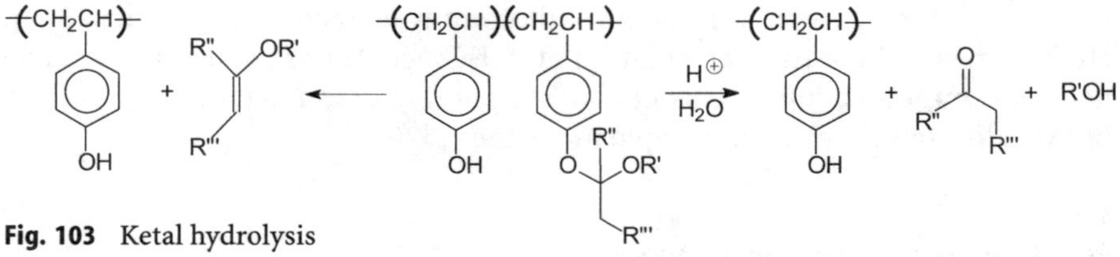

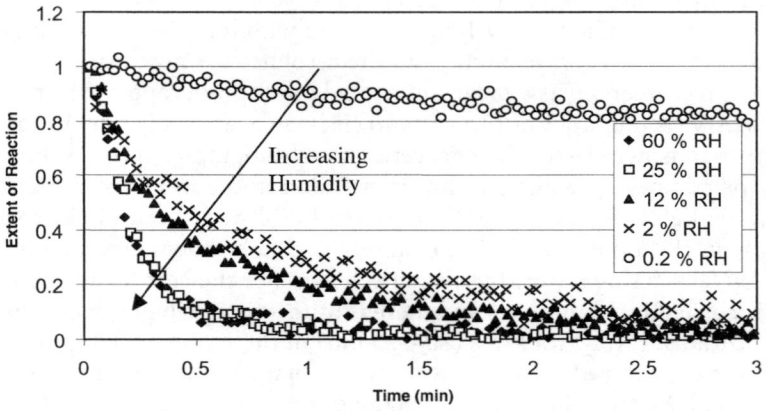

Fig. 103 Ketal hydrolysis

Fig. 104 Effect of humidity on ketal deprotection [222]

real-time infrared reflectance-absorbance spectroscopy [222]. In addition, the ketal group is selected so as to generate low volatility deprotection products. Another important application of the e-beam technology is photomask printing, which has its own specific requirements such as stability after coating on thick chrome-on-glass (COG) substrates. The aforementioned KRS resist has been employed in this application also because it does not require PEB for deprotection to proceed, which is advantageous as the COG substrates are not particularly efficient conductors of thermal energy. Fluctuation in the PEB temperature could results in loss of the line width control. Figure 105 presents high resolution 75 nm line/space images printed in KRS by e-beam exposure. An interesting approach has been proposed recently for the acetal-based e-beam resist system. High molecular weight blocking groups were employed to prevent outgassing during exposure and upon acid-catalyzed hydrolysis these compounds were designed to produce phenolic or carboxylic hydroxyl groups capable of deprotonation during development (Fig. 106) [323].

While SCALPEL and PREVAIL utilize high energy electrons and fairly thick resist films, requiring high sensitivity, a low energy electron beam projection lithography (LEEPL) approach [324] using thin film imaging has some advantages including less constraint on sensitivity and a small proximity effect. Thin film imaging using a bilayer system alleviates the line collapse in high aspect

Chemical Amplification Resists for Microlithography

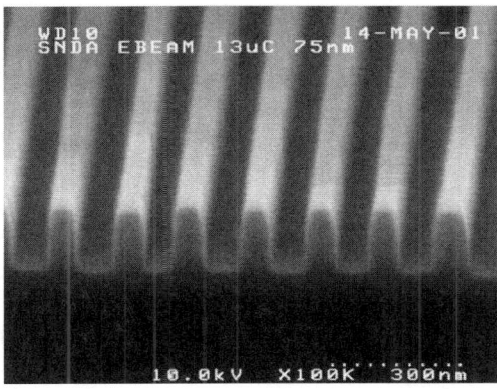

Fig. 105 High resolution images (75 nm line/space) printed in KRS by EPL (courtesy of D.M. Medeiros, IBM)

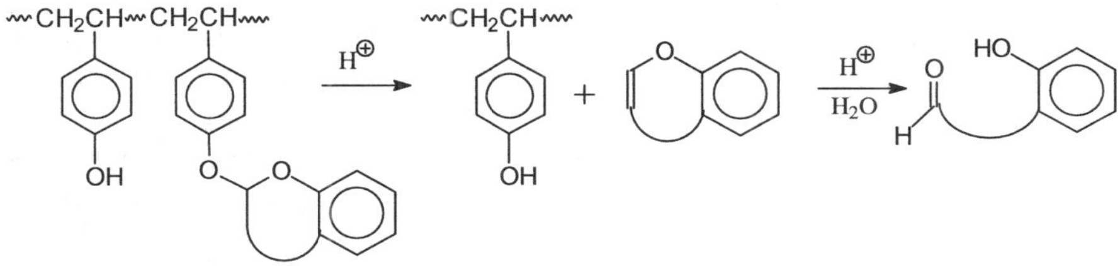

Fig. 106 Multiple anion nonvolatile acetal (MANA)

ratio patterning of single layer resists. The bilayer resist systems will be described in detail later in this chapter.

4.2 Depolymerization

PMMA was the very first e-beam [325] and deep UV [326] resist with an extremely good resolution, which functions on the basis of radiation-induced main chain scission; low molecular weight polymers are more soluble in a developer solvent. A large effort was directed toward enhancement of sensitivity of these polymethacrylates and related polymers till the early 1980s but the increase in G_s values (the number of main chain scission events per 100 eV of absorbed dose) or quantum yields achieved was marginal [327]. The ultimate form of polymer main chain degradation is depolymerization to starting monomer. Polymers characterized with low ceiling temperatures (T_c) could undergo depolymerization upon scission of one bond. The initial bond scission could be acid-catalyzed. Repeated catalytic main chain scission of acid-labile backbones could constitute another mechanism of depolymerization, which is essentially the same reaction mechanism as the acid-catalyzed deprotection discussed earlier.

Although the acid-catalyzed deprotection discussed in the earlier chapter is the paradigm for essentially all the advanced positive resists, polymers that undergo depolymerization might provide lower LER and/or higher resolution and are currently being revisited.

4.2.1
Thermodynamically-Driven Depolymerization

There are three mechanisms for initial backbone cleavage that cascades into complete depolymerization (Fig. 107) [110, 328]:

- Protonation of a backbone heteroatom followed by scission
- Protonation and cleavage of a pendant group
- Protonation and cleavage of a polymer end group

Polyaldehydes can be prepared by anionic or cationic polymerization and are known to have low T_c in many cases [329]. Since aliphatic polyaldehydes are crystalline and rather intractable materials, lithographic evaluation was performed on non-crystalline amorphous polyphthalaldehyde, which can be synthesized by anionic or cationic cyclopolymerization at cryogenic temperatures [330]. The polymer must be end-capped (with an acetyl group, for example) at the end of the polymerization before the mixture is warmed up for polymer isolation. Once properly end-capped, the acetal polymer is stable thermally to ~200 °C. However, the acetal linkage is very labile toward acid. The C-O bond of polyphthalaldehyde is catalytically cleaved with a photochemically-generated acid, which triggers depolymerization of the entire polymer chain and therefore results in spontaneous formation of positive relief images by irradiation alone (self-development) (Fig. 108) [24, 25]. The acid-catalyzed depolymerization of polyphthalaldehyde leading to self-development was one of the very first chemical amplification mechanisms proposed and later has been subjected to molecular orbital calculation [331]. Single-component non-catalyzed positive e-beam resists employing aliphatic aldehyde copolymers were reported [332].

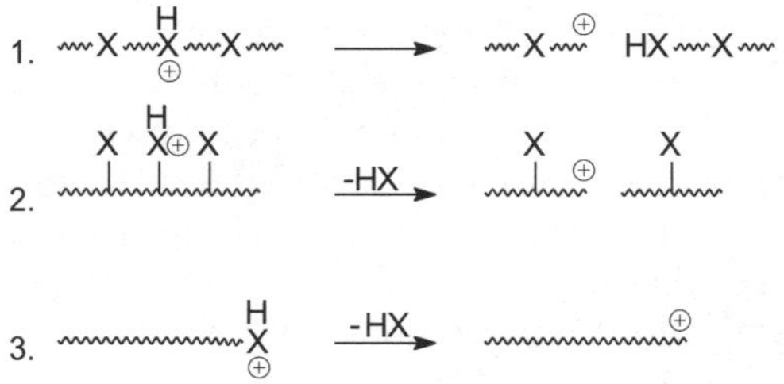

Fig. 107 Three modes of chain cleavage for depolymerization

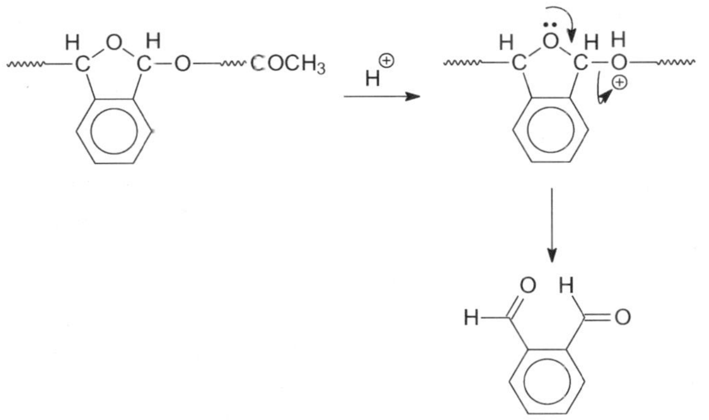

Fig. 108 Acid-catalyzed depolymerization of polyphthalaldehyde for self-development

The self-development involves evaporation of materials during exposure and thus could contaminate expensive exposure tools (cf. low activation energy protecting groups discussed earlier). The potential tool contamination problem has been alleviated by suppressing the depolymerization at ambient temperatures. In this approach complete depolymerization is accomplished by heating the exposed film after exposure (thermal development) to accelerate the unzipping process and to facilitate evaporation of the monomer (crystalline at room temperatures). Poly(4-chlorophthalaldehyde) prepared by cationic polymerization with BF_3OEt_2 was imaged by this thermal development technique [333, 334].

However, polymers that degrade readily cannot be resistant to dry etching and thus cannot function as a resist. Two approaches have been pursued to render the highly sensitive polyphthalaldehyde system more dry etch resistant and practical. One approach was to incorporate silicon into the polyphthalaldehyde structure for use as a thermally-developable, oxygen RIE barrier resist in bilayer lithography [334–336], which will be described later. The other approach was to utilize polyphthalaldehyde as a polymeric dissolution inhibitor of a novolac resin [89, 337]. Unsubstituted polyphthalaldehyde blends homogeneously with a cresol-formaldehyde novolac resin while it is not miscible with PHOST. The lipophilic polyaldehyde dispersed in the novolac film retards the dissolution of the phenolic resin film in an aqueous base solution. The polymeric dissolution inhibitor is completely reverted to the starting monomer upon PEB by photochemically-induced acid-catalyzed depolymerization and thus removed from the exposed area of the film through evaporation of the monomer produced. As a consequence, the exposed region that lacks the dissolution inhibitor dissolves rapidly in the aqueous base developer, providing positive-tone images.

Poly(α-acetoxystyrene) is an acetophenone enol ester and its acidolysis involves the polymer backbone in contrast to the side chain deprotection discussed earlier (Fig. 30). Protonation of the side chain carbonyl oxygen results in scission of the C-O bond to form a highly stable, tertiary benzylic carbocation in the backbone and acetic acid (Fig. 109) [338]. Polyphenylacetylene is formed upon β-proton elimination. However, main chain scission could compete to generate an α-acetoxystyrene terminal carbocation. Unzipping propagates from the scission point due to the low T_c of 47 °C. The degree of depolymerization amounts to 80% in solution at room temperature. The solubility change by the structural alteration and the molecular weight reduction through depolymerization bring about positive-tone imaging with xylenes as a developer [338]. In sharp contrast to other α-substituted styrenes, α-acetoxystyrene readily undergoes radical polymerization, which is characterized with a low T_c [339]. The radical polymerization of α-acetoxystyrene is an equilibrium process with T_c of 47 °C at 1 mol/L of monomer. The rate of polymerization (R_p) can be expressed by $R_p=k[AIBN]^{0.45}([M]-[M]_e)^{1.0}$, considering the depolymerization process with an equilibrium monomer concentration of $[M]_e$ [339]. The overall activation energy has been determined to be 116 kJ/mol. The enthalpy and entropy of polymerization obtained are −26.5 kJ/mol and −82.6 J/(K*mol), respectively. The T_c is low because of the low ΔH_p but high enough to allow radical homopolymerization owing to the low ΔS_p, which is significantly lower than the typical value (−100 to −125 J/(K*)) for vinyl monomers [339]. Monomer reactivity ratios of bulk copolymerization of α-acetoxystyrene with MMA (M_1) are $r_1=0.644$ and $r_2=0.754$ at 60 °C. The temperature dependence (50–80 °C) of the reactivity ratios has revealed that the copolymerization mechanism can be adequately described by the Mayo-Lewis model; a depolymerization process does not have to be taken into consideration in copolymerization. Q and e values of α-acetoxystyrene can be calculated as 0.82 and −0.45, respectively, from the reactivity ratios. Poly(α-acetoxystyrene) is transparent in the 250 nm range and as durable in CF_4 plasma as polystyrene or novolac [338]. Because

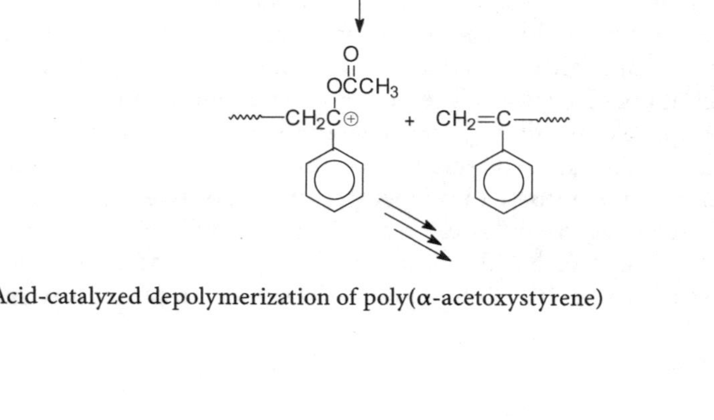

Fig. 109 Acid-catalyzed depolymerization of poly(α-acetoxystyrene)

of the rigid backbone, the polymer has a high T_g (no glass transition observed below 200 °C) and is resistant to common solvents but its film tends to crack during development due to solvent-induced stress [338].

Poly(4-hydroxy-α-methylstyrene), prepared by cationic polymerization, undergoes depolymerization from the polymer end [100, 110, 328]. A photochemically-generated acid attacks a polymer end group to yield a terminal carbocation, from which depolymerization propagates (Fig. 110). This phenolic polymer can be prepared by cationic polymerization of 4-*tert*-butoxycarbonyloxy-α-methylstyrene with BF_3OEt_2 in liquid *sulfur dioxide* with the acid-labile *t*BOC group intact [105] followed by heating the polymer powder to ~200 °C (Fig. 110). While the thermal process produces the phenolic polymer without main chain cleavage, deprotection with an acid in solution results in significant backbone scission even at room temperature. In contrast, poly(4-hydroxy-α-methylstyrene) synthesized by living anionic polymerization of 4-dimethyl(*tert*-butyl)silyloxy-α-methylstyrene followed by desilylation with HCl (Fig. 110) is very much inert to acidolysis [100, 110]. The difference between the

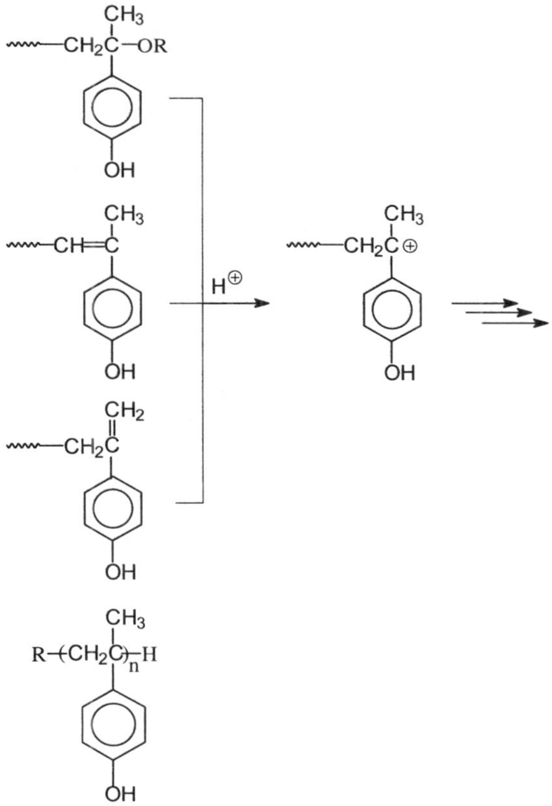

Fig. 110 Acid-catalyzed depolymerization of poly(4-hydroxy-α-methylstyrene)

cationic and anionic polymers suggests that the depolymerization proceeds from the polymer end. The anionic polymers are expected to have acid-stable end groups such as alkyl groups derived from an alkyllithium initiator and hydrogen atoms introduced by protonation of the growing anions at termination (Fig. 111). The cationically-prepared polymers contain acid-labile end groups. Accidental and intentional termination with hydroxyl compounds introduces OH or OCH$_3$ groups to form *tertiary* benzylic alcohols or ethers that are highly susceptible to acidolysis (Fig. 110). α-Methylstyrenic end groups introduced by deprotonation during polymerization can readily react with photochemically-generated acid to form terminal carbocations (Fig. 110). Thus, cationically-prepared polymers undergo depolymerization from the end due to the low T_c. The electron-donating *p*-OH group stabilizes the carbocation, rendering the depolymerization very facile.

The depolymerization mechanism from the polymer end has been recently revisited in the design of positive electron beam resists. 2-Phenylallyl-terminated poly(α-methylstyrene) was prepared by living anionic polymerization, which exhibited a significantly lower depolymerization temperature on TGA than the H-terminated counterpart [340]. The 2-phenylallyl-terminated polymer depolymerized completely when treated with *n*-BuLi in THF at room temperature. A single-component resist (without PAG) formulated with the 2-phenylallyl-terminated poly(α-methylstyrene) demonstrated a higher e-beam sensitivity (500 µC/cm^2 at 20 keV) than the one based on the H-terminated polymer when developed with methanol/methyl isobutyl ketone (2/3 vol/vol) [340]. However, the sensitivity of the non-catalyzed single-component system

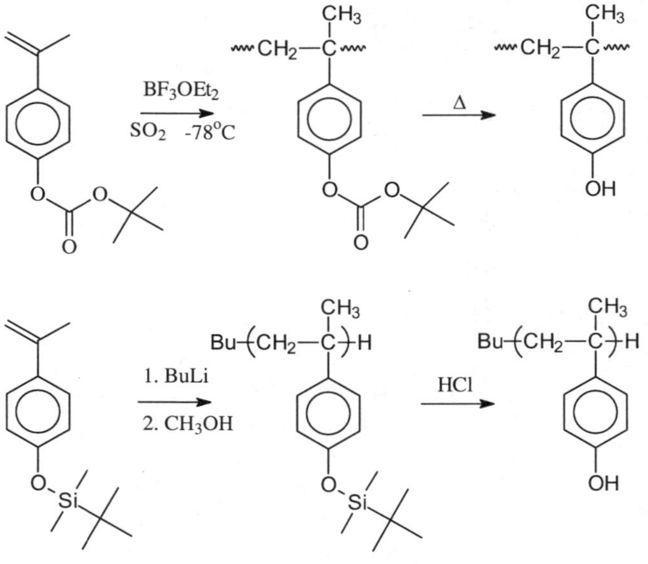

Fig. 111 Synthesis of poly(*p*-hydroxy-α-methylstyrene)

was a couple of magnitudes lower than those of acid-catalyzed chemical amplification systems.

4.2.2
Repeated Catalytic Main Chain Scission

If the reactions described in 4.1 occur in the polymer backbone, depolymerization results. Polymers consisting of *tertiary, secondary* allylic, or *secondary* benzylic carbonates, esters, or ethers in the backbone yield stable carbocations upon mild heating through acid-catalyzed cleavage of the C-O bond (Fig. 112) [341, 342]. The carbocations eliminate β-protons to form olefins. The reaction is repeated on the polymer chain and eventually results in complete degradation. If the reaction fragments evaporate out of the resist film during PEB, this degradation mechanism could provide thermal development.

Polycarbonates based on 2-cyclohexen-1,4-diol and a dihydroxy compound liberate benzene through aromatization, a dihydroxyl compound, and carbon dioxide, upon acidolysis [343]. The low volatility of the dihydroxyl compound hampers complete development by heating alone and necessitates wet development.

Polyformals containing *secondary* allylic and benzylic groups undergo complete acidolytic decomposition to volatile products such as an aromatic com-

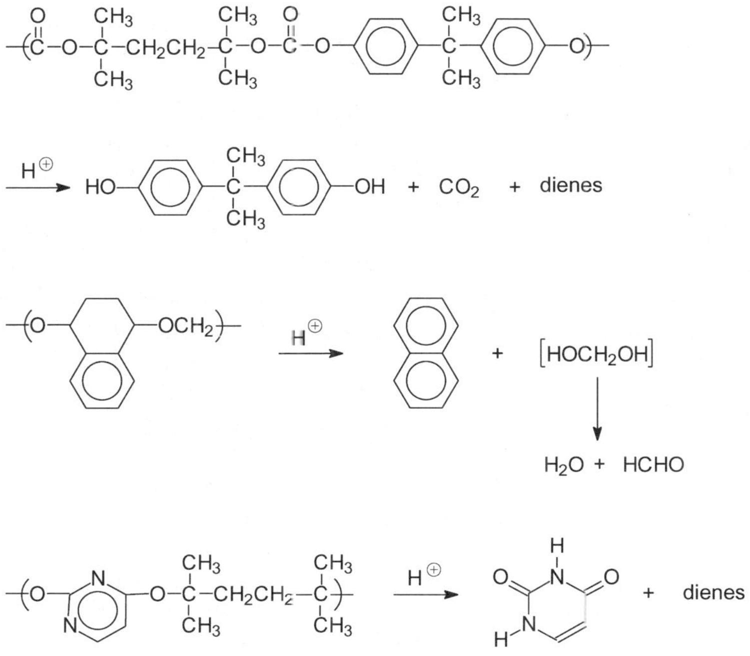

Fig. 112 Acid-catalyzed main chain cleavage

pound, formaldehyde, and water (Fig. 112) and thus a resist based on such polyformals develops thermally without use of a developer solvent [343]. However, positive images obtained by thermal development are very much rounded due to the low T_g of the polymers and perhaps also due to plasticization by the aromatic degradation product.

This catalytic main chain scission has been revisited recently in an attempt to reduce LER and outgassing but high resolution imaging or the validity of the concept has not been demonstrated [344].

Polyethers consisting of alkoxypyrimidine units undergo an acid-catalyzed tautomeric change from the alkoxypyrimidine to pyrimidone, releasing dienes (Fig. 112) [345].

4.3
Rearrangement

Acid-catalyzed rearrangement reactions are also very useful in the design of chemical amplification resists. However, the potential of this imaging mechanism has been exploited only partially so far.

4.3.1
Polarity Reversal

The acid-catalyzed deprotection of polymer pendant groups results in a change of the polarity from a nonpolar to a polar state. However, there are instances where a postexposure treatment renders the unexposed areas more polar than the exposed, resulting in polarity reversal.

Replacement of one of the methyl groups of poly(*tert*-butyl 4-vinylbenzoate) with a cyclopropyl group lowers its thermal deprotection temperature by as much as 80 °C [127]. However, the thermolysis of poly(2-cyclopropyl-2-propyl 4-vinylbenzoate) at ~160 °C does not convert the ester polymer cleanly to poly(methacrylic acid) but is accompanied by ca. 10% rearrangement of the *tertiary* dimethyl cyclopropyl carbinol ester to a *primary* 4-methyl-3-pentenyl ester (Fig. 113) [127]. Acid promotes the rearrangement to as much as 66%. Thus, when the resist film containing triphenylsulfonium hexafluoroantimonate is postbaked at 100–130 °C, the exposed areas are insoluble in nonpolar organic solvents such as anisole due to the presence of 4-vinylbenzoic acid units (33%), resulting in negative imaging with anisole as a developer (based on the polarity change). When the PEB temperature is raised to 160 °C, deesterification takes place in the unexposed area to the extent of ca. 90% and therefore the unexposed polymer film becomes soluble in aqueous base. The exposed area contains only 33% of the acidic units and is therefore insoluble in the base developer. The high temperature PEB results in negative imaging with aqueous base due to polarity reversal (Fig. 113).

Poly(methacrylic acid) and methacrylic acid-methacrylate copolymers undergo anhydride formation losing water and/or alcohol at relatively low tem-

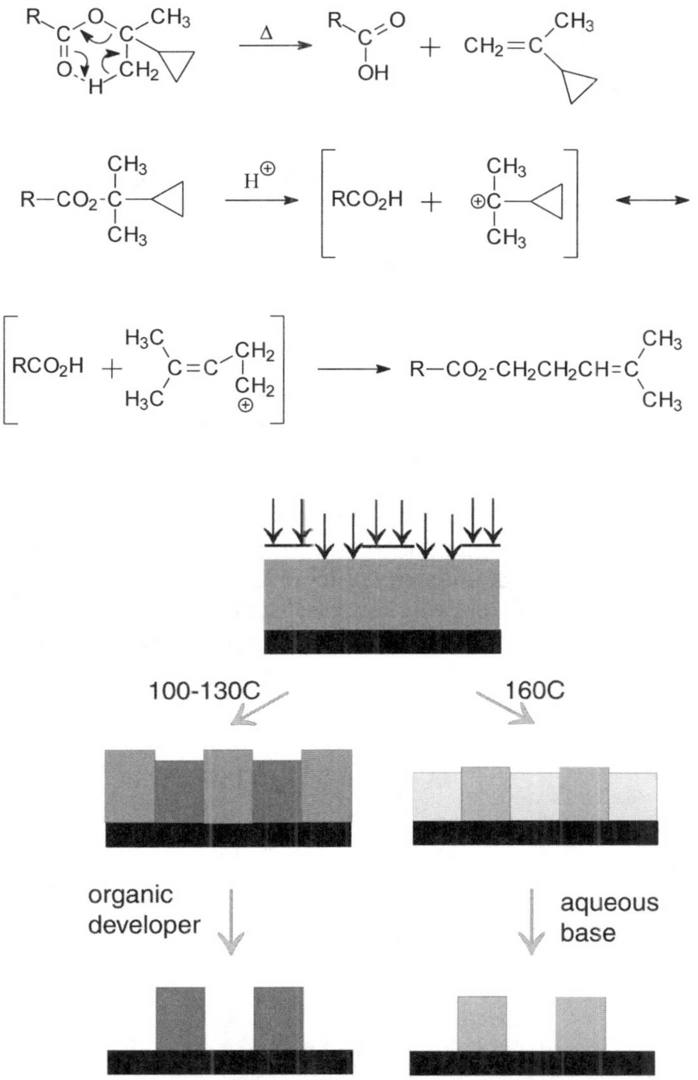

Fig. 113 Thermal deesterification and acid-catalyzed rearrangement of cyclopropyl carbinol ester; negative imaging processes based on polarity change and polarity reversal

peratures of ~180 °C while poly(*tert*-butyl methacrylate) is stable to >200 °C. The tone of the PTBMA resist can be reversed by utilizing the anhydride formation (destruction of carboxylic acid) [121]. When the TBMA resist is exposed, baked at temperatures below 150 °C, and developed with aqueous base, a positive image results (a negative image with use of a nonpolar organic solvent). However, when the exposed PTBMA resist is baked at a high temperature

(180 °C), anhydride formation proceeds in the exposed area while PTBMA in the unexposed regions remains intact. After this high temperature process, the PTBMA resist film is flood-exposed and baked at <150 °C, which converts PTBMA to poly (methacrylic acid) in the initially unexposed areas. This entire process renders the unexposed area more polar (carboxylic acid) than the exposed area (anhydride), allowing negative imaging with aqueous base [121]. This concept has been later employed in the design of water-castable and water-developable resist [346].

4.3.2
Claisen Rearrangement

Poly(4-*tert*-butoxystyrene) is converted to PHOST through acid-catalyzed deprotection, releasing isobutene, but its model compound, 4-*tert*-butoxytoluene, undergoes significant realkylation onto the *ortho* position as well as deblocking in solution [125]. When the *tert*-butyl group is replaced with a cyclohexenyl group, acid-catalyzed deprotection to yield PHOST and 1,3-hexadienene is only partial and realkylation onto the *ortho* position competes through Claisen rearrangement (Fig. 110) [347]. The net result of Claisen rearrangement is still a polarity change from a nonpolar to polar state. Poly(4-phenoxymethylstyrene) is isomerized in a similar fashion with an acid as a catalyst to a *C*-alkylated phenolic structure (Fig. 114).

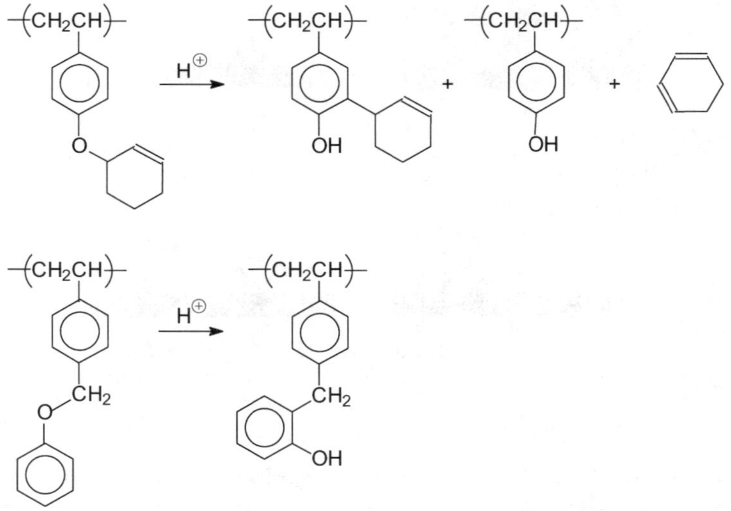

Fig. 114 Claisen rearrangement for polarity change

4.3.3
Pinacol Rearrangement

A change of a polarity from a polar to nonpolar state (reverse polarity change) can be accomplished by the pinacol-pinacolone rearrangement and has been exploited in chemically amplified lithographic imaging [151, 348–350]. The pinacol rearrangement involves conversion of *vic*-diols to ketones or aldehydes with an acid as a catalyst (Fig. 115).

A polymeric pinacol, poly[3-methyl-2-(4-vinylphenyl)-2,3-butanediol], has been prepared by radical polymerization of the styrenic diol monomer and shown to be cleanly and quantitatively converted to a non-conjugated ketone in the solid state by reaction with a photochemically-generated acid [151, 348, 350]. The rearrangement reaction can be readily monitored by IR spectroscopy as the disappearance of the hydroxyl OH absorption is accompanied by appearance of a new ketone carbonyl absorption (Fig. 116). Since a polar alcohol (isopropanol) dissolves the polar diol polymer in the unexposed regions but cannot dissolve the less polar ketone polymer produced in the exposed regions, the resist functions as a negative system with alcohol as a developer. The diol polymer is stable thermally to 225 °C in the absence of acid.

Aqueous base developable two- and three-component negative resists have been designed on the basis of pinacol rearrangement [151, 348–351]. In the two-component design, the styrenic pinacol monomer was copolymerized with 4-acetoxystyrene with AIBN as the initiator and the resulting copolymer was hydrolyzed with base to yield an aqueous base soluble copolymer (Fig. 117) [351]. The diol groups are converted by reaction with a photochemically-generated acid in the exposed areas to ketones, which form hydrogen bonds with the surrounding phenolic OH groups and thus inhibit the dissolution of the copolymer film in an aqueous base solution. The three-component approach involves the use of a small diol such as benzopinacole or *meso*-hydrobenzoin in a phenolic matrix resin together with a PAG [151, 348–350]. In these resist systems small ketone or aldehyde generated in the exposed areas through pinacol rearrangement functions as a dissolution inhibitor while the unexposed

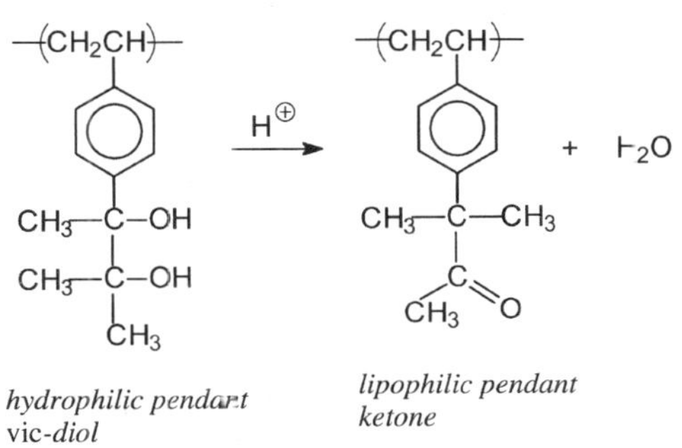

Fig. 115 Pinacol rearrangement of polymeric *vic*-diol for reverse polarity change

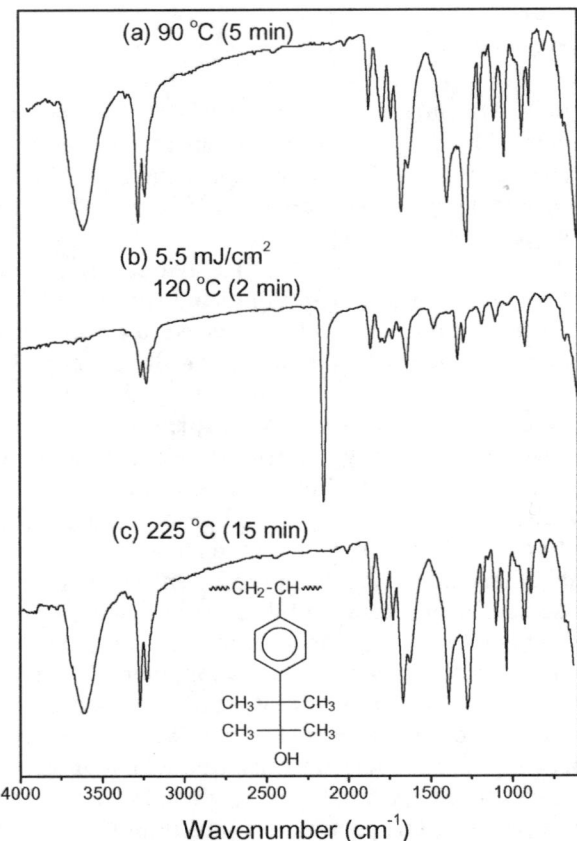

Fig. 116 IR spectra of polymeric pinacol demonstrating its thermal stability and acid-catalyzed rearrangement to ketone [151]

film containing polar diol is highly soluble in aqueous base. The dissolution promotion/inhibition effect of added diols must be taken into consideration [151, 348, 350]. Phenyl trityl ketone produced from benzopinacole is a very strong dissolution inhibitor of phenolic resins. The IR carbonyl absorption of the ketone product is shifted to a lower wavenumber due to hydrogen-bonding with the phenolic OH group.

An aqueous base developable negative resist for use in 193 nm lithography was developed by combining NBHFA and norbornene bearing a pendant *vic*-diol (Fig. 118) [352]. One problem in the design of 193 nm resists is that the benzylic stabilization effect cannot be utilized to drive the acidolysis reactions and two kinds of ketone can be formed in this case.

The reverse polarity change mechanism to convert polar polymer to nonpolar polymer could be an excellent basis to design a resist that could provide positive-tone images upon development with supercritical CO_2.

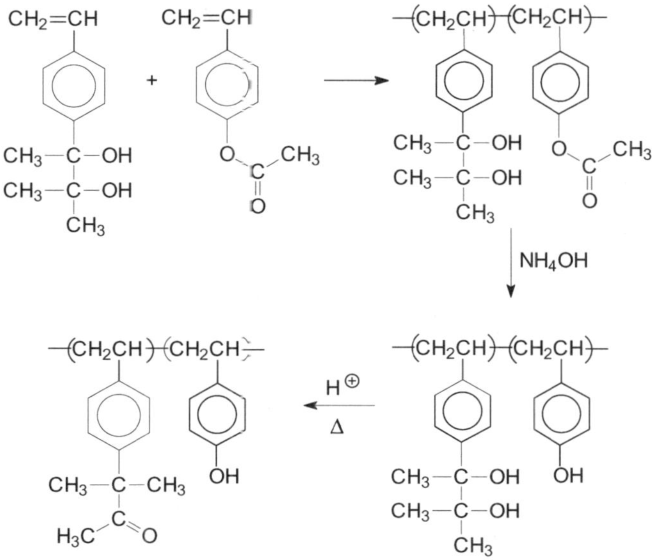

Fig. 117 Preparation of aqueous base soluble pinacol copolymer

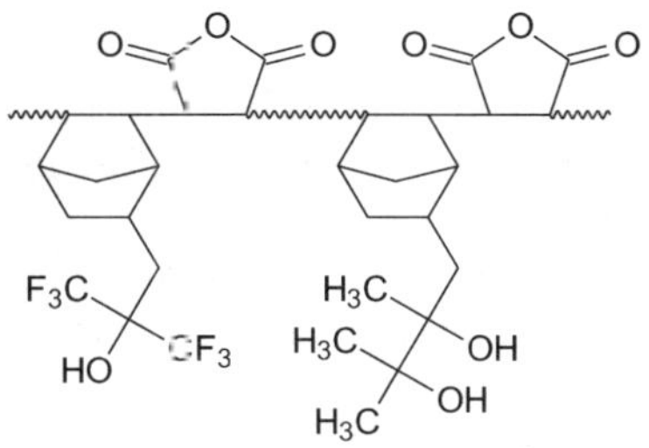

Fig. 118 Pinacol polymer for negative tone 193 nm imaging

4.4
Intramolecular Dehydration

Dehydration is the first step of pinacol rearrangement of *vic*-diol. *Tertiary* alcohols can dehydrate intramolecularly with an acid as a catalyst to form olefins, which provides another mechanism of a reverse polarity change from a polar to nonpolar state [353].

Poly[4-(2-hydroxy-2-propyl)styrene] undergoes acid-catalyzed dehydration to yield a stable *tertiary* benzylic carbocation, which then eliminates a β-proton to form a pendant olefinic structure (Fig. 119) [353]. This intramolecular dehydration reaction converts the hydrophilic alcohol to a highly lipophilic olefin and allows negative tone imaging with a polar alcohol as a developer.

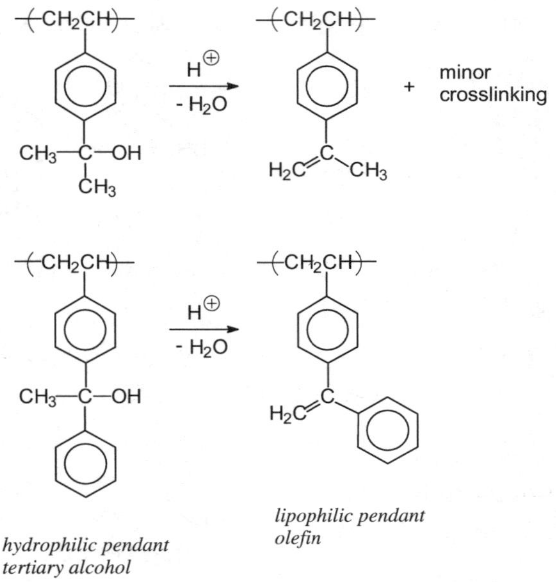

Fig. 119 Intramolecular dehydration for reverse polarity change

However, the α-methylstyrene structure produced by the dehydration further undergoes acid-catalyzed linear and cyclic dimerizations to a small degree, resulting in concomitant crosslinking and therefore precluding positive imaging with a nonpolar organic solvent [353].

The dimerization pathway can be eliminated by replacing one of the methyl groups with a phenyl ring to generate a 1,1-diphenylethylene structure (Fig. 119) [351]. The resist system based on this polymer can be developed in a positive mode with use of a nonpolar solvent such as xylene and in a negative mode with a polar alcohol as a developer. This is the first and only example of dual tone imaging systems based on the reverse polarity change from a polar to nonpolar state and is the most sensitive resist currently known. Generation of a carbocation through dehydration is extremely facile in this case due to an additional stabilizing phenyl ring.

4.5
Condensation/Intermolecular Dehydration

Acid-catalyzed condensation has been the primary and dominant foundation for aqueous base developable negative resist systems [354–363]. The first commercial chemical amplification resist was built on this mechanism. The condensation resists are typically three-component systems comprising a base soluble binder resin bearing reaction sites for crosslinking (phenolic resin), a radiation-sensitive acid generator, and an acid-sensitive latent electrophile (crosslinking agent). The very first system designed on the basis of acid-catalyzed condensation employed a novolac/DNQ positive photoresist and a N-methoxymethylated melamine crosslinker [354]. As mentioned earlier (Fig. 5), photolysis of DNQ produces indenecarboxylic acid, which is strong enough to react with the melamine crosslinker and to generate a carbonium ion, releasing methanol. The N-carbonium ion undergoes electrophilic substitution onto the electron-rich benzene ring of the novolac resin, regenerating a proton (Fig. 120). Since the melamine compound is multifunctional, crosslinking results. While this chemically amplified negative tone imaging requires baking after exposure, the resist develops in a positive mode without PEB (no amplification) [354].

The novolac resin and DNQ have been replaced with PHOST and chloromethyltriazine as a HCl generator, respectively, for deep UV applications [356–358, 361, 362]. Furthermore, various crosslinkers have been evaluated. The PEB conditions (temperature and time) strongly influence the crosslinking efficiency (sensitivity and contrast) and resolution of these resists, as is typically the case with chemical amplification resists. The rate-determining step for crosslinking in these resists is the formation of a carbocation from the protonated ether moiety. In addition to the C-alkylation initially proposed, O-alkylation is responsible for crosslinking and also results in destruction of the base-solubilizing phenolic OH groups [362, 363]. Furthermore, it has been demonstrated that lower molecular weight PHOST provides a higher resolution [364]. The condensation resists have provided excellent lithographic performance in deep UV, e-beam, and X-ray exposures. However, the first generation negative resists employed a rather weak 0.14 N TMAH developer. As mentioned in the deprotection positive resist section, the new resist systems had to be compatible with the industry standard 0.26 N TMAH developer optimized for the DNQ/novolac resist. Use of the stronger developer with the first generation negative resists

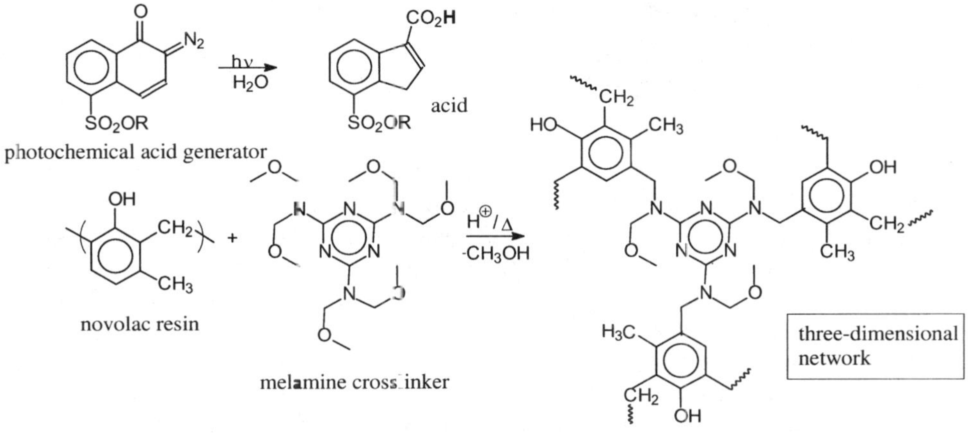

Fig. 120 Acid-catalyzed crosslinking of novolac through condensation

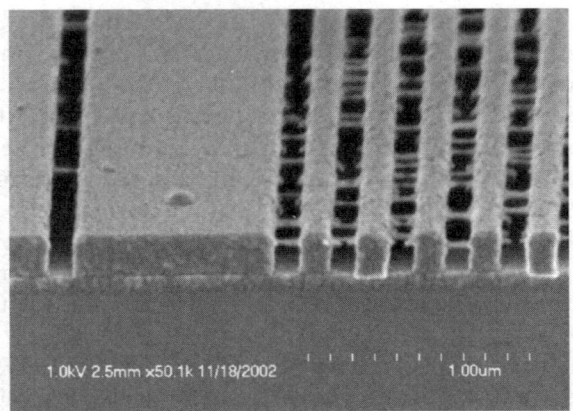

Fig. 121 Microbridging in crosslinking negative resist based on condensation (courtesy of C.E. Larson, IBM)

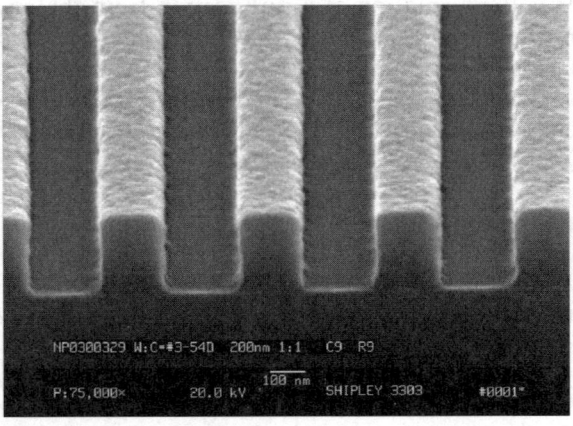

Fig. 122 Negative 200 nm line/space patterns printed at 248 nm in a condensation resist with 0.26 N TMAH developer (courtesy of J.W. Thackeray, Shipley)

resulted in microbridging, which limited the resolution (Fig. 121) [364]. Partial blocking of PHOST to lower its base solubility has resulted in successful formulation of a 0.26 N-compatible negative resist free from microbridging [364]. Figure 122 presents a scanning electron micrograph of 200 nm features printed in such a negative resist on a KrF excimer laser stepper using a 0.26 N TMAH solution. While positive resists are needed for memory device fabrication because of their contact hole capability, negative resists can print isolated lines nicely and therefore could find use as logic resists. Furthermore, a recent simulation study showing an advantage of negative resists in trench imaging [365] has renewed an interest in developing high performance 193 nm negative resists.

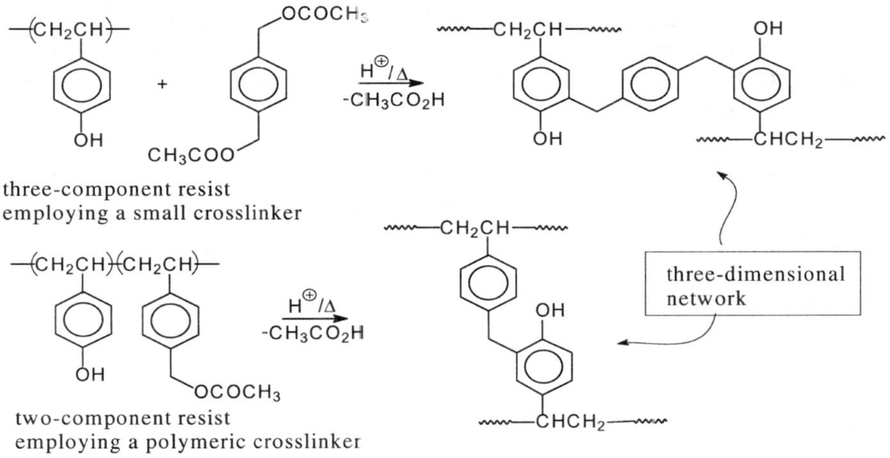

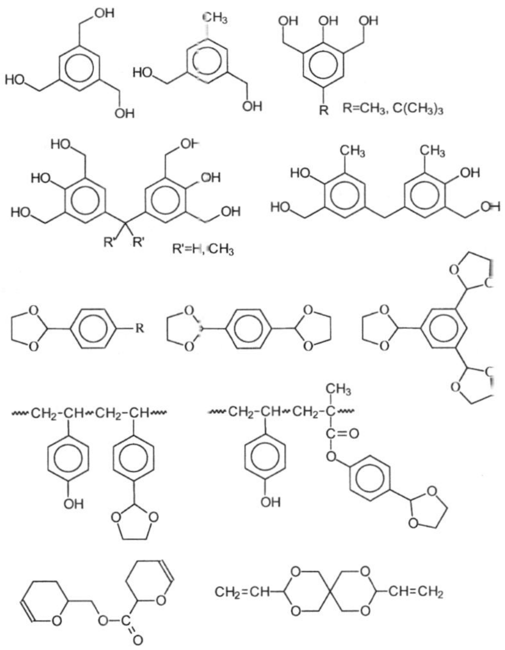

Fig. 123 Latent electrophiles

Other crosslinkers have been also extensively studied (Fig. 123). Benzyl acetate derivatives are such a latent electrophile, which yields a stable benzylic carbocation, releasing acetic acid, upon acid treatment. The carbocation undergoes electrophilic substitution reactions onto the electron-rich benzene ring and crosslinks the phenolic resin when the latent electrophile is multifunctional [366–370]. The crosslinker can be an additive or incorporated into a phenolic resin through copolymerization. In the former (three-component) approach, 1,4-di(acetoxymethyl)benzene is added to PHOST (or novolac) along with a PAG. In the latter (two-component) approach, 4-vinylbenzyl acetate is copolymerized with BOCST, followed by selective removal of the *t*BOC group in refluxing glacial acetic acid, to afford a copolymer bearing both the latent electrophile and the crosslinking site on the same polymer chain. While the base solubility is controlled by adjusting the copolymer composition, only less than 5% of the acetoxy group is removed at an imaging dose of ~2 mJ/cm^2 [370]. Mechanistic studies by NMR employing 4-isopropyphenol and a substituted benzyl acetate as model compounds in chloroform revealed that the rate-determining step was the formation of a benzylic carbocation [367, 368]. In an early stage of the reaction, some *O*-alkylation products were observed, which underwent acid-catalyzed rearrangement to *C*-alkylation products [367, 368]. The three-component design has been extended to include a number of latent electrophiles such as benzyl alcohol derivatives [371] (Fig. 123). Vinyl cyclic acetals (Fig. 123) have also been reported as a cationic crosslinker of PHOST [372, 373].

Calixarenes and dendrimers (see above) have been employed as a matrix for negative imaging through condensation [374].

When a bulky substituent is attached to the phenolic group through *C*- or *O*-alkylation, the dissolution rate of the phenolic resin in aqueous base becomes slower (Fig. 124). Thus, such condensation can provide negative imaging even without involving crosslinking. *N*-Hydroxy- and acetoxymethylimides, which are monofunctional latent electrophiles, undergo acid-catalyzed condensation with a phenolic resin, reducing the dissolution rate in the exposed regions and therefore providing negative images upon aqueous base development. The model reactions have indicated that the *C*- vs *O*-alkylation depends on the latent electrophile structure [150].

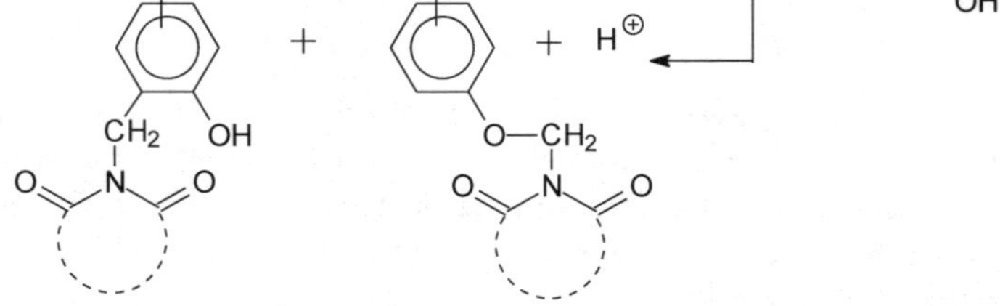

Fig. 124 *C*- and *O*-Alkylation of phenol with a monofunctional electrophile

Aldehydes can methylolate phenols with an acid as a catalyst and therefore function as latent electrophiles in the negative resist design based on condensation (Fig. 125) [150]. The methylolated phenolic resin is expected to dissolve more slowly in aqueous base than its precursor resin and therefore this process could be exploited in negative imaging. Furthermore, the methylolated phenol can undergo further condensation with phenol. If the phenol is polymeric, the second reaction results in crosslinking, lowering the dissolution rate even further.

In the above condensation resist designs, the phenolic resin offers a reaction site as well as base solubility. Self-condensation of polymeric furan derivatives has been utilized as an alternative crosslinking mechanism for aqueous base development (Fig. 126) [375]. The copolymer resist is based on poly[4-hydroxystyrene-*co*-4-(3-furyl-3-hydroxypropyl)styrene], which was prepared by radical copolymerization of the acetyl-protected furan monomer with BOCST followed by base hydrolysis. The furan methanol residue, highly reactive toward electrophiles due to a mesomeric electron release from oxygen that facilitates the attack on the ring carbons, readily yields a stable carbocation upon acid treatment. Thus, the pendant furfuryl groups serve as both the latent electrophile and the nucleophile. Model reactions indicated that the furfuryl carbocation reacts more preferentially with the furan nucleus than the phenolic functionality.

Poly[4-(1-hydroxyethyl)styrene] bears *secondary* alcohol as a pendant group and crosslinks through acid-catalyzed interchain dehydration to form ether linkages. Development with alcohol produces negative images [353]. The self-condensation induces a polarity change from a polar to a nonpolar state, which

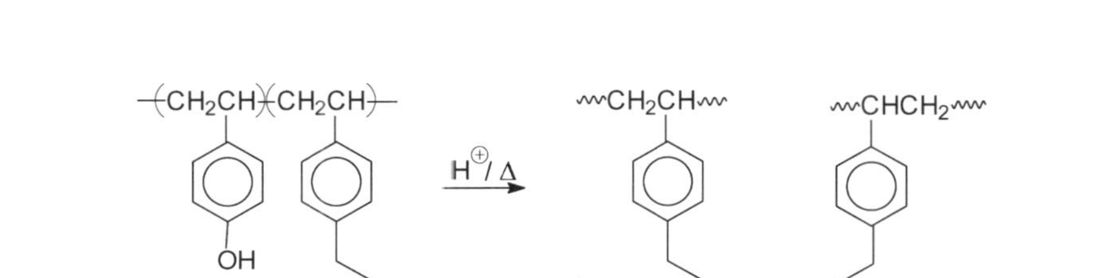

Fig. 125 Condensation of phenol with aldehyde

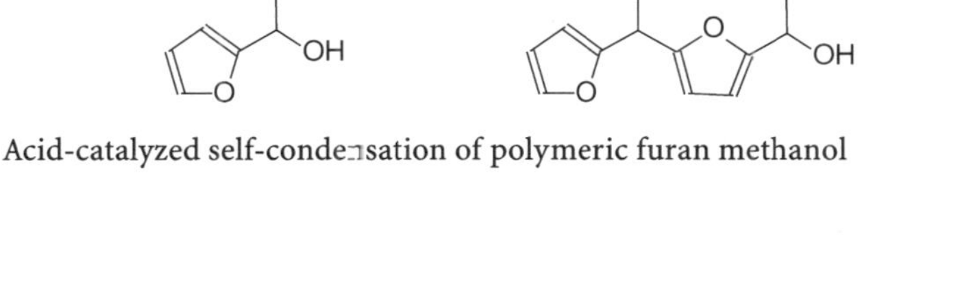

Fig. 126 Acid-catalyzed self-condensation of polymeric furan methanol

is responsible for the negative imaging in conjunction with crosslinking. This negative resist system has been modified for aqueous base development by copolymerizing the styrenic *secondary* alcohol monomer with 4-acetoxystyrene, followed by base hydrolysis (Fig. 127) [351]. Model studies indicated that the acid-catalyzed intermolecular dehydration results in self-condensation to di(α-methylbenzyl) ether, O-alkylation to α-methylbenzyl phenyl ether, and C-alkylation to o-(α-methylbenzyl)phenol (Fig. 127). All these condensation reactions contribute to crosslinking and therefore to negative aqueous base development. In addition, the ether formation is a reverse polarity change which also contributes to the negative imaging. In a fashion similar to the pinacol rearrangement discussed above, small phenylcarbinols were incorporated into PHOST along with a PAG to formulate an aqueous base developable negative e-beam resist [376]. Aromatic *tertiary* diols and triols presented in Fig. 128 were evaluated as a dissolution inhibitor precursor and a water generator. The dimethylcarbinol of benzene was selected as the most suitable material because it gave the highest sensitivity and promoted the dissolution of PHOST in aqueous base whereas the carbinols with two benzene rings functioned as a dissolution inhibitor. The insolubilization mechanism has been reported to be primarily O-alkylation of phenol and it has been proposed that water generated through acid-catalyzed dehydration accelerates acid diffusion (anisotropic acid diffusion; acid diffu-

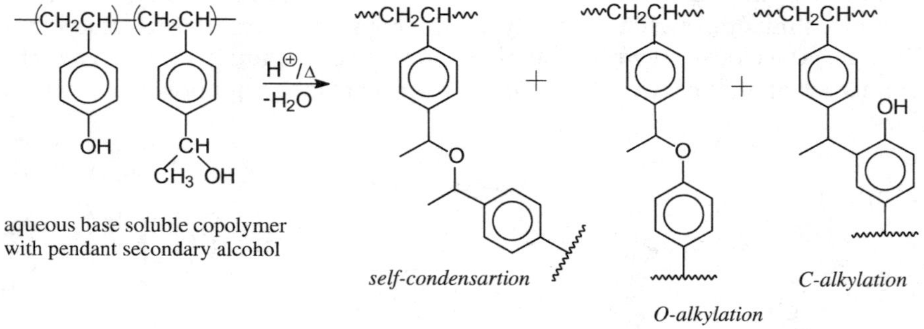

Fig. 127 Acid-catalyzed dehydration for negative tone aqueous base development

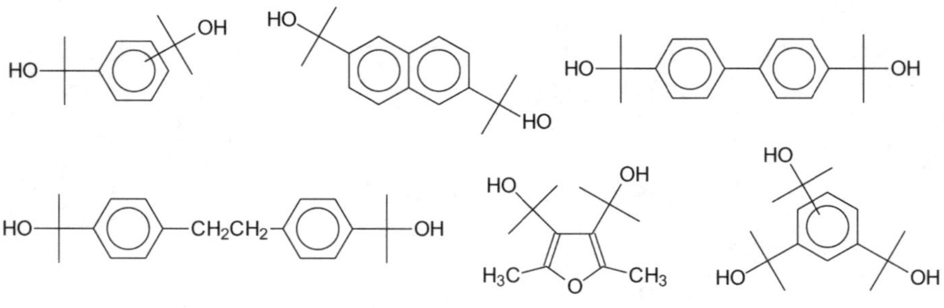

Fig. 128 Aromatic *tertiary* diols and triols for acid-catalyzed dehydration

sion is faster in the exposed regions where water is present). The dehydration resist employing anionically-prepared PHOST as a phenolic matrix resin, diphenyliodonium triflate as a PAG, and 1,3,5-tris-(2-(2-hydroxypropyl)benzene) as a water generator/dissolution inhibitor precursor exhibited a smaller PEB temperature dependence of the line width (6.2 nm/°C) than the acid-hardening resist (44 nm/°C), high e-beam sensitivity (5 µC/cm^2 at 50 kV), and high resolution (<100 nm) using a 2.38% TMAH developer [377].

A high contrast negative resist, called a contrast boosted resist, using a water-repellent compound that is converted to a hydrophilic compound during aqueous base development, has been developed for e-beam lithography [378]. The water-repellent compound such as 1,3,5-tris(bromoacetyl)benzene in a three component resist consisting of a novolac resin, hexamethoxymelamine, and 1,3,5-tris(trichloromethyl)triazine PAG acts as a strong dissolution inhibitor in the exposed regions and is converted to a hydrophilic dissolution promoting compound in the unexposed region, resulting in contrast enhancement. This function is similar to that of anhydrides and acetoxystyrene in positive tone aqueous base development discussed above.

Self-condensation of silanol compounds in a phenolic matrix resin has found use in formulation of negative resists that develop with aqueous base (Fig. 129) [379]. This system is similar to pinacol rearrangement of small *vic*-diol to a dissolution inhibiting compound and thus functions on the basis of a polarity change instead of crosslinking. Base-soluble silanol compounds such as diphenylsilanediol are dissolution promoters of phenolic resins (a novolac resin in this case) but are converted to polysiloxanes upon PEB through acid-catalyzed condensation, providing negative images in aqueous base development. However, siloxane oligomers do not inhibit the dissolution of the novolac resin at all when blended, presumably because the hydrophobic siloxane compounds are located apart from the hydrophilic phenolic OH groups. In contrast, the hydrophilic silanol compounds associate themselves with the phenolic OH groups and are then converted to siloxane oligomers, maintaining their initial orientation relative to the phenolic OH groups. Thus, the siloxane oligomers generated from silanols in a phenolic resin can inhibit dissolution of the resin in aqueous base due to the hydrophobic barrier surrounding the phenolic OH groups. Silsesquioxanes have been also employed in a similar negative resist formulation [380]. Application of silanol condensation to bilayer lithography will be discussed later.

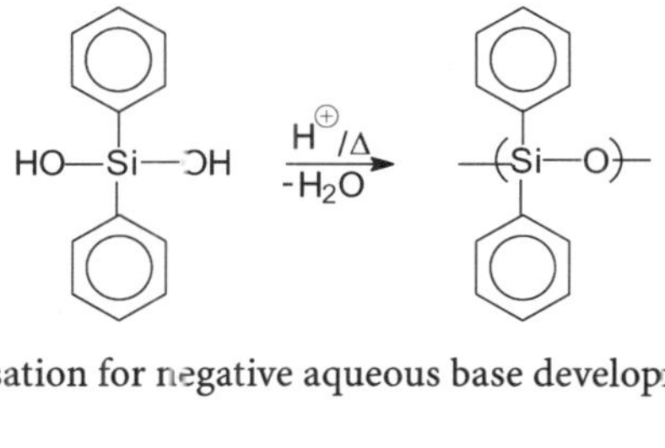

Fig. 129 Silanol condensation for negative aqueous base development

Exclusion of aromatic structures in 193 nm resists makes the design of ArF negative resists difficult because the benzylic stabilization of carbocations cannot be utilized and electron-rich phenol which serves as a reaction site cannot be employed in the condensation reaction. For negative 193 nm applications aliphatic polymers bearing pendant OH or epoxide groups (and CO_2H for aqueous base development) were reported to undergo crosslinking using urea/melamine latent electrophiles (Fig. 130) [381]. ^{1}H NMR investigation of model reactions of hydroxytricyclo[5.2.1.0^{2,6}]decene and carboxytetracyclo[4.4.0.1^{2,5}.1^{7,10}]dodecene with 1,3,4,6-tetrakis(methoxymethyl)glicoluril in the presence of p-toluenesulfonic acid in $CDCl_3$ at 50 °C has revealed that the carboxylic acid did not react at all and that the OH group exhibited high reactivity toward the carbocation generated from the uril compound. Cyclic urea crosslinkers showed higher transparency than the linear urea or melamine

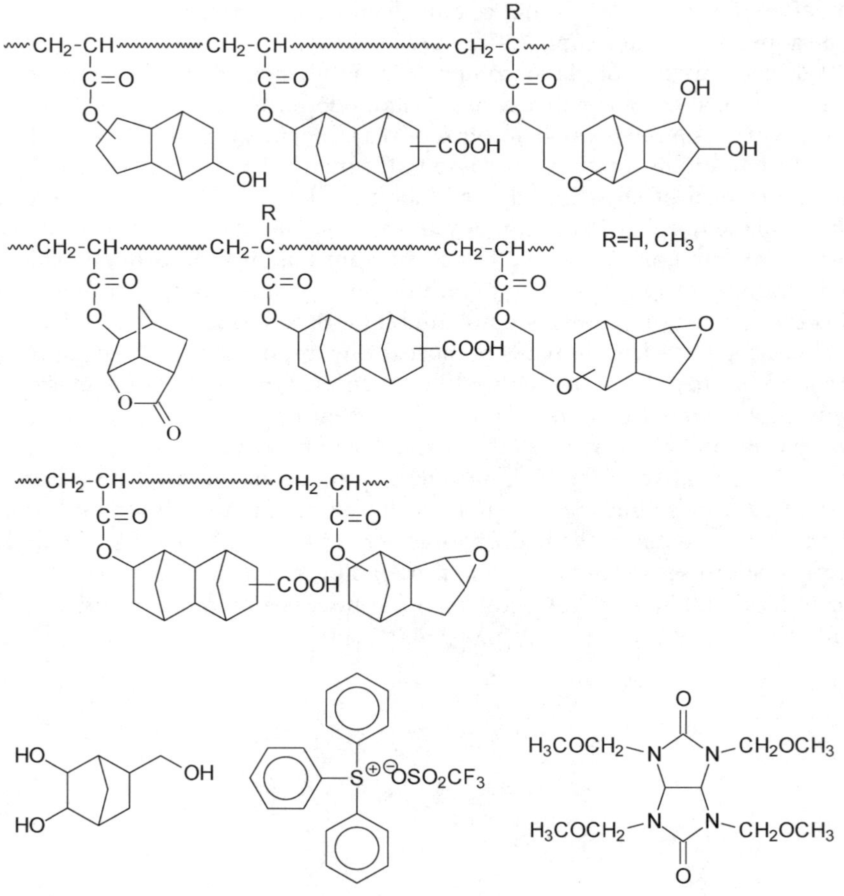

Fig. 130 ArF resists based on condensation

at 193 nm. A resin containing 68 mol% OH was selected as a resist polymer, the glicoluril as the crosslinker, and triphenylsulfonium triflate as a PAG for 193 nm negative imaging using 2.38 wt% TMAH developer. Addition of 2,3-dihydroxy-5-hydroxymethylnorbornan to the resist formulation improved the resolution [381].

In a similar approach polyacrylates bearing alcoholic hydroxyl groups (hydroxyethyl methacrylate) was mixed with a small molecule functionalized with two or three epoxides and a PAG [382].

4.6
Esterification

Esterification of carboxylic acid results in a reverse polarity change and thus can be exploited in the design of aqueous base developable negative resists especially for 193 nm lithography. As mentioned earlier, 193 nm resists are predominantly based on the use of carboxylic acid as a polar group for good transparency.

A methacrylate terpolymer containing 35 mol% methacrylic acid units was mixed with diepoxides (Fig. 131) and triphenylsulfonium triflate and imaged in a negative mode by exposure to ArF excimer laser followed by development with 0.06% (weak) TMAH solution [383]. The aromatic epoxide was selected for imaging because the more transparent aliphatic epoxide was liquid and thus lowered the T_g of the resist film to an unacceptable level. The insolubilization mechanism was speculated to be crosslinking through acid-catalyzed esterification between the carboxylic acid and epoxide.

The MA copolymers of NB involving cyclopolymerization of dienes (Fig. 80) have been employed in negative 193 nm imaging [384]. In this application the anhydride was converted to γ-hydroxy acid using $NaBH_4$, with concomitant formation of γ-lactone (Fig. 132). The γ-hydroxy acid moiety undergoes acid-catalyzed intramolecular esterification (ring-closure) to form γ-lactone, resulting in a reverse polarity change for negative imaging with aqueous base (Fig. 132) [384]. Because carboxylic acid was not completely destroyed to form less polar lactone, aqueous base penetrated into the exposed regions, causing swelling-induced distortion. Furthermore, the 193 nm transparency was not high enough and the shelf life of polymers bearing γ-hydroxy acid was quite short [384].

In an attempt to overcome the above-mentioned problems, acrylates bearing pendant androsterone with δ-hydroxy acid (Fig. 133), 3-β-acryloyloxyandrosterone, was polymerized with a radical initiator in THF to 80–90% yield [385]. The polymer was oxidized with CH_3CO_2H/H_2O_2 to form δ-lactone in 95% yield. The lactone ring was hydrolyzed in THF with 0.2 N aqueous NaOH in 85–90% yield. The final polymer contained 25 mol% unreacted lactone, had a good transparency of 0.21/μm at 193 nm, and a much better shelf life than the γ-hydroxy acid system. ^{13}C NMR of the polymer treated with methanesulfonic acid in THF and IR studies of the resist film after exposure indicated that the

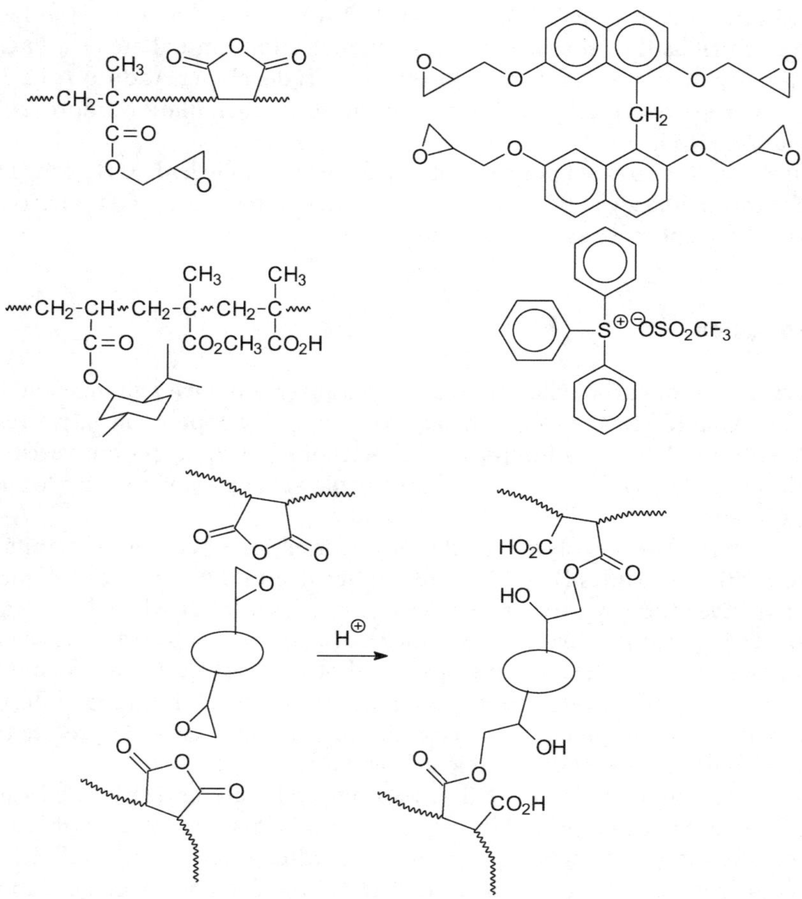

Fig. 131 Acid-catalyzed esterification involving epoxide and anhydride (or carboxylic acid)

intramolecular esterification indeed occurred and was responsible for the negative aqueous base development without concomitant crosslinking as evidenced by clean and fast dissolution of highly exposed films in THF. The resist produced 120–100 nm dense line/space patterns when developed with a weak developer (0.048 wt% TMAH) after exposure on an ArF excimer laser stepper (NA=0.60) with a phase shift mask [385]. Hydrolysis of several lactone-containing 193 nm resist polymers was investigated and acid-catalyzed esterification of γ-hydroxy acid produced from α-acryloyloxy-β,β-dimethyl-γ-butyrolactone units was utilized in the development of a high performance 193 nm negative resist [386].

The negative behavior of 193 nm methacrylate resists based on methacrylic acid and hydroxyethyl methacrylate has been investigated and transesterification involving the hydroxyethyl ester has been proposed as a new mechanism of negative imaging (Fig. 134) [387, 388].

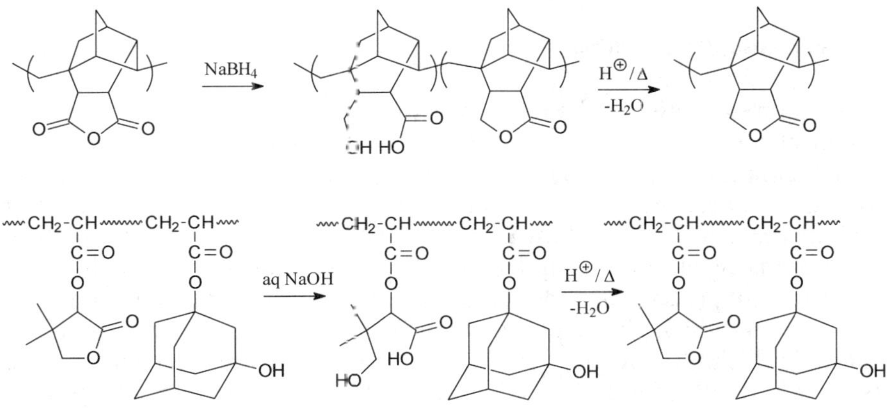

Chemical Amplification Resists for Microlithography

Fig. 132 Intramolecular esterification of γ-hydroxy acid for a reverse polarity change

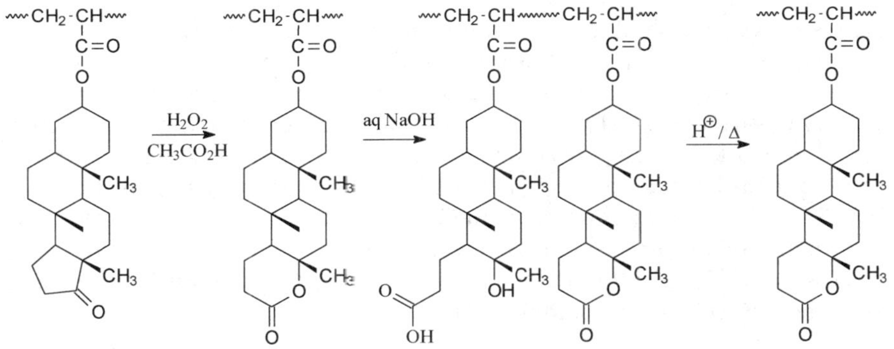

Fig. 133 Intramolecular esterification of δ-hydroxy acid

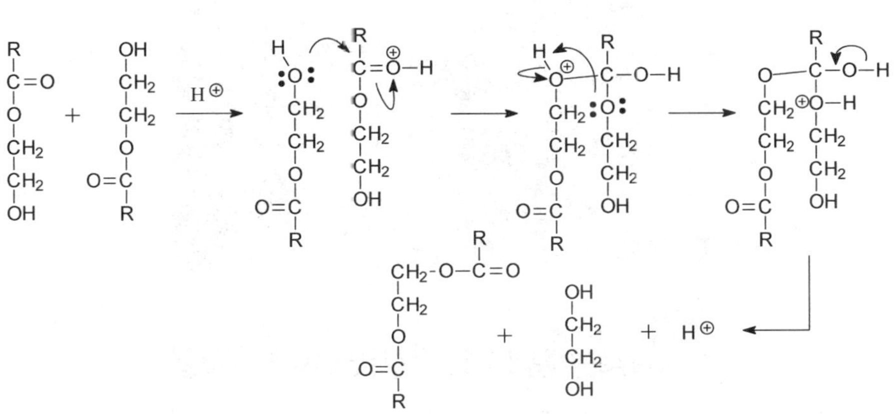

Fig. 134 Transesterification involving hydroxyethyl ester for negative imaging

4.7
Polymerization/Crosslinking

The onium salt cationic photoinitiators were originally developed for photochemical curing of epoxy resins [33]. Non-nucleophilic gegenanions such as hexafluoroantimonate and hexafluoroarsenate were chosen for this application to minimize premature termination by coupling between the growing cation and the gegenanion. One of the first chemical amplification resists was based on crosslinking of epoxy resins (Epi-Rez SU-8) via cationic ring-opening polymerization of pendant epoxide groups [25]. This first imaging was carried out using a rather thick film (6 µm) and the SU-8 resist is currently a material of choice in thick film imaging (50–100 µm) [389] for micro-electromechanical systems (MEMS) (Fig. 135) [390]. Copolymers of styrene with allyl glycidyl ether have been prepared for deep UV applications (Fig. 136) [391]. Pendant episulfides have been also employed as cationically-polymerizable groups for crosslinking in negative resist formulations (Fig. 136) [392, 393].

However, the crosslinking mechanism does not generally offer high resolution because swelling during development with organic solvents distorts developed images in the form of bridging and/or snaking. The developer solvent which can dissolve the uncrosslinked polymer in the unexposed regions has an affinity toward the crosslinked polymer and penetrates into the three-dimensional network, inducing swelling. The epoxy chemistry has been combined with aqueous base development, which is less prone to swelling, by blending PHOST with an epoxy-novolac resin (15 wt%) and triphenylsulfonium hexafluoroantimonate (10 wt%) [394]. Another blend formulation employed a novolac resin, triphenylsulfonium hexafluoroantimonate, and a bifunctional small epoxide, bis(cyclohexene oxide) [394].

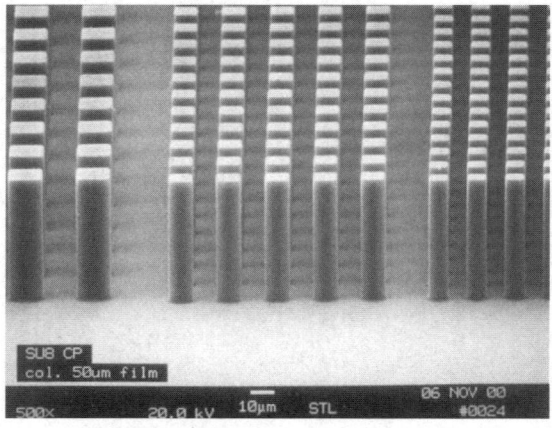

Fig. 135 Thick SU-8 negative resists patterns (5, 10, and 20 µm posts in 50 µm thick film) (courtesy of D. Johnson, MicroChem Inc.)

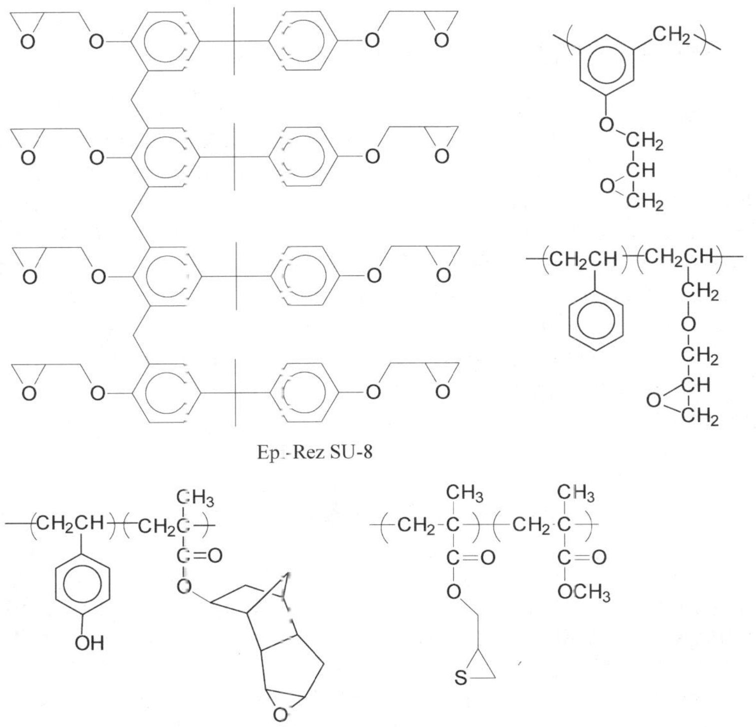

Fig. 136 Epoxy resins for cross-linking by cationic ring-opening polymerization

A copolymerization technique has been also employed to render the epoxy crosslinking system developable with aqueous base. Dicyclopentyloxy methacrylate was copolymerized with BOCST and the copolymer solution was heated to reflux (145 °C) in a casting solvent (PGMEA) to remove the tBOC group to afford the desired base-soluble copolymer (Fig. 136) [395]. The two-component negative resist containing a PAG provided high lithographic performance in deep UV and e-beam imaging using aqueous TMAH solution as a developer [395].

Cationically reactive vinyl ethers have been utilized also in the design of crosslinking negative phenolic resists. Since the pendant vinyl ether reacts with the phenolic unit in the polymer to form acetal crosslinks during high temperature PAB, acid-catalyzed deprotection of the acetal group results in de-crosslinking, providing positive images. Thus, the vinyl ether resist systems exhibit interesting and complex dual mode imaging, depending on the composition of the polymer, bake conditions, and exposure doses (Fig. 36) [161, 396].

5
Environmentally Friendly Processes

Aqueous base has completely replaced organic solvents in the development process of semiconductor devices while organic developers are still being used in a much smaller volume in mask making, MEMS, etc. However, a great deal of interest in environmentally friendly lithographic processes, as evidenced by establishment of NSF/SRC Engineering Research Center for Environmentally Benign Semiconductor Manufacturing [397], has spawned research activities directed toward designing resist materials that can be cast from and/or developed with water or other environmentally benign solvents such as supercritical carbon dioxide to reduce further volatile organic compound (VOC) emission from lithographic fabrication facilities and to facilitate cost saving in treatment of waste streams.

5.1
Water-Processable Resists (Casting and Development)

Attempts have been made to design chemical amplification resists that can be cast from and/or developed with pure water, although casting from a solvent with a high surface tension (such as water) might not produce high quality thin films needed for microlithography.

An early effort to design a chemical amplification resist that could be developed with water involved synthesis of acrylamide-based water-soluble homo- and copolymers [398]. Negative images were generated by crosslinking through acid-catalyzed self-condensation. In this case the casting solvent was still organic cyclohexanone and the resolution was reportedly limited to 2 µm, presumably due to swelling.

This work led later to a design of a resist that can be cast and developed with water, which consisted of three water soluble components; poly(methyl methoxy[(1-oxo-2-propenyl)amino]acetate) [poly(methyl acrylamidoglycolate methyl ether)], 2,4-dihydroxyphenyldimethylsulfonium triflate, and 1,4-butanediol (Fig. 137). The crosslinking mechanism responsible for the negative imaging (ca. 30 mJ/cm^2 at 248 nm) was investigated by using model compounds and reported to be via transesterification/transetherification at low temperatures and via alcoholysis of the amide at high temperatures [399].

Commercially available water soluble copolymers of maleic anhydride with ethylene and methyl vinyl ether (presumably partially hydrolyzed to vicinal carboxylic acids) and triphenylsulfonium triflate were dissolved in water as a casting solvent. Spin-cast films were baked at 130 °C for 10 s, exposed to 254 nm radiation, and postexposure-baked at 130 °C for 40 s. Development with pure water provided negative tone images, presumably due to acid-catalyzed dehydration between vicinal carboxylic acids to form less polar anhydride, as was demonstrated for polarity reversal (4.3.1). However, a copolymer of maleic acid with methyl vinyl ether failed to provide any negative images,

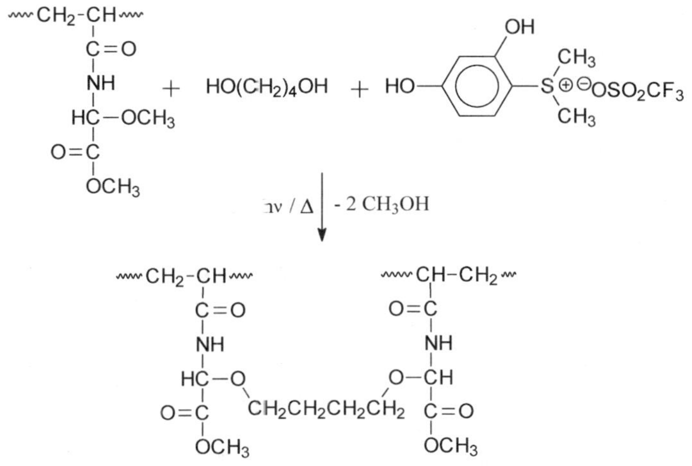

Fig. 137 Water-processable negative resists

presumably due to the extremely high water solubility of the carboxylic acid copolymer [346].

Poly(vinyl alcohol), another classical and commercially available water soluble polymer, was utilized in conjunction with the aforementioned water-soluble PAG and hexamethoxymethylmelamine as a latent electrophile [400]. Poly(vinyl alcohol) prepared by 80–90% hydrolysis of poly(vinyl acetate) was used as a matrix polymer (completely hydrolyzed polymer is insoluble in water due to extreme levels of hydrogen-bonding). Even at a high loading (20 wt%) of PAG and crosslinker, the imaging dose was quite high (>200 mJ/cm^2) and thermal crosslinking occurred at >130 °C. The negative tone imaging based on crosslinking of a water-castable resist with water as a developer can suffer from swelling-induced resolution limits as the exposed/crosslinked regions still have an affinity toward water (Fig. 138) [400]. Another attempt to utilize poly(vinyl alcohol) in negative imaging with water was based on the fact that while pure poly(vinyl alcohol) is insoluble in water due to extensive hydrogen-bonding, a small amount of substitution by other functionalities is enough to disrupt hydrogen-bonding and to give aqueous solubility. Poly(vinyl alcohol) was pro-

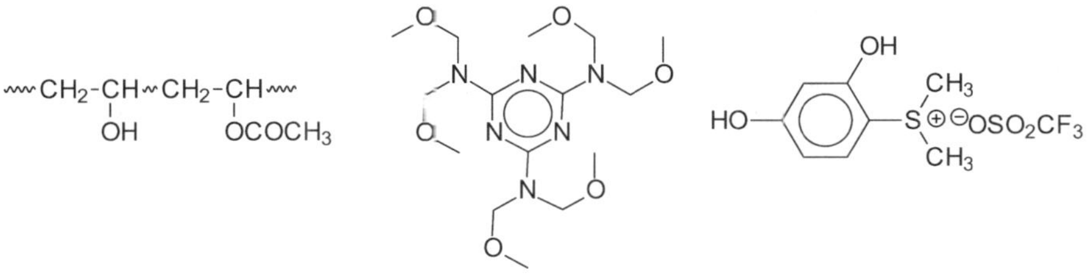

Fig. 138 Poly(vinyl alcohol)-based water processable resists

tected with 3–4 mol% *tert*-butyl carbonate, rendering the polymer soluble in water (Fig. 138). A formulation based on the copolymer and 4-methoxyphenyl-dimethylsulfonium triflate reportedly gave negative images upon development with pure water [401].

Poly(3-*O*-methacryloyl-D-glucopyranose) was prepared through deprotection of poly(1,2:5,6-di-*O*-isopropylidene-3-*O*-methacryloyl-α-D-glucofuranose) and used as a water-soluble resin that could be crosslinked by generating triflic acid from water soluble (4-methoxyphenyl)dimethylsulfonium triflate (Fig. 139) [402]. The presence of some remaining acetonide rings greatly improved the sensitivity and contrast of these water-soluble resist systems. The resist based on a polymer still containing some isopropylidene rings may not be fully water-soluble but rather is water-dispersible. However, such materials were able to resolve 0.45 μm dense patterns and 0.225–0.200 μm isolated lines at 248 nm, though swelling was still observed [402].

2-Isopropenyl-2-oxazoline was homo- and copolymerized with styrene via radical mechanism to prepare two-component negative resists (Fig. 140) [403]. Polymers containing at least 80 mol% oxazoline afforded casting from and developing with water. However, higher quality imaging was achieved when the resist was cast from 2-methoxyethanol, presumably because side reactions such

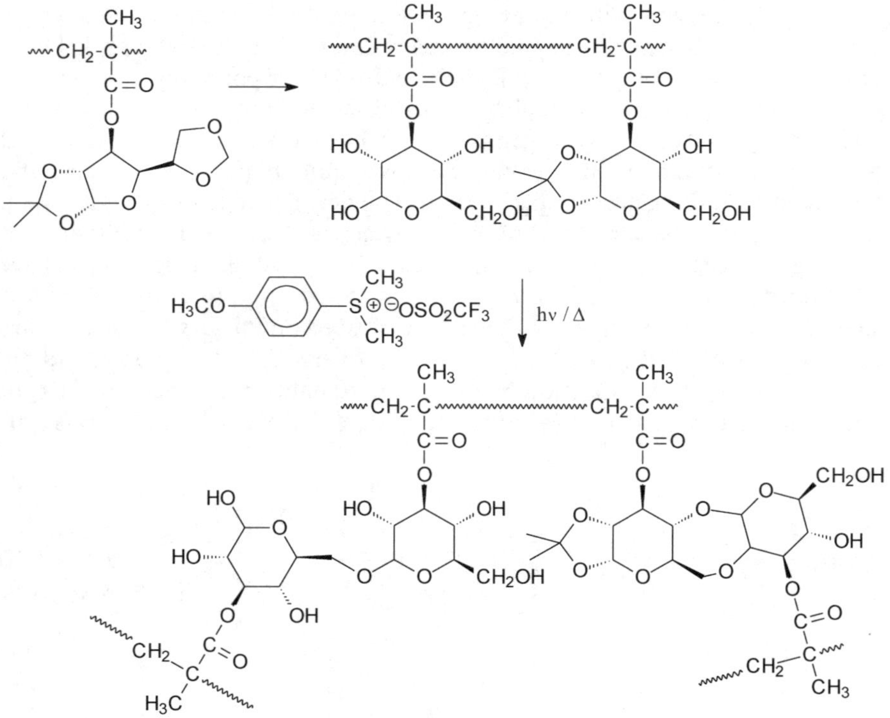

Fig. 139 Sugar-based water processable resists

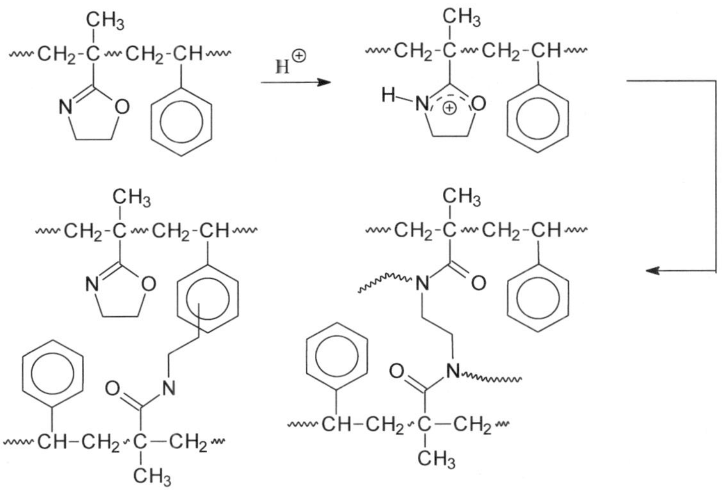

Fig. 140 Oxazoline-based water processable negative resists

as partial hydrolysis of the pendant oxazoline rings in aqueous environments were avoided. In fact, radical polymerization of the oxazoline monomer in water resulted in extensive hydrolysis [403].

As aqueous base development of crosslinking negative resists suffers from swelling during development, the concept of a reverse polarity change through pinacol rearrangement [151, 350, 351] (4.3.3) from a polar to nonpolar state has been also incorporated in the design of water processable resists. Since polystyrene with a pendant *vic*-diol and copolymers with 4-hydroxystyrene are insoluble in pure water, the styrenic diol monomer was copolymerized with methyl styrenesulfonate and the resulting copolymer was converted to a polyanion with a tetramethylammonium cation [404]. Copolymers containing 55–70 mol% diol were targeted as water-soluble resist polymers. Because the styrenesulfonate unit has a strong absorption in the 248 nm region, imaging employing a deep UV PAG such as 4-methoxyphenyldimethylsulfonium triflate did not yield high sensitivity, high contrast, and high resolution. A PAG extending its absorption to the 365 nm region was prepared and added to the copolymer for i-line imaging. Alternately, a naphthalene chromophore pendant from styrene was terpolymerized with the diol monomer and styrenesulfonate as a sensitizer of the deep UV sulfonium salt for i-line exposure [404].

The design of a positive tone resist that can be cast from and developed with water is much less straightforward because the exposed areas must remain soluble in water and the unexposed regions must somehow become insoluble in water. A dual solubility change must be incorporated; an original water-soluble polymer film is insolubilized thermally during the postapply bake step and the exposed regions are rendered soluble in water through an acid-catalyzed process. In an early investigation, the aforementioned poly(2-isopropenyl-2-oxazoline), 2-phenyl-1,3-dioxane-5-yl-carbonic acid, and 4-methoxyphenyl-

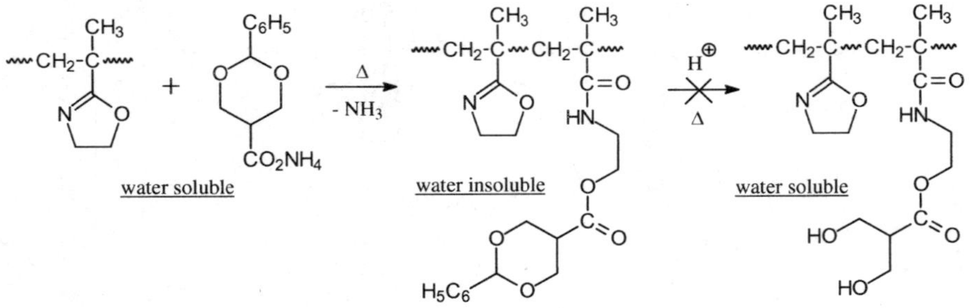

Fig. 141 Water-castable and -developable positive resist – an attempt

dimethylsulfonium triflate dissolved in water was cast into a film, which was insolubilized through crosslinking by heating. Formation of an amide-ester was expected to append an acetal functionality for subsequent acid-catalyzed deprotection. However, the crosslinking was irreversible and the desired positive images were not obtained after exposure and PEB presumably because the oxazoline rings were prone to acid-catalyzed crosslinking or so basic that the acid-catalyzed acetal cleavage was prevented (Fig. 141) [401, 405].

Thermal crosslinking of acidic polymers with use of di(vinyl ethers) and acid-catalyzed de-crosslinking through cleavage of the acetal crosslinks [161, 396] were attempted in water development (Fig. 142). Commercial poly(acrylic acid), 40 mol% (relative to COOH) tetra(ethylene glycol) divinyl ether, and 5 mol% (relative to COOH) 2,4-dihydroxyphenyldimethylsulfonium triflate were dissolved in methanol (not H_2O) [405]. Spin-cast films were baked at 110 °C for 3 min, exposed to UV, postexposure-baked at 90 °C for 1 min, and then developed with water for 20 s to generate low resolution positive images [405]. The positive resist design based on the vinyl ether chemistry has been then modified by introducing vinyl ether as a pendant group on poly(acrylic acid). Since carboxylic acids readily add to vinyl ethers in aqueous solution, the reactive carboxylic acids were protected as their ammonium salts. The ammonium salt copolymer was highly soluble in water. Upon PAB ammonia is volatilized from the films and the corresponding free carboxylic acid is formed. The free acid quickly reacts with the vinyl ether groups to form acetal crosslinks, rendering the film aqueous insoluble. Upon exposure, photochemically generated acid and water hydrolyze the acetal linkages to de-crosslink the exposed film. Thus, development with pure water generates positive images. In addition to the expected poor dry etch stability of the aliphatic resist, a slow dark reaction limited its shelf life (less than six days even at 8 °C).

To address these shortcomings the carboxylic acid was replaced with weaker phenol for a better shelf life and dry etch stability and the vinyl ether functionality was separated from the acidic functionality employing a three component design. A water soluble phenolic copolymer was prepared by radical copolymerization of 4-acetoxystyrene and sodium styrenesulfonate, followed by deacetylation with ammonium hydroxide. A water soluble bis(vinyl ether)

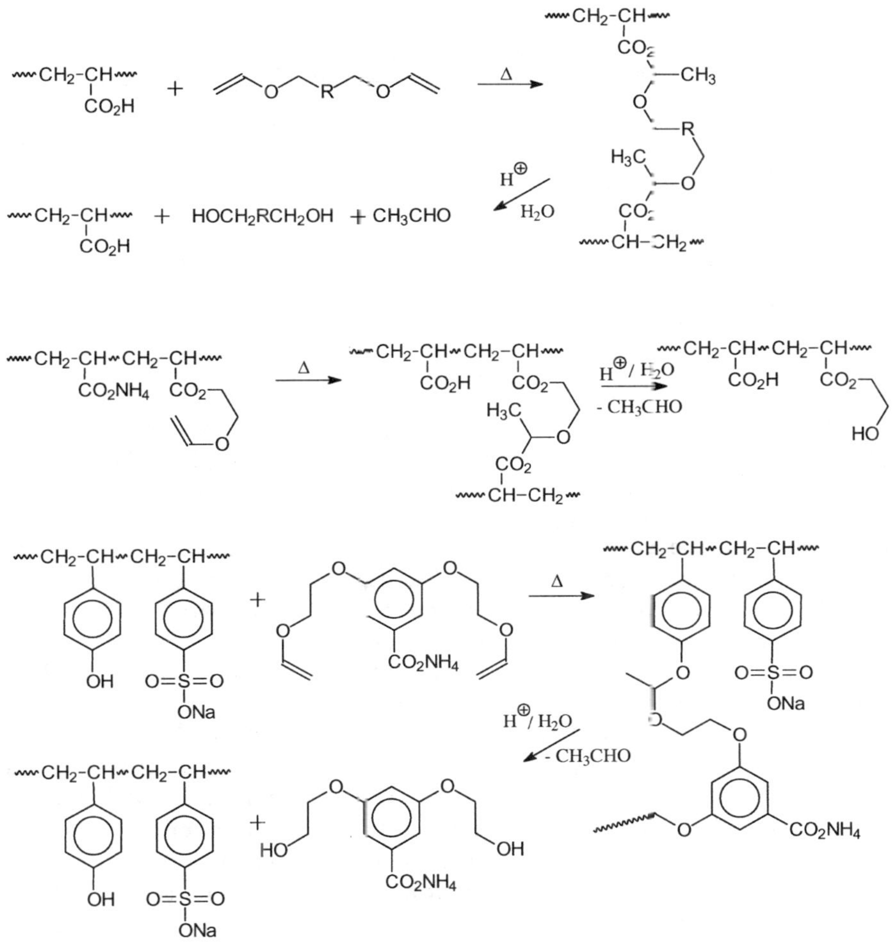

Fig. 142 Vinyl ether-based water processable positive resist

was synthesized from 3,5-dihydroxybenzoic acid, which was first protected as the methyl ester and reacted with 2-iodoethyl vinyl ether. The water insoluble bis(vinyl ether) was hydrolyzed with NaOH to give a water soluble sodium salt, which was then converted to the ammonium salt. A three component resist was formulated by dissolving the phenolic copolymer, bis(divinyl ether), and 4-methoxyphenyldimethylsulfonium triflate in water. Spin-cast films were insolubilized in water by baking above 110 °C for 10 min and finally re-solubilized in water by exposure to 254 nm radiation and PEB at 90 °C for 1 min (Fig. 142) [404].

However, no aqueous resist system has been reported that has both high resolution and dry etch resistance. In order to design an aqueous processable positive resists with more practical performance, pure water was replaced with

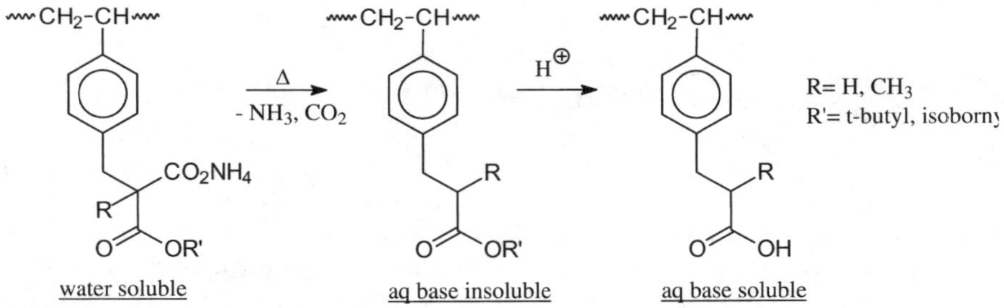

Fig. 143 Water-castable and aqueous base developable positive resist based on decarboxylation

widely accepted aqueous TMAH solution in development [406]. β-Keto carboxylic acids, such as malonic acid, undergoes thermal decarboxylation to produce propionic acids and carbon dioxide. The ammonium salts of the half esters of malonic acids provide not only the initial solubility for the polymer but also the insolubilization mechanism at PAB via sequential volatilization of ammonia and decarboxylation. Acid-catalyzed deprotection renders the exposed area soluble in aqueous base. The polarity switching group was attached to polystyrene for good dry etch resistance (Fig. 143) [406]. In the case of *tert*-butyl ester concomitant occurrence of decarboxylation and deprotection during PAB reduced the developer selectivity and contrast significantly. More stable isobornyl ester provided a better selectivity between decarboxylation and deprotection, and thus performed better lithographically. The polymer film cast from water had to be baked at 155 °C for at least 5 min to be insolubilized in 2.38 wt% TMAH aqueous solution but exhibited significant swelling. The solubility threshold existed around 4 min at 165 °C and the swelling of the film was reduced as the bake time was increased. Imaging experiments were carried out at 248 nm, using an aqueous solution of the ammonium salt of the isobornyl-protected polymer and triphenylsulfonium nonaflate. Spin cast films were baked at 165 °C for 5 min, exposed to 20 mJ/cm^2 of 248 nm radiation, postbaked at 140 °C for 1 min, and developed with 0.26 N TMAH for 30 s, providing 1.0 µm line/space patterns. Smaller features suffered from pattern deformation due to swelling and adhesion failure. With longer PAB time of 10 min no image was obtained after development. The decarboxylation was found by IR to occur only slowly even at 165 °C and prolonged heating at this temperature induced crosslinking [406].

5.2
CO_2-Processable Resists (Casting, Development, Rinse, and Strip)

Use of liquid or supercritical carbon dioxide in microlithography (and also in other areas such as separation and cleaning) has attracted a great deal of attention recently, which stems from the environmental/health consideration and from practical lithographic process issues. The 1997 Semiconductor Industry

Association National Technology Roadmap for Semiconductors (and its 1999 update) and the 1996 Electronics Industry Environmental Roadmap both emphasize the importance of reducing wet organic and aqueous consumption in the industry. Among supercritical fluids (SCF), carbon dioxide has been touted as the solvent of choice because it is nonhazardous and inexpensive and has been proposed to be employed in the lithographic process to replace one or two or entire wet process. SCF CO_2 has high diffusivity, comparable to gas, which may aid in rapid and effective dissolution. It has no surface tension since liquid and vapor states are not present simultaneously, and produces no lateral forces to damage high aspect ratio resist patterns. SCF CO_2 has density higher than other supercritical fluids (up to 900 g/L), near that of its liquid state. The most useful trait is the supercritical solvating capability that can be tuned by minor adjustment of temperature and pressure.

Resist systems that can be cast from and/or developed with environmentally-benign CO_2 have been sought. As polar polymers are not soluble in CO_2 in general, protected fluorine-containing nonpolar polymers were employed, which were converted to polar polymers by acid-catalyzed deprotection, producing negative images upon development with SCF CO_2. The first example of the use of SCF CO_2 in lithographic imaging was disclosed by Allen and Wallraff, who developed negative images using SCF CO_2 in a polymethacrylates resist containing an acid-labile group and fluorine or siloxane esters [407]. In this case the casting solvent was not CO_2 but conventional organic solvents. Similarly, TBMA copolymers with siloxane ester of methacrylate were prepared by random radical copolymerization and by group transfer polymerization to form block copolymers. The block polymer and triphenylsulfonium hexafluoroantimonate were dissolved in an organic solvent (PGMEA) and developed in a negative mode after exposure to 193 nm radiation (Fig. 81) [408]. Block copolymers were prepared by group transfer polymerization of tetrahydropyranyl methacrylate and 1H,1H-perfluorobutyl or 1H,1H-perfluorooctyl methacrylate, dissolved in an organic solvent (PGMEA) together with a PAG, and developed in a negative mode with SCF CO_2 after exposure to 193 nm radiation [409].

An interesting all-CO_2 process has been reported [410]. Radical copolymerization of TBMA and the aforementioned 1H,1H-perfluorooctyl methacrylate was carried out using AIBN as the initiator in SCF CO_2, the isolated copolymer and a newly prepared CO_2 soluble PAG were dissolved in CO_2 (Fig. 144). The films cast from CO_2 were developed, after exposure to 193 nm radiation and PEB, with CO_2, producing negative images of the mask. Coatings from liquid CO_2 can be extremely uniform, smooth, and thin on large surface areas because CO_2 has a low surface tension and viscosity and it can wet practically any surface. No residual solvent is found. The rate of AIBN decomposition in SCF CO_2 has been found to be 2.5 times lower than that observed in benzene, which has a higher dielectric strength and stabilizes the transition state of the primary scission reaction more than SCF CO_2. Furthermore, a high radical efficiency has been observed in SCF CO_2, which is due to the low viscosity of CO_2 (no cage effects to promote recombination of primary radicals).

Fig. 144 All-CO_2 process (polymerization, casting, and development)

Thus, significant efforts have been directed toward designing resist systems that can be cast from and/or developed with SCF CO_2. However, the use of SCF CO_2 in microlithography is likely to be implemented first in the rinse step after aqueous base development as it could provide a solution to the line collapse problem first reported by X-ray lithographers in high aspect ratio imaging and then frequently observed in the current water rinse process. The mechanism of resist pattern collapse has been elucidated by Tanaka et al. by observing resist patterns in a rinse liquid (H_2O) using an atomic force microscope (AFM) [411]. The cause of resist pattern collapse is related to the surface tension (γ) and the

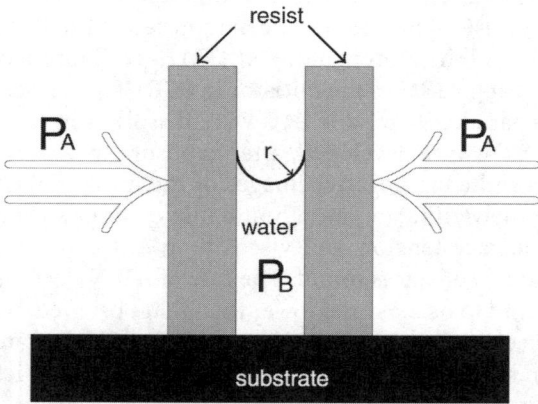

Fig. 145 Line collapse

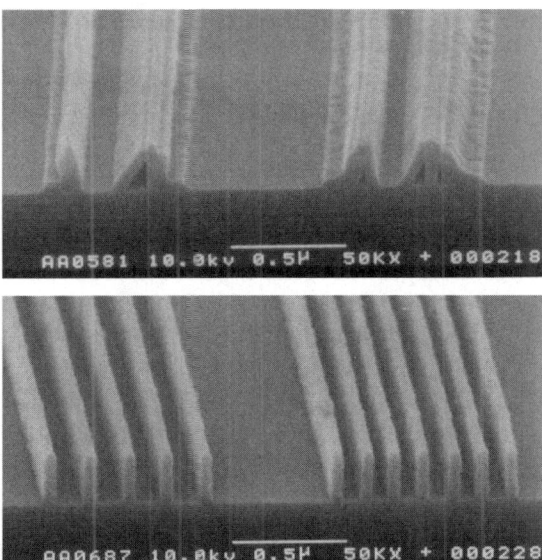

Fig. 146 High aspect ratio images (60 nm lines in 300 nm thick resist) dried with N_2 blow after H_2O rinse (*top*) and with SCF CO_2 (*bottom*) [412b]

resulting capillary forces acting on the resist walls. The capillary force (P) can be expressed as $P=\gamma/r$, where r is a radius of the curvature of the water-air interface and γ is a surface tension of water (Fig. 145). These researchers suggested the use of low surface tension rinse liquids or a surface tension free drying process, such as freeze-drying or supercritical drying, as promising methods for preventing pattern collapse. Namatsu et al. [412] and then other research groups [413] have demonstrated that the low surface tension of SCF CO_2 eliminates image collapse during development and rinse. Figure 146 presents high aspect ratio images (60 nm wide/300 nm high) dried with N_2 blow after H_2O rinse and with SCF CO_2.

6
Bilayer Lithography and Top-Surface Imaging

The semiconductor manufacturing prefers single layer resists over multilayer systems from the viewpoint of simplicity and cost and therefore strives to improve single layer aqueous base development to meet the ever-demanding challenges. However, bilayer lithography and dry development schemes can reduce some of the burdens placed on resist materials and imaging technologies. Due to the paramount difficulties associated with single layer lithography, bilayer and dry imaging techniques were heavily investigated a decade ago and have re-surfaced recently as viable/necessary techniques as the semiconductor technology moves to the minimum feature size of <130 nm. Use of a bilayer process can extend the life of the current workhorse technology before a new shorter wavelength technology becomes available and/or improve process latitudes and yields of cutting edge product lines.

The linewidth control over steps and wafer topography becomes extremely challenging and difficult as the feature size shrinks. Reflection from the topographic features on the wafer results in linewidth variation. Higher resolution is achieved in optical lithography at the expense of the depth-of-focus (DOF) because DOF is proportional to the exposure wavelength and inversely proportional to NA^2 while the resolution is proportional to the wavelength and inversely proportional to NA. The multilayer scheme discussed in this section can provide solutions to these difficult problems encountered in optical lithography. Furthermore, the use of a thin imaging layer can improve resolution and linewidth control and a thick bottom layer offers superb etch resistance for etching of substrate, thus imaging and etch performances do not have to reside in resist. The dry development process is more suited to high aspect ratio imaging while thin tall lines tend to fall down in wet development due to a capillary effect as mentioned earlier, which has become increasingly serious.

The more directional dry etching process employing plasmas has long replaced the isotropic wet etching technique in substrate fabrication, which allows a tighter control of the image transfer step. The advent of the dry etch technique motivated resist chemists and engineers to design resist systems that can be developed with a plasma without use of a developer solvent. Generation of primary resist images with use of a plasma was first reported in 1979. In this plasma-developable photoresist (PDP) process [414], the resist is coated in the usual fashion and exposed. The exposed film is then heated to produce a relief image of a negative tone, which is then transferred into the underlying layer by RIE. The PDP concept was extended to the design of an X-ray resist system, which is based on poly(2,3-dichloropropyl acrylate) as a host polymer and a silicon-containing monomer [415]. X-ray exposure of the resist film initiates radical crosslinking of the polymer, and hence it incorporates the silicon-containing monomer into the network (fixing). The exposed film is then heated to remove the unreacted organometallic monomer from the unexposed areas. Development is accomplished by treating the baked film with an O_2 plasma.

Two general schemes are available in the use of plasmas, especially oxygen reactive ion etching (RIE), in the resist development processes, which typically employ organometallic polymers that form refractory oxide upon treatment with oxygen plasma. Organosilicon polymers are of particular interest, which are converted to a thin (ca. 50 Å) layer of silicon oxide in oxygen plasma. It is this thin oxide layer that is impervious to further etching. The first approach, which is bilayer lithography, involves coating a substrate with a thick (0.5–2 µm) planarizing layer of an organic (aromatic) polymer, followed by application of a thin (<0.5 µm) organosilicon resist (two coating steps). The silicon species is selectively removed from either the exposed or the unexposed areas by development, which provides a protective stencil for transfer of the resist image to the planarizing layer by oxygen RIE. The second approach to patterning with oxygen RIE is selective incorporation of silicon in an organic resist film after image-wise exposure, which provides single layer all-dry processes without use of a developer solvent. Many variations of these general schemes have been developed as described in this section.

6.1
Bilayer Lithography with Organosilicon Resists

The bilayer lithography technique employing an organometallic resist has evolved from a trilayer imaging scheme involving a top organic imaging layer, a thin middle layer of silicon oxide as an oxygen RIE barrier, and a thick bottom organic polymer layer for planarization of a topographic substrate (Fig. 147). In this trilevel scheme, the top resist layer is imaged with a developer in a conventional fashion and the three-dimensional resist pattern thus generated is then used as a protective mask to etch the middle oxide layer with a fluorocarbon plasma. The imaged oxide layer now protects the underlying organic polymer

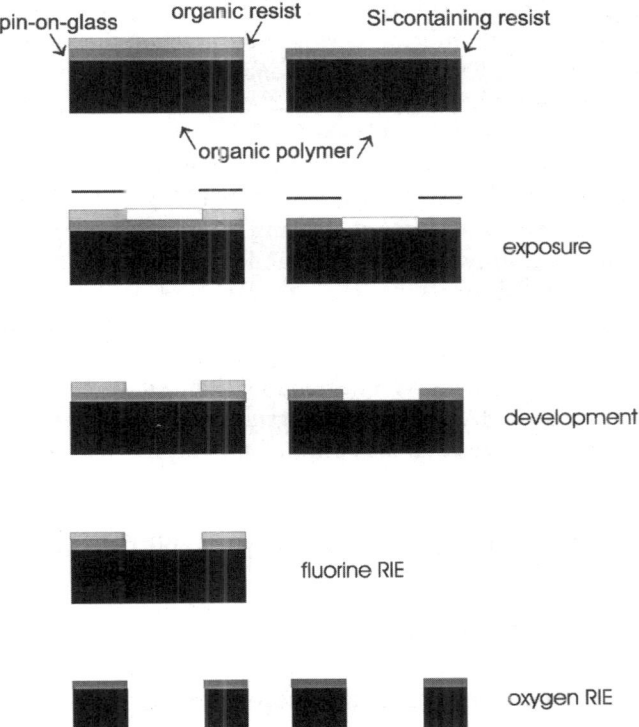

Fig. 147 Trilayer (*left*) and bilayer (*right*) lithography

film in the subsequent oxygen RIE pattern transfer step. The bilayer lithography technique (Fig. 147) combines the functions of the top imaging resist and the middle oxide film in one layer through use of an organometallic resist and thus simplifies the time-consuming and therefore costly trilayer process. In general organosilicon polymers for use in the oxygen RIE bilayer process contain >10 wt% Si for good etch resistance.

The bottom planarizing layer plays a non-trivial role in the bilayer scheme and thus must satisfy a number of criteria:

- Suitable for spin-coating
- Resistant to common casting solvents with no interfacial mixing with the top resist layer
- Highly opaque at the exposing wavelength to eliminate the topography effect
- Not detrimental to the top resist
- Etched rapidly in oxygen plasma
- Thermally stable
- Resistant to fluorocarbon plasmas
- Easy stripping after substrate fabrication

The polymers most commonly employed as a planarizing material in the oxygen RIE bilayer lithography are hard-baked (crosslinked) novolac resists and cured polyimide. These materials possess all the good attributes as a bottom layer but lack the strippability. To avoid interfacial mixing with the imaging resist layer, the bottom layer is typically rendered insoluble by crosslinking or curing, which enhances its thermal stability but makes clean stripping extremely difficult. Poly(4-vinylbenzoic acid) has been reported to be suitable as a strippable bottom layer material for use in all-dry bilayer lithography with poly(4-timethylsilylphthalaldehyde) (discussed later) [336]. This acidic polymer is soluble in 2-ethoxyethanol for spin-coating but resistant to common casting solvents such as PGMEA and cyclohexanone even without crosslinking. It is highly opaque in the deep UV region with OD of 3.4/µm at 248 nm. The polymer has a high T_g of 250 °C and remains soluble even after prolonged heating at 230 °C. It is resistant to fluorocarbon plasmas due to its aromatic nature but etches rapidly in oxygen plasma. The polymer film can be readily stripped with commonly available TMAH solution after substrate etching because of its high solubility in aqueous base even after high temperature treatment.

6.1.1
Semi-Dry Bilayer Lithography (Wet Development/O_2 RIE Pattern Transfer)

The first organometallic polymers employed in the bilayer imaging were commercial polysiloxanes, which function as crosslinking negative resists (not chemically amplified) [416]. Poly(phenylmethylsilsesquioxane), which undergoes acid-catalyzed silanol condensation and crosslinks upon PEB, was evaluated as a chemically-amplified bilayer resist using an organic developer (4-methyl-2-butanone/2-propanol=1/1) [417]. A low molecular weight ladder

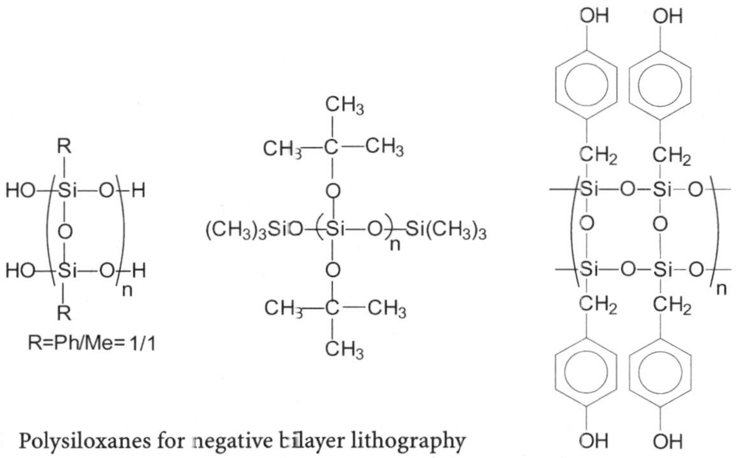

Fig. 148 Polysiloxanes for negative bilayer lithography

polymer (M_n=1700, M_w/M_n=1.56) (Fig. 148) was selected because condensation occurs between OH end groups.

Poly(di-*tert*-butoxysiloxane), which combines acid-catalyzed deprotection and subsequent silanol condensation to produce an O_2 RIE resistant glass (SiO_2), has been evaluated as an e-beam resist resin [418]. The polymer was synthesized by reacting diacetoxy(di-*tert*-butoxy)silane with triethylamine and water and subsequent end-capping with trimethylsilyl chloride in the presence of triethylamine. The polymer in the unexposed area is soluble in an organic solvent such as methyl isobutyl ketone while the three-dimensional inorganic network produced in the exposed areas is insoluble. Thus, development with the organic solvent results in negative tone imaging.

Acid-catalyzed silanol condensation to form insoluble networks has been also utilized in the design of aqueous base developable negative resist systems [419]. An aqueous base soluble silicone polymer was synthesized by a sol-gel reaction of a mixture of phenyltrimethoxysilane and 2-(3,4-epoxycyclohexyl)-ethyltrimethoxysilane. This polymer contained a high concentration of silanol OH groups and thus was soluble in aqueous base.

Poly(hydroxybenzylsilsesquioxane) is soluble in aqueous base. The polymer was synthesized by treating poly(methoxybenzylsilsesquioxane) with BBr_3 and then partially protected with the *t*BOC group to design an aqueous base developable positive resist [420]. This partially-protected polymer with T_g of 90 °C is highly transparent at 248 nm. Acid-catalyzed deprotection renders the exposed region highly soluble in an aqueous TMAH solution, while the unexposed film dissolves insignificantly. A similar silsesquioxane homopolymer with T_g of 115 °C and a Si content of 17 wt% was employed in the design of aqueous base developable 248 nm negative resists based on acid-catalyzed condensation using melamine and uril derivatives (see 4.5) [421]. The O_2 etch selectivity of the resist layer vs a hard-baked novolac resist was greater than 25:1. The underlying novolac resist was selected so as to closely match the reflective index to minimize the reflectivity from the interface between the top resist layer and the underlayer. High resolutions of 137.5 nm dense line/space patterns and 130 nm isolated lines were printed by annular illumination on a 248 nm step-and-scan tool (NA=0.60) using a chrome-on-glass mask and 0.14 N TMAH. The shelf life stability of such aqueous base soluble silsesquioxane polymers is a concern.

As is the case with the single layer resists, use of aqueous base as a developer has been mandated in the bilayer scheme involving wet development, thus necessitating incorporation of a polar functionality along with an acid-cleavable group and Si. Each functionality must be present in resist in a large enough concentration to perform its function; a polar group for aqueous development and adhesion, a protected group for high development contrast, and Si for etch resistance. It is sometimes difficult to incorporate the three functional groups in one polymer in sufficient concentrations. This dichotomy has been overcome by designing a Si-containing acid-labile group [421b, 422]. The versatility of silicon in organic chemistry can in part be attributed to the electronic effects of silicon substituents on chemical reactivity. One of the most interesting and widely studied characteristics of silicon is its ability to stabilize β-carbocations ($R_3SiCH_2CH_2^+$) [423]. This stabilization is believed to be due to both the electropositive nature of silicon and to hyperconjugation involving donation of the C-Si σ electrons into the empty carbocation p orbital. The effective stabilization requires the C-Si bond to be coplanar with the empty p-orbital. The β-silyl carbocations of this type are significantly more stable than the parent ethyl cation with stabilization energies similar to that of a *tert*-butyl cation. Thus, β-silyl

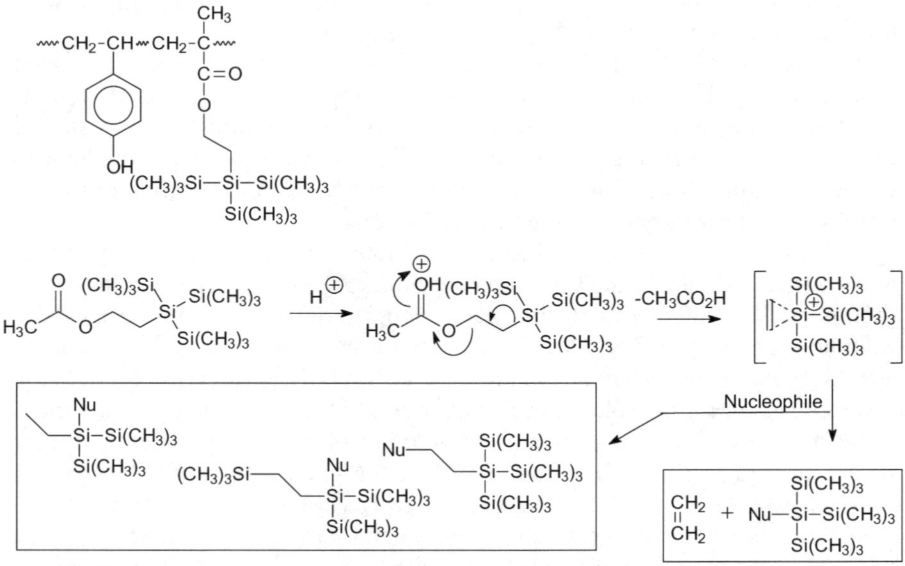

Fig. 149 EIRIS polymer and β-silicon elimination

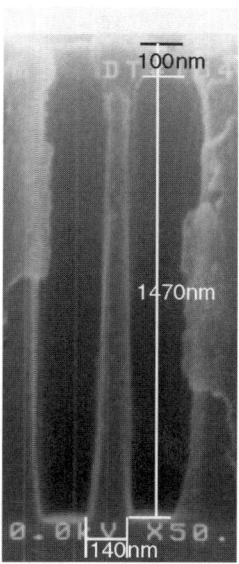

Fig. 150 High aspect ratio image printed in 248 nm bilayer resist by oxygen RIE [422c]

groups have been employed as an acid-labile protecting group in the design of aqueous base developable 248 nm positive bilayer resist. Methacryloxyethyltris(trimethylsilyl)silane and methacryloxyethyltrimethylsilane were synthesized and copolymerized with 4-hydroxystyrene using AIBN, copying the 248 nm ESCAP design (Fig. 149). The deprotection mechanism was elucidated by NMR analysis of model reactions between $AcOCH_2CH_2SiR_3$ and triflic acid in $CHCl_3$ and also by ^{29}Si NMR analysis of exposed resist films (Fig. 149). When R was CH_3, only a single Si-containing product (hexamethyldisiloxane) was detected in addition to acetic acid and ethylene. When R was $Si(CH_3)_3$, a number of Si-containing species were detected and significant silylation of the phenolic OH group was observed. This 248 nm bilayer positive resist (EIRIS™) developed at IBM is now commercially available from JSR (Fig. 150).

Bilayer resist systems have been actively pursued also in 193 nm lithography. One example is based on polymethacrylate. 3-[Tris(trimethylsilyloxy)silyl]-propyl methacrylate, methacrylic acid, and methacrylate bearing acid-labile ester (*tert*-butyl, tetrahydropyranyl, and 1-ethoxyethyl) were terpolymerized using AIBN in THF [424]. In order to ensure sufficient dry etch stability, the Si monomer fraction in feed was kept between 0.15 and 0.3, leading to polymers with Si contents of 10–15 wt%. An increase in the Si monomer concentration results in lower T_g and a higher methacrylic acid concentration results in higher T_g and faster dissolution rates in aqueous base. A terpolymer made from methacrylic acid/THP methacrylate/Si-methacrylate=3/5/2 was employed in 193 nm imaging using triphenylsulfonium triflate as a PAG. The strength of the TMAH developer employed was only 0.003 N and the thermal flow stability of

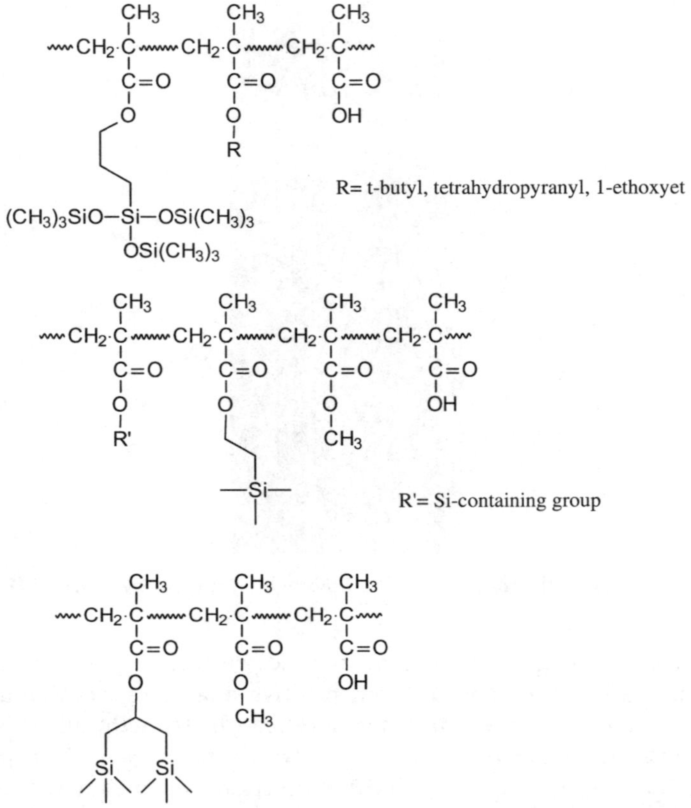

Fig. 151 Si-containing polymethacrylates for 193 nm bilayer lithography

the resist pattern was about 120 °C [424]. The same Si-containing methacrylate was later terpolymerized with acetal-protected methacrylate and property enhancing methacrylate (Fig. 151) [425, 426]. Block terpolymers were also prepared by the nitroxide (TEMPO) procedure for comparison. Importance of the optimization of the underlying layer in bilayer lithography has been described [426] and detailed evaluation of the methacrylate bilayer resist on a 193 nm step and scan exposure tool can be found in the literature [427].

The cleavable Si group as described above has been also incorporated in a polymethacrylate resist for 193 nm bilayer lithography (Fig. 151) [422b]. In a similar fashion 1,3-bis(trimethylsilyl)isopropyl methacrylate was terpolymerized with methacrylic acid and methyl methacrylate with AIBN in THF [428]. The COMA system has been modified to be used as a bilayer positive resist by incorporating norbornenes bearing a passive Si group and an acid-cleavable Si group (Fig. 152) [429].

Polysilsesquioxanes have found a use in the design of 193 nm bilayer resist. The phenol structure has been replaced with cyclohexyl carboxylic acid for

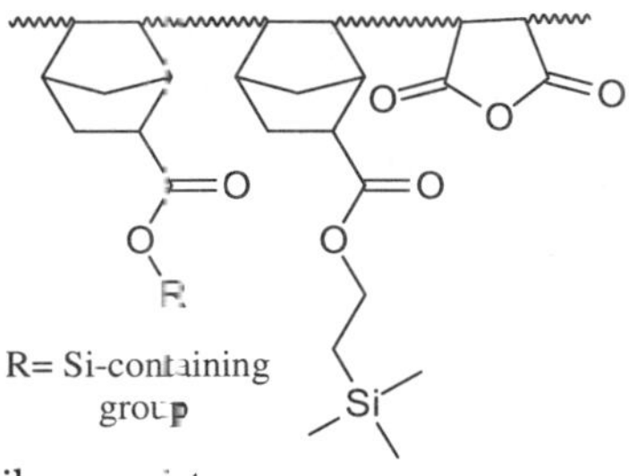

Fig. 152 COMA-based Si bilayer resist

base solubility and 193 nm transparency. The carboxylic acid was partially protected with an acid-labile group for chemically amplified positive imaging [430, 431]. The TMAH developer employed was 0.12 wt% (0.04 wt% in some cases). The cyclohexylcarboxylic acid polymer in CASUAL (Chemically Amplified Si-contained Resist Using Silsesquioxane for ArF Lithography) has a low calculated T_g of 13 °C and the resist based on an acetal-protected polymer did not provide high enough development contrasts due to its excessive dissolution (even in a weak developer). In an attempt to improve the contrast (preferably in a stronger developer), the remaining carboxylic acid was converted to alcohol through an amide linkage, which increased the contrast dramatically even in the standard 2.38 wt% developer but decreased the 193 nm transparency significantly. Another attempt to minimize the unexposed dissolution rate without sacrificing the transparency was to employ a copolymer of cyclohexylcarboxylic acid silsesquioxane and tricyclodecylalcohol silsesquioxane (Fig. 153) [432]. A blend of polysilsesquioxane and the aforementioned methacrylate terpolymer (see above) was employed as a positive bilayer 193 nm resist, which could be developed with 2.38 wt% TMAH [433].

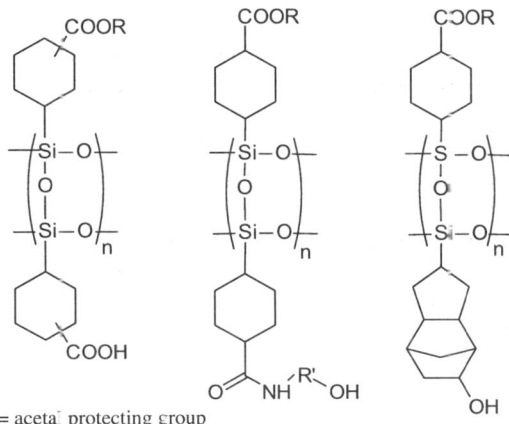

Fig. 153 Polysilsesquioxanes for 193 nm bilayer lithography

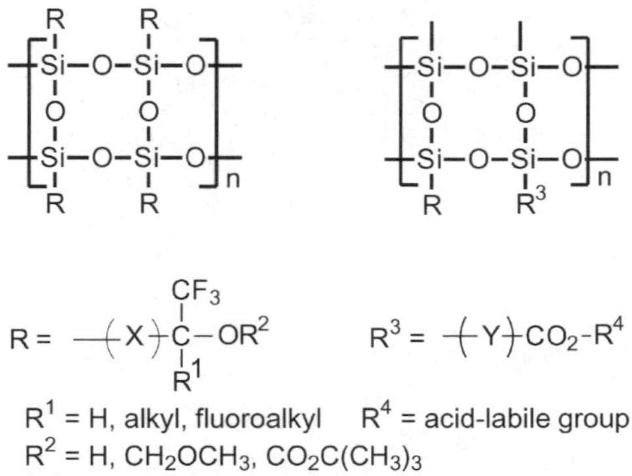

Fig. 154 Fluoroalcohol-containing polysilsesquioxanes for 157 nm bilayer lithography

Silsesquioxane polymers are a logical choice for the design of 157 nm bilayer resists because of their good transparency. Since carboxylic acids and phenols absorb strongly at 157 nm, the fluoroalcohol functionality has been incorporated into the peripheral Si-H bonds by hydrosilylation (Fig. 154) [434–436] 1,3,5,7,9,11,13,15-Octakis(dimethylsilyloxy)pentacyclo-[9.5.1.1^{3,9}.1^{5,15}.1^{7,13}]octasiloxane was reacted with 1,1-di(trifluoromethyl)-3-butene-1-ol (4.1.1) protected with methoxymethyl or tBOC, in THF using a Karsted's catalyst at room temperature. The protected polysilsesquioxanes thus obtained were a viscous liquid or wax. When unprotected 1-methyl-1-trifluoromethyl-3-butene-1-ol was employed instead, a solid product exhibiting a glass transition at 5.4 °C and melting at 62 °C was obtained, which cast a clear transparent film from a PGMEA solution. These silsesquioxane oligomers were developed as low molecular amorphous materials (see 4.1.8). Matrix-assisted laser desorption ionization-time of flight mass spectroscopy (MALDI-TOF MS) analysis of these materials exhibited only one peak [434]. The 157 nm absorption of the silsesquioxane resist polymers ranged from 3.5 to 2.0/µm.

Copolymers of 4-silyloxy-α-methylstyrene or silyl methacrylate with methacrylonitrile were prepared by radical copolymerization of silyl monomer or by silylation of acidic copolymer and evaluated as 157 nm positive bilayer resists [437].

LER and resist outgassing during exposure are major issues in bilayer lithography [438, 439]. Si species generated from a bilayer resist during exposure could badly contaminate lenses. Thus, formation of gaseous products must be minimized by proper selection of deprotection chemistry and resist components.

6.1.2
All-Dry Bilayer Lithography

The very first chemical amplification resist designed for use in bilayer lithography employed Si-containing polyphthalaldehyde as a top layer resin,

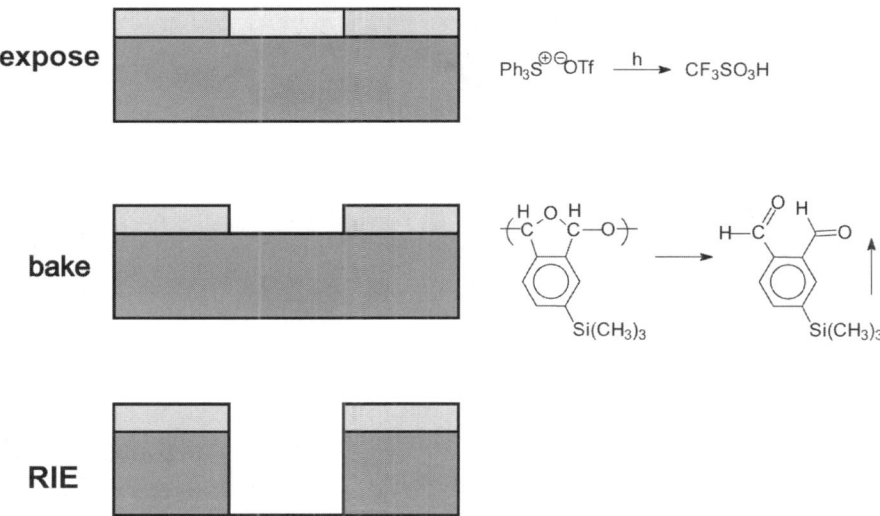

Fig. 155 All-dry bilayer lithography (thermal development-O_2 RIE pattern transfer)

which undergoes acid-catalyzed depolymerization to provide positive images [333–336]. Poly(4-trimethylsilylphthalaldehyde) was prepared by anionic cyclopolymerization at cryogenic temperatures and end-capped with acetyl groups by adding acetic anhydride to the cold polymerization mixture. The end-capped polymer is stable thermally to ~200 °C but undergoes clean depolymerization at ~100 °C in the presence of a photochemically-generated acid. The resist consisting of polysilylphthalaldehyde and triphenylsulfonium triflate (1.2 wt%) develops completely by PEB alone (thermal development) and the positive relief image functions as a stencil for oxygen plasma etching of a 2-μm-thick hard-baked novolac resist film (Fig. 155). Deep UV sensitivity curves for the thermal development of the resist are presented in Fig. 156. Scum-free clean development is critical in bilayer lithography because residual silicon remaining in the exposed regions after development can interfere with clean etching of the planarizing layer. Acid-catalyzed and thermodynamically-driven depolymerization completely reverts the Si-containing polymer to the starting monomer, which evaporates out during the PEB step (which is also the development step), providing clean complete removal of the Si species from the exposed areas simply by moderate heating after a low dose UV exposure, as Fig. 156 demonstrates. The sulfonium triflate is completely decomposed in oxygen plasma to volatile fragments due to its non-metallic nature. The thermally-developable resist has demonstrated sub-half-micrometer resolution in e-beam and KrF excimer laser exposures. Figure 157 shows a scanning electron micrograph of sub-half-micrometer positive images printed on a KrF excimer laser stepper in a bilayer structure by thermal development of the polysilylphthalaldehyde resist and subsequent oxygen RIE pattern transfer [336]. As men-

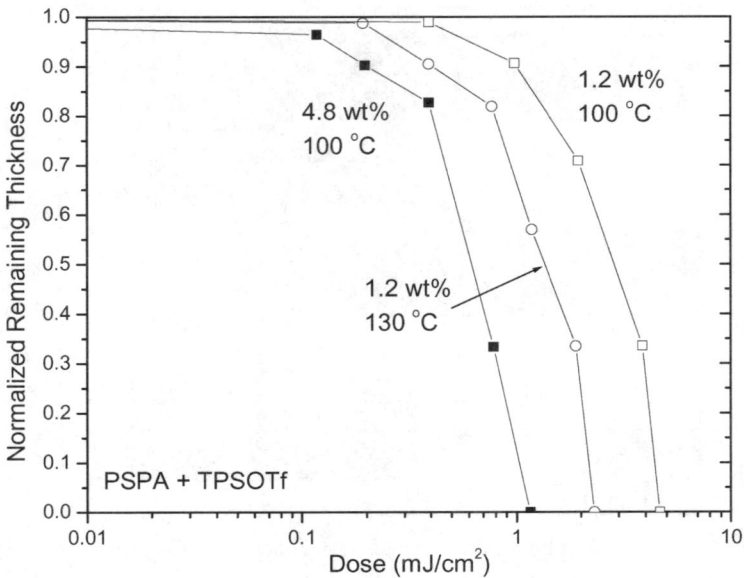

Fig. 156 Deep UV sensitivity curves of thermal development of poly(4-trimethylsilyl)-phthalaldehyde containing 1.2 and 4.8 wt% TPSOTf [334]

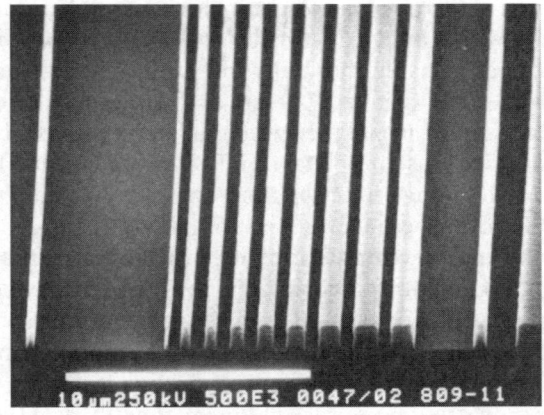

Fig. 157 Sub-0.5 μm line/space patterns printed by thermal development and subsequent O_2 RIE pattern transfer [336]

tioned earlier, poly(4-vinylbenzoic acid) was employed as a strippable bottom layer specifically for the polysilylphthalaldehyde resist [336].

Poly[4,5-bis(trimethylsilyl)phthalaldehyde] has been used in a similar all-dry process to etch a ~2 μm-thick polyimide film [440].

Acid-catalyzed hydrolysis of poly(4-trimethylsilyloxystyrene) was employed in the design of an aqueous base developable positive resist as described earlier

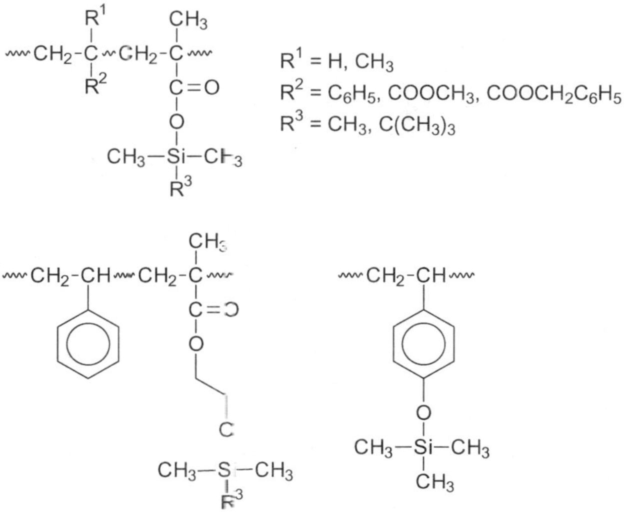

Fig. 158 Acid-catalyzed desilylation for positive bilayer lithography

but Si-containing single layer resists were not widely accepted because of difficult stripping with O_2 plasma. Desilylation can be useful in positive all-dry bilayer lithography, if the Si-containing fragments generated by acidolysis can be cleanly removed from the exposed areas without use of a developer solvent (by heating, for example). This concept was first reported in non-amplified photochemical cleavage of Si-containing side group pendant from poly(methyl methacrylate) [441]. A system consisting of poly(4-trimethylsilyloxystyrene) and triphenylsulfonium hexafluoroarsenate was studied in this context. In a similar fashion copolymers of trimethylsilyl methacrylate with styrene, methyl methacrylate, and benzyl methacrylate were evaluated in conjunction with 1,2,3,4-tetrahydronaphthylideneamino p-toluenesulfonate as the acid generator in terms of acid-catalyzed desilylation (Fig. 158) [442]. The silyl ester hydrolysis reached a limited value of 85% at 390 mJ/cm² of 254 nm radiation and only 45% of trimethylsilanol generated at this high dose escaped from the film. It was necessary to treat the resist film, after UV exposure, with an organic vapor such as acetone, n-hexane, or alcohol at room temperatures to completely remove the silanol. The treatment with acetone vapor not only improved the removal of the silanol but also enhanced the acid-catalyzed hydrolysis. This study indicates that removal of Si-containing fragments produced by acidolysis is not a trivial task. As mentioned earlier, a small amount of Si species remaining in the exposed areas after development can result in residue formation after oxygen RIE.

6.2
Silylation of Organic Resists

Instead of selectively removing Si species from an organosilicon resist film, one can selectively introduce Si species into an organic resist film, which could then provide a potential for all-dry development with oxygen RIE. Two mechanisms for selective silylation are available: reactivity-controlled silylation and diffusivity-controlled silylation. Silylation can be carried out either in a gas phase or in solution. If the penetration depth of the exposing radiation is shallow, due to a strong absorption in photolithography, for example, the photochemical reaction occurs only in the top surface of the film, freeing the latent image formation from the challenging substrate and topography effects as discussed earlier. Wet development would not work adequately in this case but selective silylation can produce a quasi-bilayer structure that is suitable for oxygen RIE development. The process is called "top-surface imaging (TSI)." The selectivity of silylation is a key factor in high resolution imaging.

6.2.1
Reaction-Controlled Silylation

In this scheme, a functionality that does not react with a silylating reagent is converted to a reactive group to incorporate Si selectively in the exposed area, or conversely, a reactive functionality is rendered inert by a radiation-induced process so that Si is selectively introduced in the unexposed region.

The dual-tone resists based on acid-catalyzed deprotection (see 4.1) are ideal for selective silylation because the polarity change is equivalent to the alteration of reactivity (Figs. 12 and 22). In these resist systems, reactive phenolic hydroxyl or carboxylic acid groups are unmasked during PEB through acid-catalyzed deprotection. When the exposed/postbaked resist film is treated with a silylating reagent such as hexamethyldisilazane or dimethylamino-(trimethyl)silane, Si is incorporated only into the exposed areas through covalent bonding but not into the unexposed areas consisting of carbonate or ester which is inert to silylation. Thus, the silylation selectivity is high. The silylated resist film is then subjected to oxygen RIE to provide negative tone images without use of a developer solvent (Fig. 159) [443]. Figure 160 presents IR spectra of the tBOC resist after PAB (a), after UV exposure (10 mJ/cm^2 at 254 nm) and PEB (b), and after treatment with gaseous dimethylamino(trimethyl)silane for 5 min at 100 °C and 200 torr (c). The IR study clearly indicates that PHOST generated by acid-catalyzed deprotection can be completely silylated in the solid state under mild conditions. The process can be extended to all-dry bilayer lithography simply by using a wafer coated with an organic polymer layer.

The tone of this imaging process can be reversed according to the scheme in Fig. 161 [443b, 444]; the phenolic OH group in the imagewise-exposed area generated by acid-catalyzed deprotection is blocked first with a non-metallic group which is unreactive toward silylation, the tBOC group in the initially

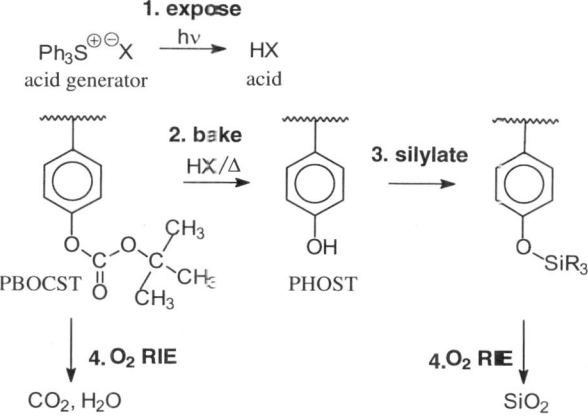

Fig. 159 Negative tone O_2 RIE development of *t*BOC resist by acid-catalyzed deprotection followed by selective silylation

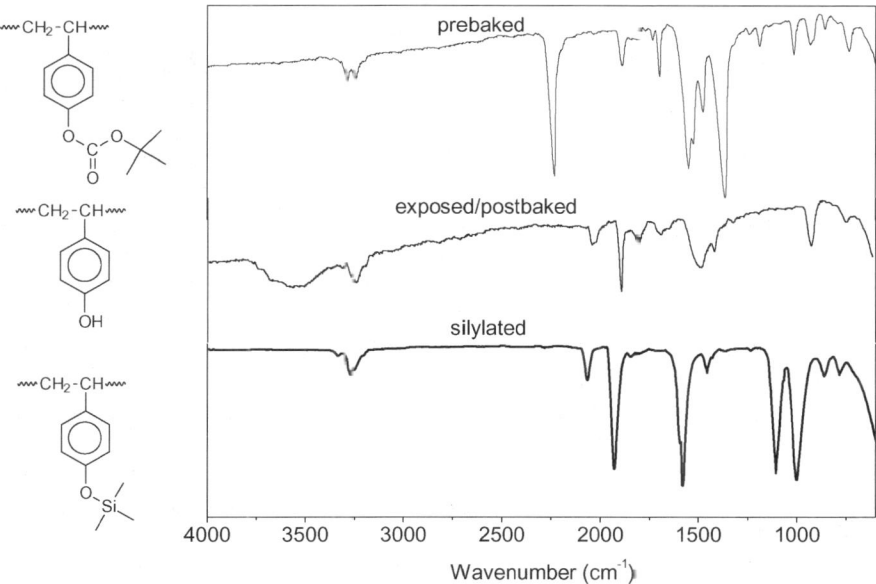

Fig. 160 IR spectra of *t*BOC resist after prebake, after exposure to 10 mJ/cm^2 and postbake, and after treatment with gaseous dimethylaminotrimethylsilane for 5 min at 100 °C and 27.7 kPa [443]

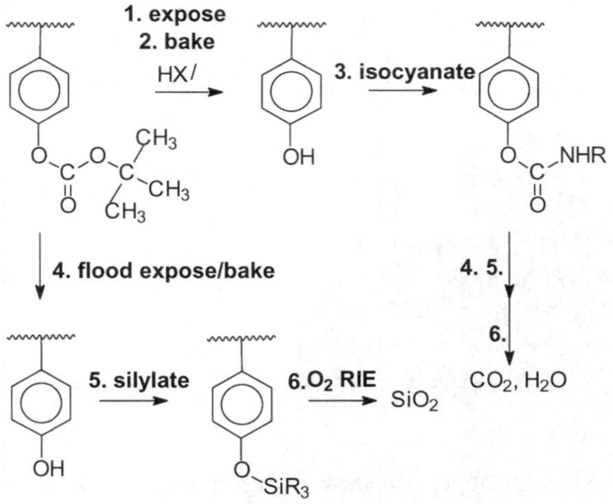

Fig. 161 Positive tone O_2 RIE development of *t*BOC resist by selective silylation through polarity reversal

unexposed area is removed by flood deep UV exposure and baking to generate reactive phenolic OH groups, and since silylation results in incorporation of Si in the initially unexposed area; this process provides positive tone images of the mask upon oxygen RIE. The masking reagent useful in the first gas phase functionalization is isocyanate, which reacts with phenol in the presence of triethylamine to form a urethane.

As mentioned earlier, however, deprotection and subsequent silylation can take place throughout the entire thickness of the *t*BOC resist film because of its high transmission in the 250 nm regions. Addition of an appropriate dye to the *t*BOC resist formulation is required to render the system opaque at 248 nm and suitable for TSI. Use of the 248 nm *t*BOC resist at 193 or 157 nm for silylation is TSI. Chemical amplification resists that undergo acid-catalyzed deprotection to form reactive acidic repeat units shrink in the exposed areas. The silylation process incorporates a mass into the exposed areas of the film and causes the exposed film to swell. In order to avoid distortion, the film shrinkage must equal the film swelling. To achieve a zero volume change in the film, it is possible to silylate with an appropriate mixture of two silylating agents having an average molecular weight equivalent to the mass of the gaseous products liberated upon exposure and bake [445]. In contrast, the diffusion-controlled silylation described in the next section always results in an increase in film thickness. The *t*BOC resist exposed to 248 nm radiation and silylated to a nearly zero net thickness change produced well-defined 200 nm dense patterns. However, when the *t*BOC resist was exposed to 193 nm radiation (TSI), extraordinary LER and residues were observed after O_2 RIE regardless of the silylation parameters. This was attributed to the shallow penetration of the 193 nm light (TSI) and the shallow silylated profile, casting a serious doubt on the usefulness of TSI, especially near the resolution limit of the tool [445]. Nevertheless, the negative tone TSI with the *t*BOC resist at 193 nm (NA=0.6) produced 100 nm L/S patterns by off-axis illumination using a binary mask [446].

Poly(*tert*-butyl 4-vinylbenzoate) is useful in deep UV TSI because of its high absorption below 300 nm and high T_g. The ester polymer is converted to a carboxylic acid polymer in the exposed area, which can be silylated in a fashion similar to the *t*BOC resist [336]. Silylation of methacrylate 193 nm resists has been investigated also. Carboxylic acids can be more readily silylated than phenol but desilylation of silyl ester is faster than silyl ether, rendering the silylated image less stable. TBMA and TBA were copolymerized by changing the feed ratio to prepare copolymers with varying T_g (and composition). Lowering T_g resulted in faster silylation and desilylation. In order to improve the stability of the silylated state, the T_g was increased by incorporating bulky alicyclic groups such as isobornyl and adamantyl and a bulky silylating agent (trimethylsilyl dimethylamine instead of dimethylsilyl dimethylamine) was employed [447].

In an attempt to overcome the LER problem associated with the use of PHOST-based resists (*t*BOC and diffusion-controlled systems discussed below) in 193 nm lithography, which is TSI, alicyclic polymers with higher transparency and higher T_g were employed in silylation [448]. As mentioned earlier, silylation of an opaque resist in a TSI scheme produces a non-square silylation profile very similar to the light energy deposition profile, resulting in LER. A second contribution to LER is the silylation contrast of the system. The silylation contrast can be measured by plotting the extent of silylation monitored by IR (Si-H stretch at 2100 cm^{-1}) as a function of exposure dose as presented for the *t*BOC resist exposed to 248 nm illumination and silylated with 30 Torr of dimethylsilyldimethylamine for 60 s at 90 °C in Fig. 162 [448]. The silylation response is not a step function and not as nonlinear as the dissolution response to dose in wet development steps. Since the silylation contrast is rather low and the response is linear, the edge of the silylated feature is somewhat indistinct without being able to sharpen the pseudo-Gaussian shape of the aerial image and this grayscale response manifests itself as LER. Another factor is T_g of silylated polymers. Although PHOST has a fairly high T_g of 150–180 °C, silylation reduces T_g drastically. In fact, the fully deprotected and silylated (with dimethylsilyldimethylamine) *t*BOC resist film has a low T_g of 70 °C. Thus, silylation temperatures in excess of 70 °C flow silylated polymer, resulting in problematic imaging with LER. Therefore, alicyclic systems that are characterized with high 193 nm transparency and high T_g have been tested for 193 nm silylation (not TSI). The poly(norbornene sulfone) bearing a pendant hexafluoralcohol protected with *t*BOC (see Figs. 21 and 77) proposed as a single layer 193 nm positive resist [115] has been shown to be selectively silylated in the exposed area where acidic hexafluoroisopropanol is generated. Although the silylation response was not of high contrast, the deep 193 nm light penetration produced a much more square silylation profile, resulting in significantly reduced LER after O_2 RIE [448]. The poly(norbornene sulfone)-based system was also evaluated as TSI at 157 nm [448c].

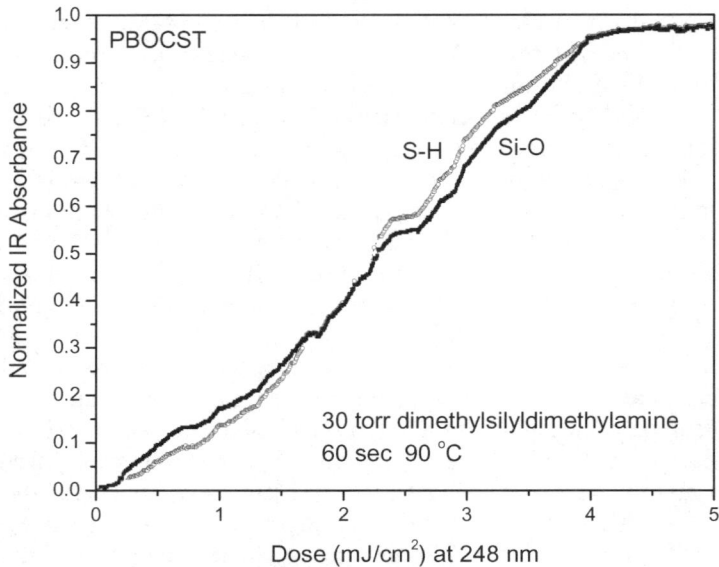

Fig. 162 Silylation contrast plots of tBOC resist exposed to 248 nm radiation [448a]

6.2.2
Diffusion-Controlled Silylation

An alternative approach to selective silylation employs a phenolic resist that can be silylated. Selective silylation is achieved in this case by crosslinking the phenolic resist so as to limit the diffusion of the silylating reagent. The "diffusion enhanced silylation resist" (DESIRE) process was originally proposed to render the commonly-available novolac/DNQ resists dry-developable with oxygen plasma for near UV lithography [449]. The novolac resist undergoes thermal crosslinking in the unexposed areas upon PEB, which prevents diffusion of a silylating reagent, while the gaseous silyl compound diffuses at a much faster rate into the uncrosslinked exposed regions and reacts with the surrounding phenolic OH groups. The process yields negative tone images upon oxygen RIE (Fig. 163). Since the thermally crosslinked phenolic resin in the unexposed areas is also silylated to a minor degree, pre-burning with fluorocarbon plasma prior to oxygen RIE is employed to remove a thin uniform layer of organosilicon polymer. The DESIRE process with the novolac/DNQ resist has been applied to deep UV lithography, which is TSI due to the high opacity of the resist below 300 nm [450]. Certain formulation adjustments were needed to prevent crosslinking by deep UV exposure in this case. The deep UV crosslinking of the resist was combined with near UV flood exposure, followed by silylation, to reverse the tone, which was called PRIME (Positive Resist Imaging by dry Etching) [451]. The novolac/DNQ resist and novolac resin by itself have been evaluated for 193 nm TSI through silylation [452].

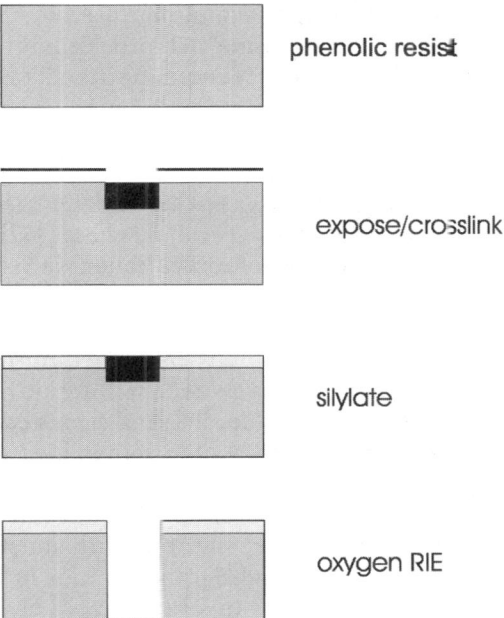

Fig. 163 DESIRE process

PHOST crosslinks efficiently by exposure to ArF excimer laser radiation even without a crosslinking reagent perhaps through photo-oxidation and thus can be selectively silylated in the unexposed regions for positive tone TSI [452, 453]. TSI utilizing oxygen RIE was thought to be a very logical approach to ArF lithography because of excessive absorption of most organic polymers at 193 nm. The DESIRE process can be combined with the chemical amplification concept as crosslinking of a phenolic resin can be readily achieved through acid-catalyzed condensation. Since crosslinking occurs in the exposed areas, silylation proceeds more rapidly in the unexposed regions and therefore this system provides positive images upon oxygen RIE. Thus, this process was named SUPER (Sub-micron Positive dry Etch Resist) [454]. The novolac-based acid-hardening resist containing melamine crosslinker and PAG (Shipley SAL601, Fig. 120) was employed in 193 nm and X-ray positive TSI. Although the pure PHOST performs very well lithographically in the TSI scheme at 193 nm, the chemical amplification scheme (acid-catalyzed condensation for crosslinking) is likely to be needed to increase the sensitivity to an acceptable range [455]. The chemically-amplified TSI process for use in 193 nm lithography has been developed, focusing on three critical areas, sensitivity, contrast, and LER. It has been reported that the use of a chemically amplified resist can improve the sensitivity by a factor of 1.5–2 without compromising LER. While the silylation contrast of this chemical amplification resist was poor ($\gamma<1$), the RIE contrast was excellent ($\gamma\gg10$). The use of disilanes as a gaseous silylating agent further reduced the dose-to-size and increased the contrast. Furthermore, using sulfur dioxide in the plasma etch process improved the sidewall passivation of the resist lines and thus reduced the overall LER [455b]. The TSI process employing an acid-hardening resist SAL601 was evaluated for X-ray lithography in an attempt to circumvent the line collapse problem encountered in the common wet development step [456].

In addition to the aforementioned gas phase silylation, incorporation of Si into exposed resist films was performed in a liquid phase [457]. The liquid phase silylation reportedly incorporates higher concentrations of Si in the PHOST film [458]. Since a phenolic resin is an imaging layer, a nonpolar solvent such as xylene is used as a silylation solvent. However, a silylating agent cannot diffuse into a phenolic film in such a nonsolvent. Therefore, a diffusion promoter such as NMP or PGMEA is added. Swelling is needed for diffusion but excess swelling deteriorates imaging. Bulky non-volatile silylating agents can be employed in silylation in solution while the volatility of Si compounds matters in the gas phase process. Thus, it has been reported that the silicon concentration in the region silylated with bis(dimethylamino)dimethylsilane in solution is about two times higher that that for gas phase silylation with dimethylsilyldimethylamine, as studied by RBS [458]. The high Si concentration in the silylation with the (bisamino)silane is reportedly due to formation of polysiloxane linkages, which is mediated by water [458]. The liquid phase silylation kinetics of SAL601 and PHOST has been investigated using 1,1,3,3,5,5-hexamethylcyclotrisilazane as a silylating agent and NMP as a diffusion promoter in xylene [459]. The extent and depth of the silylation reaction were analyzed by IR and cross-sectional SEM, respectively, as a function of the concentration of the silylating agent and NMP. When the NMP concentration is high (>2%) and the concentration of the Si agent is low (10%), the kinetics can be described as Fickian diffusion. Case II diffusion is dominant when the NMP concentration is low (<1%) and the concentration of the Si compound is high (30%). The Case II diffusion was shown to provide a higher silylation contrast than Fickian diffusion.

6.2.3
Bilayer Silylation

A discovery that a very thin silylation resist was effective in widening the process margin led to a bilayer silylation process for 193 nm lithography (Fig. 164) [460]. A thin layer (70–50 nm) of PHOST coated on a 350 nm thick film of a hard-baked novolac resist was silylated, after exposure to 193 nm radiation, with gaseous dimethylsilyldimethylamine at 80 °C. After the gas phase silylation and breakthrough etch the top and bottom organic polymer layers were etched with O_2 plasma. The bilayer silylation process demonstrated better LER, a good exposure latitude of 14% for 130 nm line/space patterns, and a high line/space resolution of 110 nm. While in the conventional single layer silylation process the depth of silylation and the degree of swelling depended on the pattern size, the bilayer process provided size-independent almost rectangular silylation profile [461].

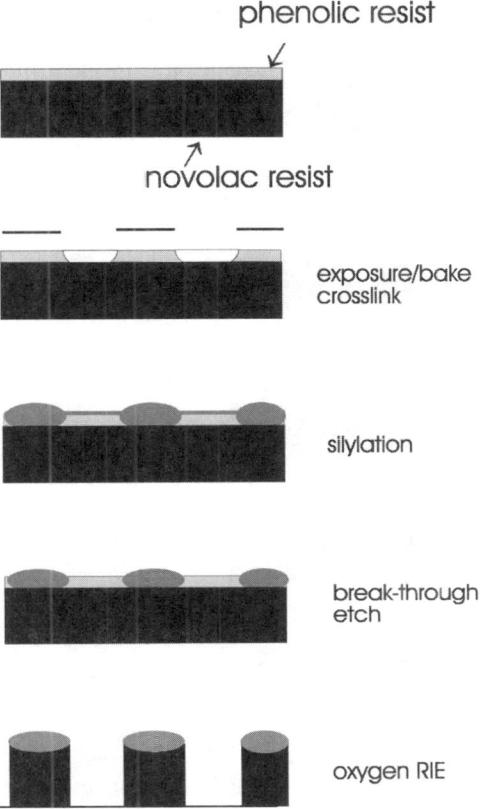

Fig. 164 Bilayer silylation process

The bilayer silylation technique was later applied to a chemically amplified negative resist, focusing on the effects of a delay between the processes [462].

6.2.4
Silylation after Wet Development

Phenolic resists can be silylated also after wet development with aqueous base [463, 464]. In the silicon-added bilayer resist (SABRE) process (Fig. 165), a diazoquinone/novolac resist is coated on a top of a planarizing layer and imaged in the conventional fashion by UV exposure and aqueous base development [463]. The phenolic polymer remaining in the unexposed area after development is silylated in a gas phase or in solution to provide O_2 RIE resistance for dry etching of the underlying layer [463, 464]. This bilayer scheme takes advantage of high contrast aqueous base development to produce square resist profiles, which are then converted to a well-defined Si mask by silylation for O_2 RIE pattern transfer. Thus, a poor silylation contrast sometimes encountered

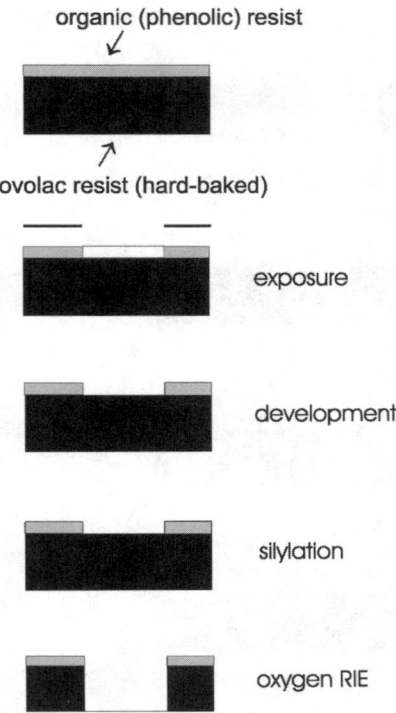

Fig. 165 Silylation after wet development

with the single layer silylation process can be overcome. While in the aforementioned Si-based bilayer resists, a balance between aqueous base development and etch resistance is critical, best available aqueous base developable resists (either positive or negative) can be employed in the silylation-after-wet-development process as a thin film. The technique has been recently adopted for 157 nm lithography by employing PHOST-based chemically amplified positive and negative resists [465]. In this silylation process after alkaline wet development (SILYAL), a thin film (70 nm) of a 248 nm chemical amplification resist (OD=5.6/µm at 157 nm) is coated on top of a hard-baked novolac resist (300 nm thick), exposed to 157 nm radiation, developed with aqueous base, silylated in a gas phase using dimethylsilyldimethylamine, and finally subjected to O_2 RIE pattern transfer. In the case of the negative resist, the best exposure dose range was quite narrow because while sufficient crosslinking was required for aqueous base development, higher exposure doses reduced Si uptake due to a higher degree of crosslinking. A major problem was a low T_g (~45 °C) after silylation (as described earlier) while the typical silylation temperature was 80 °C. The pattern deformation due to thermal flow of the silylated film was reduced either by decreasing the silylation temperature to 50 °C or by performing UV curing after development (which raised T_g to 70 °C). With use of a

50 nm-thick positive 248 nm resist on a 157 nm microstepper (NA=0.60) 110 nm line/space patterns were generated in a 300-nm-thick novolac film after wet development, gas phase silylation, and O_2/SO_2 RIE. Another interesting application of this SILYAL process is pattern size biasing based on the volume change induced by silylation (10 nm/10 s of silylation time in this case). This biasing technique is particularly useful in printing small holes. By combining underexposure of a high contrast 248 nm positive resist to open small holes and following silylation to shrink the hole size, 100 nm contact holes were etched into a 300-nm-thick novolac layer [465].

The biasing technique by silylation to print features smaller than the nominal size was first reported from Siemens in 1990 [466]. In the CARL (Chemical Amplification of Resist Lines) process an imaging layer containing anhydride is silylated in solution with bis(aminosiloxane) either after exposure (Top-CARL)

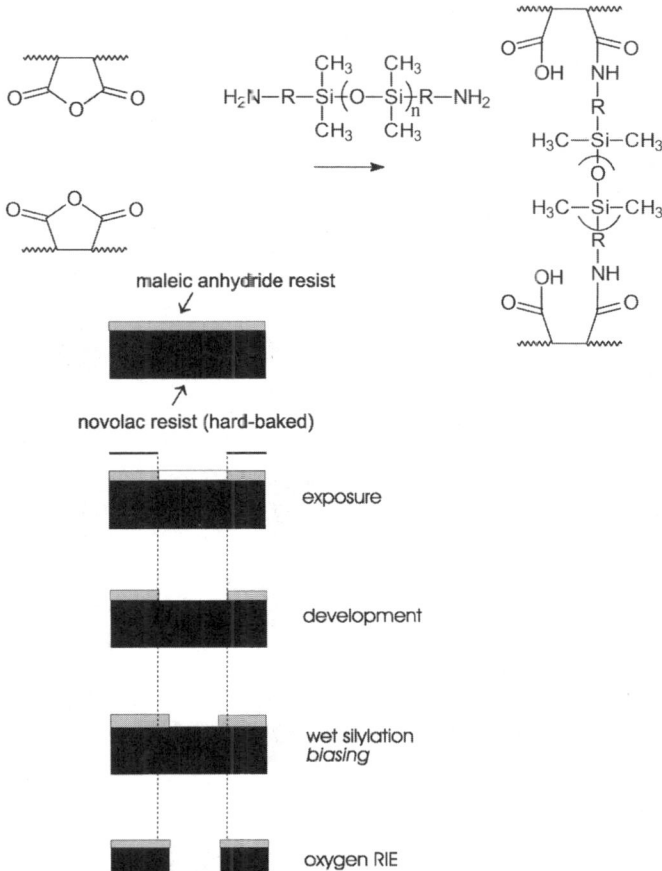

Fig. 166 Si-CARL process

[466b] in a negative mode in single- or bilayer configuration or after development (Si-CARL) [466c] in a positive bilayer mode. This solution silylation involves hydrolysis of the anhydride ring in the polymer with the amino group in the silylating agent, attachment of Si to the polymer through amide linkages, and crosslinking via the siloxane bridge due to the bifunctional aminosiloxane (Fig. 166). In the initial near and deep UV applications a non-chemically amplified resist consisting of poly(styrene-co-maleic anhydride) or poly(allyltrimethylsilane-co-maleic anhydride) and a diazonaphthoquinone was employed [466]. Maleimide was incorporated into the anhydride copolymers to improve alkaline solubility of the diazoquinone dissolution inhibition resist for the sensitivity enhancement. Then, the chemical amplification concept was incorporated in Si-CARL [467]. N-t-BOC-maleimide was terpolymerized with maleic anhydride and styrene using AIBN to prepare a chemically amplified positive resist polymer. Positive images generated by acid-catalyzed deprotection and aqueous base development were silylated with bisaminopropyl-oligomethylsiloxane (10–12 silicon units) in water/isopropanol and etched into a bottom layer by O_2 RIE [467]. For 248/193 nm dual wavelength Si-CARL application a new polymer was prepared by radical polymerization of maleic anhydride, allyltrimethylsilane, TBMA, and a plasticizing monomer [468]. The thermal properties and base solubility of the polymer were adjusted by addition of the plasticizing monomer and by partial hydrolysis of the anhydride to form pendant carboxylic acids. The polymer T_g was set to <135 °C to make use of the annealing concept for airborne contamination prevention (see 4.1.9). The process has been evaluated for 157 nm imaging as well [468].

6.2.5
Flood Silylation/Imagewise Desilylation

Instead of coating trimethylsilyl-protected PHOST containing a PAG on a planarizing layer in a bilayer scheme (see 6.1), a thick single layer of PHOST doped with a PAG is formed on a Si substrate [469]. A surface area of the PHOST film is uniformly silylated with hexamethyldisilazane in a gas phase or with hexamethylcyclotrisiloxane in solution, which essentially forms a quasi-bilayer system, allowing process steps similar to those of the bilayer resist. Exposure followed by PEB deprotects the silyl group attached to the polymer through the action of the photochemically-generated acid. However, as acid-catalyzed desilylation is hydrolysis, PEB was carried out in the presence of steam. Development with aqueous TMAH followed by O_2 RIE produced positive tone images. The system is coined "Surface-Silylated Single-Layer Resist (SSS)" [469].

6.2.6
Surface Modification

A conceptually simple example of surface modification for lithographic imaging was based on the use of focused ion beam to write a pattern onto the substrate of an organic polymer film (Fig. 167). When the ion-implanted film was subjected to O_2 RIE, the surface of the implanted regions was oxidized to indium oxide, which served as an etch barrier [470]. A bisazide/polyisoprene resist was exposed to UV light and, in a subsequent step, treated with an inorganic halide such as $SiCl_4$. Silicon was reported to be predominantly incorporated into the unexposed regions, providing positive images upon O_2 RIE [471]. This scheme has been extended to a variety of polymer systems in conjunction with $TiCl_4$, which exhibited a marked dependence of the reactivity on humidity [472]. Later, layers of water selectively absorbed in photooxidized areas of polymer films were reacted with $TiCl_4$ to generate an O_2 RIE barrier layer [473]. These early systems were not chemically-amplified.

Cationic polymerization has been utilized in surface modification to design plasma developable chemical amplification negative resist systems [474]. In this technique an onium salt cationic photoinitiator is coated directly on the substrate or with a radiation-inert binder polymer. Deep UV irradiation results in formation of strong acids in the exposed areas. Treatment of the irradiated film with an organometallic monomer that undergoes cationic polymerization in the vapor phase or in solution results in deposition of organometallic polymer only in the exposed area where the photochemically generated acid initiates the cationic polymerization. A subsequent O_2 RIE development generates negative relief images. The technique has been recently re-investigated, using vinyl ether and ethylene oxide bearing Si under the name of graft polymerization lithography. However, there is no need for the organometallic polymer to be grafted onto the host polymer in this case and such grafting is unlikely [475].

Silicon oxide formation at the near surface of UV irradiated polymers was applied to surface imaging [476]. For example, a photosensitive polymer

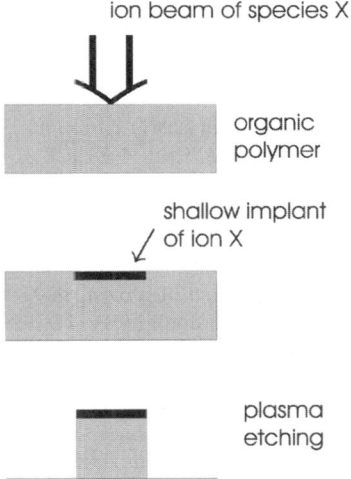

Fig. 167 Surface modification for dry imaging process

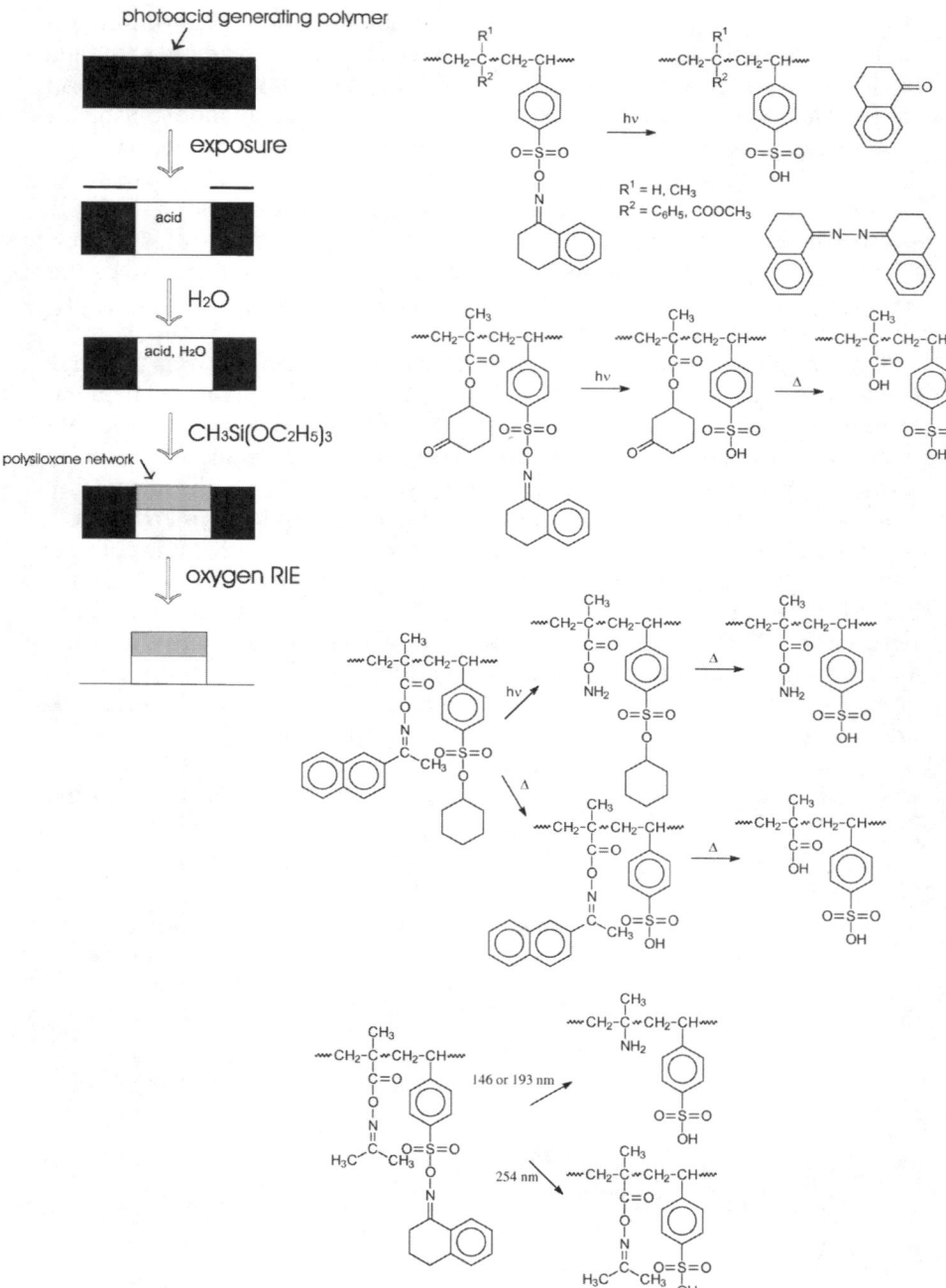

Fig. 168 Imagewise formation of polysiloxane on surface for dry process

generating a polymer-bound sulfonic acid was employed in this application. Water is absorbed from the atmosphere into the exposed regions containing the photochemically-generated sulfonic acid. When the exposed film is treated with a vapor of alkoxysilanes, silicon oxide is formed selectively only at the irradiated surface, which functions as an O_2 RIE mask to generate negative relief images (Fig. 168) [54]. The tone has been reversed by employing a polymer which generates base photochemically and acid thermally [476]. The process has been extended to incorporate chemical amplification. Thermal decomposition temperature of the cyclohexylsulfonate is lowered in the presence of a strong acid [477]. The technique was then applied to 193 nm lithography by preparing co- and terpolymers of the photosensitive styrenesulfonate with methacrylates bearing acid-cleavable ester groups. The polymer-bound sulfonic acid generated by irradiation undergoes deprotection of the ester group upon PEB. The exposed film is treated with a water vapor and then with methyltrimethoxysilane to form polysiloxane on the surface of the exposed regions. O_2 RIE etches away the unexposed areas to generate negative images [478]. Use of a polymer capable of generating sulfonic acid upon exposure to 254 nm radiation and amine by 193 (or 146) nm irradiation has produced positive tone images after surface siloxane formation and O_2 RIE [479].

6.3
Process Issues in Bilayer and Dry Lithographic Techniques

The shift from aqueous base development to the dry development employing oxygen plasma is rather drastic and therefore many issues must be addressed before implementation of the new technology in manufacturing. First of all, a new equipment (etcher) is needed and the etch uniformity (across a wafer and wafer-to-wafer) is a very important issue from a tooling point of view. The all-dry TSI technique requires a silylation equipment as an additional tool set.

There are also some lithographic process issues that must be addressed. The single layer TSI scheme is very attractive as it could solve many difficult problems associated with photolithography. However, the silylation occurs in the thin surface area of the resist film according to the latent image and therefore the silylated image does not have a square edge (Fig. 169), which could contribute to etch bias, LER, and erosion of small features. Diffusion-controlled silylation results in volume increase and T_g reduction in the silylated area. The reaction-controlled silylation based on acid-catalyzed deprotection has been reported to be free from the volume increase; the areas that are depressed due to the deprotection chemistry to eliminate small molecules are silylated, without inducing a net volume increase. Because of this potentially serious problem associated with the silylation technique, the bilayer lithography has re-surfaced as a viable alternative approach to higher resolution lithography. In the bilayer scheme, aqueous base development of the top organosilicon resist could produce a well-defined stencil with vertical wall profiles (Fig. 169) which is much more robust in oxygen RIE.

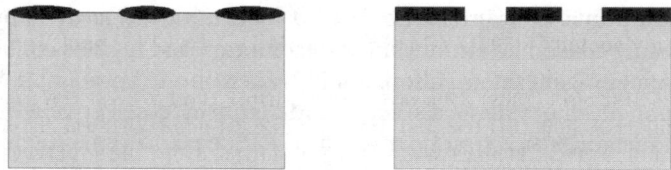

Fig. 169 Silylation profiles (*left*) and bilayer system (*right*)

Another important issue in the gas phase silylation is a control of thermal flow. The silylation is typically carried out at ~100 °C to render a liquid silylating reagent gaseous. Introduction of silicon species lowers T_g of the polymer in general. Thus, the oven temperature should not greatly exceed the T_g of the silylated polymer to afford a high silylation contrast. In a wet silylation process in solution, swelling must be well controlled. For a silylating reagent to penetrate into the resist film and to react with the polymer, swelling is necessary, but excessive swelling destroys contrast and resolution. Various silylating reagents have been evaluated in both dry and wet processes to investigate the effects of the size and reactivity of the Si compounds. Multifunctional crosslinking silylating reagents are sometimes preferred to reduce swelling.

In the case of organosilicon bilayer resists, the balance between the silicon concentration for oxygen RIE resistance and the aqueous base developability is the most important issue. The highly hydrophobic silicon groups could jeopardize aqueous base development that is an unavoidable requirement in the modern semiconductor manufacturing. An interesting solution to this dichotomy has been proposed, which employs a block copolymer of propylpentamethyldisiloxane methacrylate and TBMA synthesized by the group transfer polymerization technique, as mentioned earlier (Fig. 81) [278]. The block polymer resist dissolved much more rapidly in a 0.26 N TMAH solution after UV exposure and PEB than a resist made with a random copolymer at the same Si concentration, suggesting that the exposed block copolymer forms micelles in the aqueous base developer with the Si-containing block forming the core and the poly(methacrylic acid) block forming the corona.

While the bilayer lithography can improve the process margin and therefore yield without introducing a new exposure tool, the most serious potential problem is the resist LER after O_2 RIE, which is likely to be more devastating as the feature size moves toward <50 nm.

7
Resist Characterization

There are numerous analytical and characterization steps in resist formulation and lithographic evaluation processes. First of all, each component of a resist formulation, polymer, acid generator, and additives, must be synthesized as specialty materials in many cases. During the course of the synthesis work, common analytical and characterization techniques are employed. The purity of each ingredient and final resist formulation is an important issue, especially in terms of metal contents. The semiconductor industry is extremely sensitive to contamination of devices by trace metallic impurities and demands extremely low levels of metals in resist formulations, currently on the order of several tens of ppb, which is a formidable challenge to resist manufacturers and material suppliers. In some cases reactors and mixers coated with Teflon are employed in synthesis, blending, and formulation to reduce metallic contamination. Another impurity which is uniquely concerned about in the chemical amplification resist systems is residual acid in resist formulations, which might come from polymer, PAG, additive, or casting solvent. A trace amount of residual acid in a resist formulation could reduce the storage shelf life and induce a premature acid-catalyzed thermal reaction in the film during PAB, thereby reducing the contrast or destroying resist performance. Residual acids are typically titillated spectroscopically using a dye that bleaches on UV spectroscopy in contact with acid. Conversely, residual base impurities carried over from polymer synthesis could lower and fluctuate sensitivities for imaging.

In this section, some analysis and characterization techniques unique to resist polymers and resists are briefly described.

7.1
Molecular Weight Determination

Determination of molecular weight is one of the most fundamental characterizations unique to macromolecules. Because the molecular weight affects the dissolution rate of polymers and resists, molecular weight determination has an important position in resist chemistry and as is the case with any other fields, gel permeation chromatography (GPC) or size exclusion chromatography (SEC) is conveniently employed. In addition to the routine polymer characterization, GPC is sometimes utilized in monitoring resist chemistries involving a molecular weight change (crosslinking or main chain scission). Light scattering and membrane osmometry have been also occasionally employed. NMR has been also useful in determining degrees of polymerization by analyzing polymer end groups for low molecular weight polymers.

7.2
Thermal Analysis

As baking is an important integrated step in lithography, thermal stability of resist films and therefore resist ingredients is an indispensable requirement. In polymers, glass transition, decomposition (deprotection and/or main chain degradation), and other thermal events such as crosslinking, cyclization, rearrangement, etc. are important phenomena to be investigated. In the case of small molecules such as PAG, crosslinker, dissolution inhibitor, etc. melting, decomposition, sublimation, evaporation, and thermochemical events must be carefully studied. Thermogravimetric analysis (TGA) and differential scanning calorimetry (DSC) are the most commonly employed thermal analysis instruments. Attachment of a mass spectroscopy to TGA can provide useful information on temperature profiles of formation of volatile products associated with thermal events and acid-catalyzed reactions.

Modern advanced resists contain acidic polymers for aqueous base development, which in many cases lower the thermal stability of resist ingredients. Acid-labile protecting groups undergo thermal deprotection at much lower temperatures in the presence of an acidic functionality such as in phenolic resins (Fig. 33) and many PAGs decompose at lower temperatures in phenolic polymer films [31, 32]. Thus, it is important to study hydrolytic stability as well as thermal stability of resist ingredients. Furthermore, real resist films on wafers may have much lower T_g due to residual solvent and/or by plasticization with a PAG or additives. Therefore, it is also important to consider interactions between components. Modulating temperature DSC is highly useful in determination of T_g near decomposition (deprotection) events [480].

Thermal analyses provide guidance as to at what temperature a resist film should be baked after coating. The PAB temperature must be below a deprotection temperature in the case of positive resists based on acid-catalyzed deprotection but should be preferably higher than its T_g to minimize the free volume in the film, as discussed earlier.

Furthermore, in addition to the bulk thermal properties of polymers and resists, determination of T_g of film interfaces and of ultrathin films has become an important issue in thin film imaging (bilayer, 157 nm, and EUV). Various techniques have been employed, which include ellipsometry [481, 482], positron annihilation spectroscopy (PALS) [483], QCM [484], scanning viscoelasticity microscope (SVM) [485], x-ray reflectivity [486, 487], and thermal probe [488].

7.3
Spectroscopic Analysis

In addition to analysis and characterization of resist components, spectroscopic techniques can find use in investigation of imaging mechanisms and resist chemistries.

7.3.1
UV Spectroscopy

Because UV absorption of resist films at the exposing wavelength is a very important parameter that affects lithographic performance, UV measurements of components and resulting resists are critical in photolithography. Furthermore, since UV spectroscopy can be readily carried out on thin films coated on substrates, resist chemistries can be monitored by this technique in some cases.

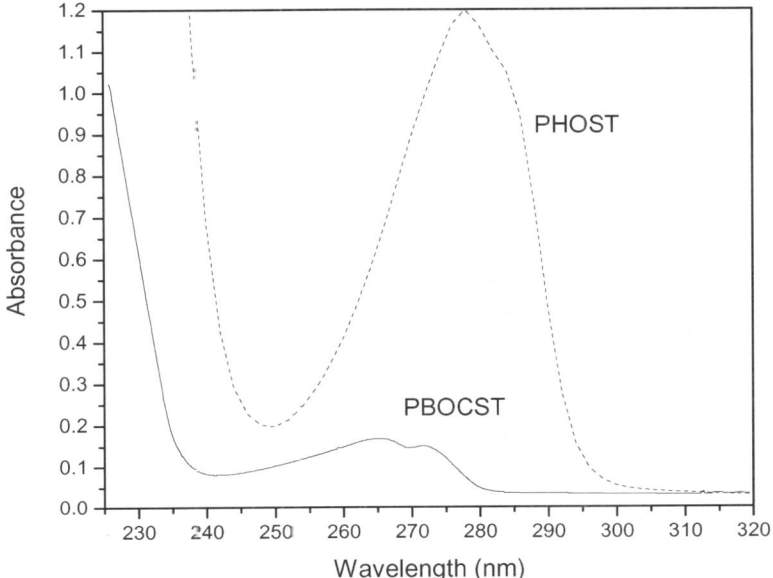

Fig. 170 UV spectra of PBOCST and PHOST [25]

For example, conversion of PBOCST to PHOST is accompanied by an increase in absorption at ca. 280 nm (Fig. 170). Furthermore, UV absorbing polymer end groups can sometimes be detected. As mentioned earlier, UV spectroscopy is highly useful in determination of acid concentration when employed in conjunction with an acid sensitive dye.

7.3.2
NMR Spectroscopy

Nuclear magnetic resonance (NMR) spectroscopy is indispensable in preparation of resist materials such as polymers and PAGs. ^{13}C NMR is particularly useful in determining copolymer compositions and in observing microstructures. The ^{13}C NMR analysis can be rendered highly quantitative by applying the inverse-gated decoupling technique in conjunction with the use of Cr(acac)$_3$ as a relaxation agent. NMR has been employed very often in elucidation of imaging mechanisms by carrying out model reactions in solution. Furthermore, it is also possible to obtain NMR data on an exposed or baked resist film. Exposed or baked films are scraped or rinsed off the wafers and then subjected to solution NMR analysis. Typically one five-inch wafer coated with a 1-μm-thick polymer film provides ~10 mg of scraped powder and therefore ten coated wafers are typically needed to obtain good ^{13}C NMR spectra using a 5 mmφ tube. Residual casting solvents in baked resin and resist films have been quantified, and side reactions and PAG decomposition analyzed by this ^{13}C NMR technique [66]. ^{19}F NMR is particularly useful in characterization of fluorine-containing 157 nm resist polymers and analysis of PAGs, many of which contain fluorine. ^{29}Si NMR has been applied to investigation of deprotection of Si-containing bilayer resists [422].

A 2D-NOESY (two dimensional nuclear Overhauser effect spectroscopy) NMR technique has been utilized in investigation of hydrogen-bonding between phenolic OH and ester groups attached to a polymer chain [489]. Solid-state ^{13}C NMR (with cross polarization and magic-angle spinning) has been used to study the chain dynamics and length scale of mixing in COMA-based 193 nm resist formulations along with 2D-WISE (two-dimensional wide line separation) NMR [248]. ^{13}C and ^{19}F NMR in solution have been utilized in investigation of interaction between phenolic and HFA OH with PAG, respectively [490, 491]. A ^{1}H NMR technique to analyze radical co- and terpolymerization kinetics in situ has been developed and applied to copolymerization of hydroxystyrenes with (meth)acrylates to prepare the 248 nm ESCAP resist [164], of norbornenes with maleic anhydride to prepare the 193 nm COMA resist [253], and of fluoroacrylates with norbornenes and vinyl ethers for 157 nm lithography [305, 492].

7.3.3
IR Spectroscopy

FT-IR is the spectroscopic technique most commonly employed in investigation of resist chemistries and imaging mechanisms because it can be easily performed on thin resist films coated on NaCl or KBr plates, or even on undoped Si wafers. IR microscopy is particularly useful in such studies as small spots on one wafer coated with a resist exposed to different doses can be automatically measured on the IR spectrometer, which can conveniently generate chemical contrast curves. This IR technique becomes particularly powerful when combined with a thermogradient plate (TGP) [493]. Acid diffusion was investigated by generating a thin layer of acid at the top of a resist film through exposure of 248 nm resists to 193 nm radiation and 193 nm resists at 157 nm and then monitoring the degree of deprotection by IR (or thickness loss) at various PEB temperatures [88, 493]. IR is also useful in isothermal kinetics studies of polymer decomposition and in investigation of hydrogen-bonding in resin and resist films. IR spectroscopy was utilized in investigation of anhydride ring opening in COMA upon contact with aqueous base [257, 258, 261].

The kinetics of the copolymerization of norbornenes and maleic anhydride has been analyzed in situ by IR spectroscopy [254]. Gas phase FT-IR has been combined with TGA to investigate outgassing from the ketal-based KRS-XE e-beam resist [222]. A flow cell which allows exposure and heating of a resist-coated wafer was attached to the IR spectrometer to study exposure- and temperature-dependent outgassing. Real time deprotection of the KRS resist has been also monitored by infrared reflectance-absorbance spectroscopy [222, 494].

7.3.4
Mass Spectroscopy

Mass spectrometry is an analytical method of choice for identification of volatile compounds and has been employed in investigation of thermolysis and acidolysis mechanisms of chemical amplification resists [96, 121, 122]. This technique has been also utilized in screening of resists systems, especially Si-containing 193 nm bilayer resists, for outgassing [438, 439]. MALDI-TOF mass spectroscopy has been applied to characterization of dendritic resist polymers.

7.4
Surface Analysis

The above spectroscopic techniques deal with solution, bulk properties, or phenomena averaged through the entire film thickness (except glazing-angle IR). However, in the solid state chemistry including the resist chemistry occurring in a glassy film, understanding the distribution of reagents such as PAG in the film and distinguishing the surfaces (air and substrate interfaces) from the bulk are extremely important. As the resist film becomes thinner and thinner, the contribution of the interfaces becomes more and more significant.

As the use of aqueous base in the development step has been mandated, the wettability of resist films is very important and therefore water contact angles of polymer and resist films are measured often. The contact angle measurement can provide important information about surface segregation of blend films.

ESCA and Auger spectroscopic studies have been carried out on organometallic resists to demonstrate that the organometallic species are converted to oxides in the thin surface layer [443]. RBS, X-ray photoelectron spectroscopy (XPS), secondary ion mass spectroscopy (SIMS) and laser ablation microprobe mass spectroscopy (LAMMS) have been employed to study the distribution of PAG and the surface of a chemically-amplified resist [90, 91, 495]. Sometimes these techniques are combined with sputtering to remove a thin surface layer to obtain a depth profile. NEXAFS (near edge X-ray absorption fine structure) spectroscopy has been utilized in probing the surface and bulk chemistry of chemical amplification resists including PAG distribution and polymer segregation [496].

Total internal reflection fluorescence spectroscopic measurements have been carried out on pyrene-doped PHOST thin films, in which pyrene serves as a fluorescence probe to study inhomogeneous distribution of small molecules and hydrophobicity of the interface layer [497].

In addition to scanning electron microscopy (SEM), scanning tunneling microscopy, (STM), atomic force microscopy (AFM), and scanning probe microscopy (SPM) have emerged as a powerful tool for investigation of surface topographies. AFM has been also applied to examination of resist patterns during rinse [411].

7.5
Neutron Scattering

Small angle neutron scattering (SANS) has been employed in investigation of chain conformation in ultrathin polymer films [144, 498].

7.6
Neutron and X-Ray Reflectometry

Thermal expansion of thin (130–5 nm) PHOST films coated on various substrates has been measured by using specular X-ray reflectivity [487]. Complementary use of neutron and X-ray reflectometry has allowed one to measure the spatial evolution of the reaction front in the tBOC resist with nanometer resolution [499]. Using a bilayer geometry with a lower layer consisting of PHOST protected with $COO(CD_3)_3$ and an upper layer consisting of PHOST containing PAG, compositional and density depth profiles were measured.

7.7
^{14}C Labeling/Scintillation

Titration of ^{14}C-labeled casting solvent during aqueous base development was employed in investigation of the effect of a residual casting solvent gradient in DNQ/novolac resists on developed image profiles [500]. Later, residual PGMEA in various baked polymer films was quantified by using the ^{14}C-labeling and scintillation technique [214]. Furthermore, as mentioned earlier, the propensity of polymer film to absorb NMP was quantitatively determined by using ^{14}C-labeled NMP, providing an insight into the formation of surface insoluble layer in positive imaging and resulting in the development of the ESCAP 248 nm resist [213].

7.8
Laser Interferometry

The development process produces final three-dimensional resist images. This is a critical process as a sinusoidal latent image generated by exposure must be converted to a step function in order to produce resist images with vertical wall profiles. Thus, it is important to understand the dissolution behavior of resist films in a developer (especially aqueous base).

The study of the dissolution behavior involves measurements of film thickness as a function of time. Although the simplest procedure requires manual measurements of the initial thickness and the thickness after development for a given period of time to generate average dissolution rates, automated in situ thickness measurement methods are much more preferred as kinetics information is much more valuable than simple average dissolution rate data. Two methodologies are available; laser interferometry and quartz crystal microbalance (QCM).

The laser end-point detection system, more commonly employed, uses a low-power He-Ne laser directed through a bifurcated fiber optic cable onto the surface of a resist film coated on a reflective substrate such as silicon. The laser light is reflected both from the resist/developer interface and from the resist/substrate interface. The two reflected lights are collected into the optical fiber and imaged onto a photodiode. The output is recorded on a strip chart. As the film dissolves, the reflected intensity goes through a series of maxima and minima due to constructive and destructive interference of the two reflected light waves. The temporal distance between maxima can be related to a change in resist thickness through $\Delta t = \lambda/2n$, where λ is the wavelength of the laser, n is the refractive index of the resist material at this wavelength, and Δt is the change in thickness occurring in the time span between maxima (or minima).

7.9
Quartz Crystal Microbalance

Piezoelectric quartz crystal oscillators function on the basis of the well-established relationship (Sauerbery equation) between the oscillation frequency of a quartz crystal and the mass of a thin film deposited on its surface [501]. QCM has been extensively used in measurements in vacuo and in the gas phase, which includes the studies on gas phase silylation for oxygen RIE development [443] (see 6.2) and on resist outgassing [439, 502]. The QCM technique has been extended to measurements in the liquid phase including aqueous media and has found powerful utility in studies of dissolution kinetics of phenolic and other acidic resists in aqueous base [503].

Roughening of the film surface during dissolution can cause such severe degradation of the reflected optical signal that calculation of the film thickness is impossible and if reflection from the resist/developer or resist/substrate interface is weak due to near-matching of the refractive indexes, detection of the desired optical interference effects can be difficult because of low signal amplitude, which restrict the use of the optical measurement technique. Furthermore, the optical technique cannot properly handle extremely fast dissolution rates of highly-exposed chemically-amplified resists. On the other hand, QCM is capable of following tremendously fast dissolution behavior (~30,000 Å/s) in minute detail [258, 490]. Figure 171 presents dissolution kinetics curves generated by using QCM of the ESCAP resist exposed above E_0 of 6 mJ/cm². Notice the time scale of 0 to 0.5 s [490]. Highly exposed films of several thousand Å thickness dissolves away completely in 0.3 s. Between E_0 and 20 mJ/cm² at which conversion of ester to carboxylic acid saturates at 95%, the dissolution rate does not change but the induction period becomes shorter with increasing dose.

The dissolution rates obtained by the either method are then plotted in many cases as a function of the dose in a log-log mode (Fig. 172). The slope of the steepest portion of the S-shaped curve is defined as a developer selectivity

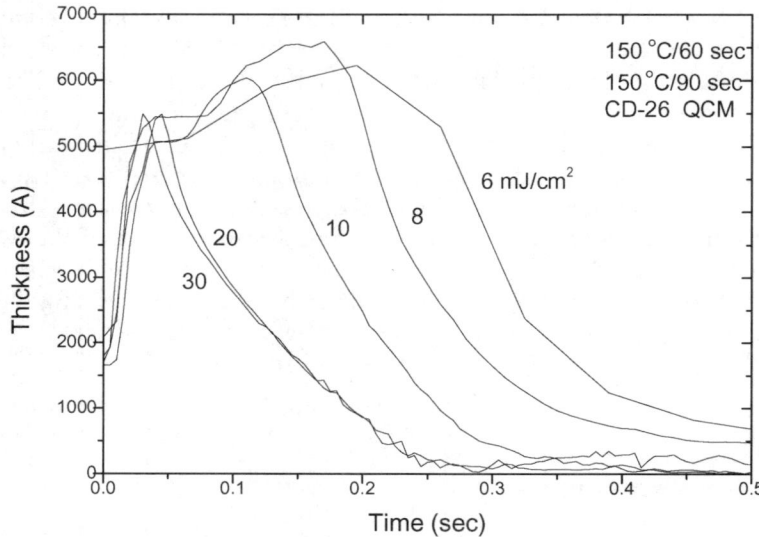

Fig. 171 Dissolution kinetics of ESCAP resist exposed above E_0 in 0.26 N TMAH as measured by QCM [490]

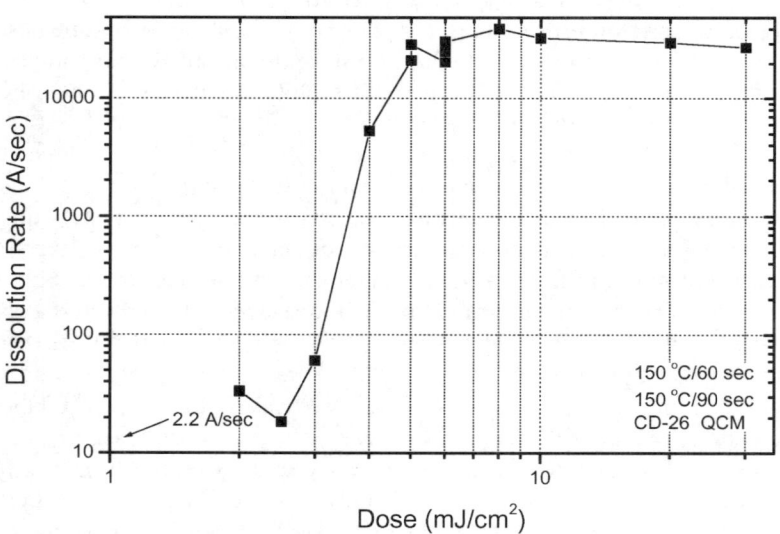

Fig. 172 Dissolution rate vs exposure dose for ESCAP [490]

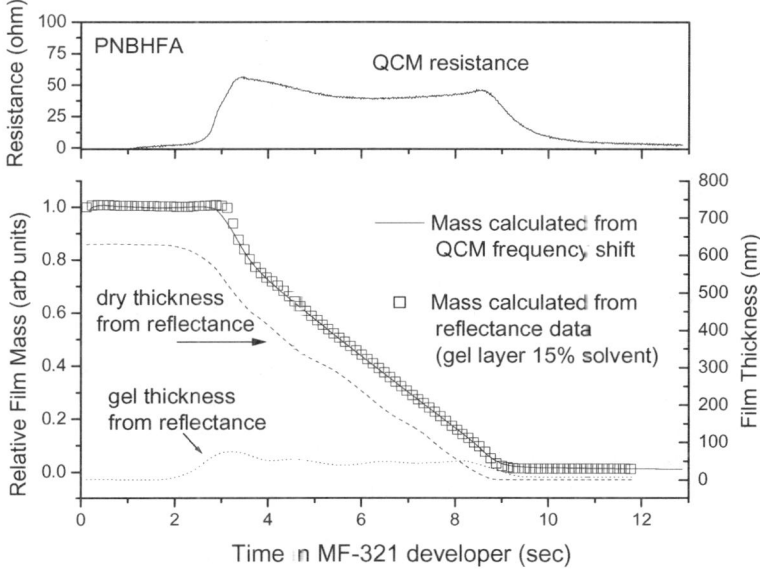

Fig. 173 Frequency, resistance, and reflectance measurements of PNBHFA in 0.21 N TMAH [491]

(n), which is perhaps more meaningful than the contrast (γ) determined from the sensitivity curve (Fig. 3). The higher n value indicates higher lithographic performance in general.

A unique QCM dissolution rate monitor has been constructed that allows changes in both oscillation frequency and resistance to be accurately recorded at high data rates during dissolution/swelling, providing more accurate description of the interfacial behavior of thin films during aqueous base dissolution [261]. Furthermore, this QCM apparatus has been coupled with a high speed visible reflectance spectrometer and with an IR spectrometer to provide information on the structure of the dissolving film and its chemical composition [261]. Figure 173 demonstrates the complex dissolution/swelling behavior of PNBHFA in 0.21 N TMAH [491]. The gel layer thickness is estimated as ca. 50 nm.

7.10
Measurements of Optical Properties of Films

Refractive indices (n) and film thicknesses of not only resist films but also underlying layers and topcoats are important parameters when developing a lithography process. Ellipsometers are employed to measure accurately film thickness, refractive index, and extinction coefficient (k) in a wide range of wavelength. The spectroscopic ellipsometry has been extended to 157 nm [504].

8
Resist Performance Parameters

Although the bulk properties such as sensitivity (E_0), contrast (γ), and developer selectivity (n) discussed earlier represent characteristics of a resist, fine line lithography provides the most meaningful and precise judgment on the performance of a resist.

8.1
Resist Sensitivity and Contrast

Radiation sensitivities of polymers are expressed simply in quantum yield, G_s, G_x, etc. The resist sensitivity is process-dependent (PAB and PEB temperature and time, developer strength, etc.). In general, the sensitivity of positive resists is defined as the dose to clear (E_0), the dose at which a large exposed pad is cleaned to the substrate with a minimal loss of the unexposed film (Fig. 3) as mentioned earlier. However, the actual imaging dose, E_{size} or E_s, which is a dose at which the target image is printed to its intended mask size, is significantly greater than E_0. In the case of negative systems, the sensitivity is the dose at which 50% of the exposed film thickness is retained after development (Fig. 3). The onset of insolubilization is called a gel dose and the imaging dose is of course much higher than the sensitivity value (E_{50}). Although there is no defined counterpart of the gel dose in positive systems, the dose at which dissolution of the exposed film begins may be called a threshold dose (E_{th}).

Resist contrasts are defined as the slopes of the linear portion of the sensitivity curves (Fig. 3) and depend on process conditions. Thus, the sensitivity (contrast) curves are constructed to semi-optimize process conditions for a given formulation. However, final optimization of a resist formulation and process conditions requires lithographic imaging of target features. A plot of a dissolution rate as a function of exposure dose (cf. Fig. 172) is very useful in assessing the developer selectivity (development contrast) as mentioned earlier.

Another useful contrast values are related to the resist chemistry in the film, which will subsequently affect the lithographic contrast. Sensitivity (contrast) curves similar to Fig. 3 can be generated by following the degree of reaction (deprotection, for example) with IR or by measuring thinning (in deprotection, for example) as the function of exposure dose. Comparison of a chemical contrast curve with a development contrast curve provides useful information on resist behavior, such as a degree of deprotection at E_0.

8.2
Linear Resolution

Resolution is perhaps the most important parameter in lithography, which drives the whole industry forward. However, the definition of resist resolution is elusive at least in part due to the fact that the resolution depends heavily on exposure tools also. When the resolution is mentioned, NA of the stepper lens and exposing wavelength should accompany in photolithography as the resolution (R) is proportional to the exposing wavelength (λ) and inversely proportional to NA of the lens; $R = k_1 \lambda / NA$.

Furthermore, small features can be printed by over- or underexposure (or by over- or underdevelopment) in many cases, which is not a true representation of resolution. Linear resolution is defined as the smallest feature printed while all other feature sizes are also delineated within ±10% of the designed mask sizes (Fig. 174).

The process-related constant k_1 (linear resolution × NA/λ) is sometimes employed in defining resist resolution, which is independent of the exposure wavelength and NA. A smaller k_1 indicates a higher resolution capability. The k_1 factor must be sufficiently large in production environments to achieve a robust process with a large latitude but the industry is learning to live with a small k_1 process as half or even quarter wavelength lithography must be implemented in production employing resolution enhancement techniques (RET).

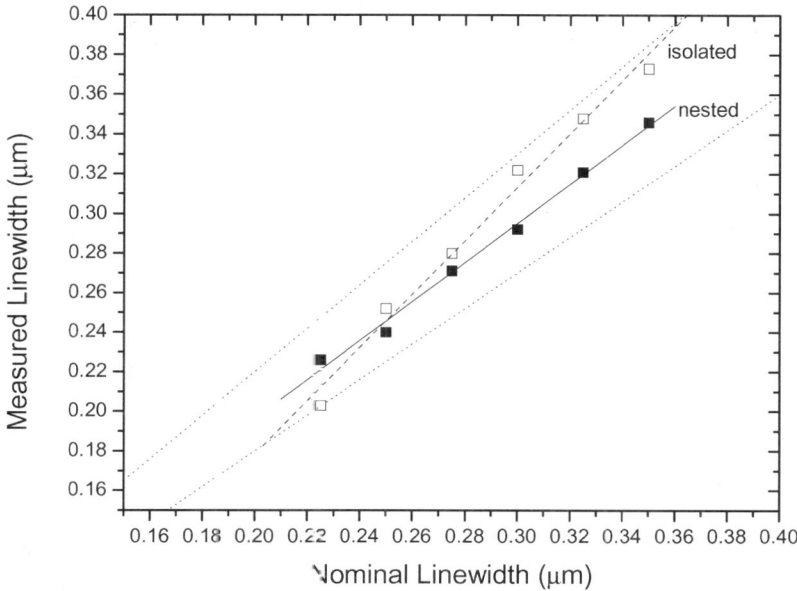

Fig. 174 Linear resolution

A number of optical techniques to enhance resolution have been proposed and have played an important role in extending the resolution and life of optical lithography, although they cannot be applied to any pattern sizes or shapes but are pattern-dependent. These techniques include annular and off-axis illumination, phase shift masks, multiple exposures, etc. [505].

8.3
Depth of Focus (DOF) or Focus Latitude

In photolithography DOF is another important parameter to describe the resist performance. Like photographic imaging, projected mask images must be sharply focused onto the resist film in photolithography. However, a resist that requires precise focusing without a margin cannot be production-worthy. Good resist systems must provide the largest margin of focus offset (DOF) as such a large focus latitude relaxes exposure tool adjustment requirements and offers a wide process latitude in the production environment. The entire resist thickness must be within the focus range and typically DOF of 1 µm would be desirable in 248 nm lithography. However, as DOF is proportional to λ and inversely proportional to NA^2, maintaining DOF becomes more and more serious challenges as the feature size continues to shrink. Thus, bilayer and TSI schemes involving thin imaging layers are attractive approaches to the solution of the DOF problem in photolithography. Resist thickness is becoming smaller as feature sizes shrink. The multiple exposure method called FLEX [506] is effective in increasing DOF for contact hole imaging.

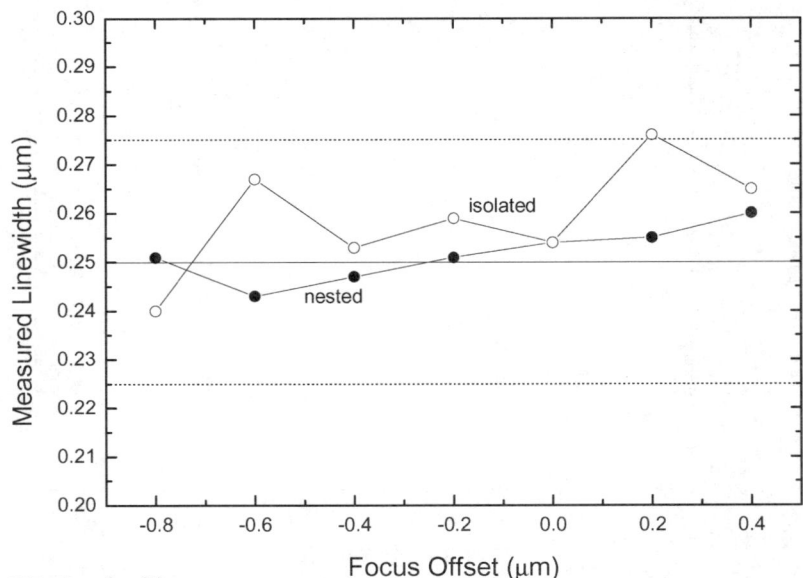

Fig. 175 Depth of focus

DOF is expressed by plotting measured linewidth as a function of focus offset and is defined as the focus range in which the linewidth is within ±10% of the nominally intended size (Fig. 175).

8.4
Exposure Latitude (Dose Latitude)

A resist that requires a precise exposure dose to provide the best image can be processed only by a well-trained engineer and cannot be implemented in manufacturing. A resist that functions equally well under over- and underexposure conditions can provide a robust manufacturing process. The dose latitude is defined as the percentage of over- and underexposure relative to the best dose in keeping the linewidth within ±10% of the nominal size. Chemical amplification resists were considered to be weak in the exposure latitude (EL) because they require only a small exposure energy. However, good chemically amplified resists can provide 30–40% dose latitude in 248 nm lithography. The exposure latitude is expressed by plotting the measured linewidth as a function of the exposure dose (Fig. 176). As DOF and exposure latitude do not go with hand-to-hand, DOF is sometimes plotted as a function of the dose latitude to find the best combination (Fig. 177).

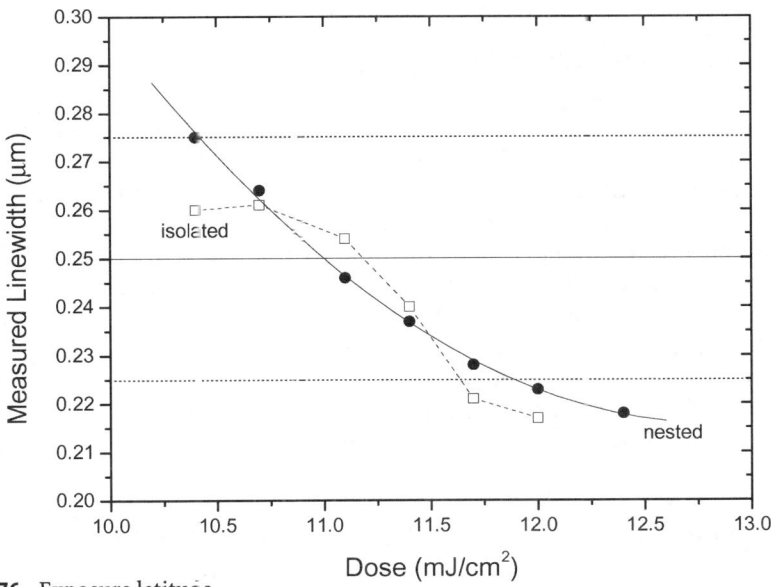

Fig. 176 Exposure latitude

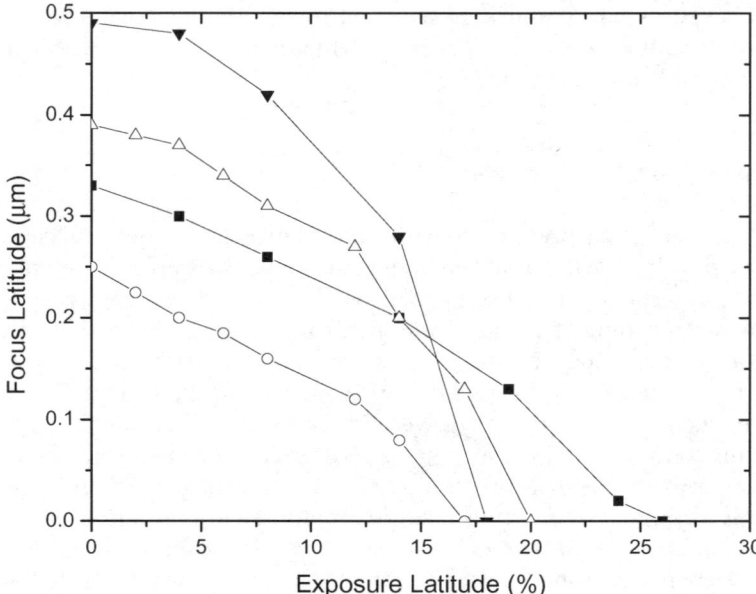

Fig. 177 Process latitude (DOF-EL)

8.5
Resist Image Profiles

The above-mentioned linewidth information is generally obtained by using a non-destructive top-down SEM. However, since the three-dimensional profile is what counts in lithography, cross-sectional SEM can provide the most important information. Chemical amplification resists tend to exhibit a "foot" in the case of positive systems and "notching" in negative systems when imaged on nitride substrates such as silicon nitride and titanium nitride, which can be examined only by cross-sectional SEM.

AFM has emerged as a non-destructive (sometimes destructive) alternative of SEM in observation of the three-dimensional relief images and surface roughness. Furthermore, AFM has been utilized in observation of pattern collapse during rinse as mentioned earlier.

8.6
Isolated-Nested Bias

Microelectronic devices have many pattern features that are different in size and also in pitch. Thus, a resist must be able to simultaneously print dense line/space patterns, relaxed pitches, and isolated lines to their intended critical dimensions as closely as possible. The difference in the sizes of isolated and dense lines printed in resist (bias) must be small, although aerial image contrasts of exposure systems are very much different for different pitches.

8.7
PEB and PED Stability

Chemical amplification resists typically require baking after exposure to drive acidolysis. However, the critical dimension of developed images, E_0, and E_s must be insensitive to a small variation of hotplate temperatures. The PEB temperature stability is desired to be <5 nm/°C.

The stability of the exposed latent image before PEB (during postexposure delay) against airborne contaminants is still an important issue to consider.

8.8
Line Edge Roughness

When the minimum dimensions of a resist pattern are <100 nm, their dimensional tolerances approach the scale of the molecular components of the film. At this level, LER, random fluctuation in the width of a resist feature, may limit the advancement of lithography. Possible contribution to LER include polymer molecular weight, molecular weight distribution, molecular structure of resist components, inhomogeneity in component distribution within the film, statistical effects influencing film dissolution, intrinsic properties of the imaging, chemistry, image contrasts of irradiation, etc.

8.9
Dry Etch Resistance

Although etch rates of resists depend on etching gasses and conditions, dry etch resistance is in general expressed as etch rates relative to representative 248 nm resists.

9
Resolution Limit – Acid Diffusion/Image Blur

When the chemical amplification concept was proposed two decades ago, engineers were skeptical about the resolution capability of such systems based on acid diffusion and chain reactions. However, the IBM *t*BOC resist was able to fabricate structures as small as 18 nm in a negative mode and <40 nm in a positive mode, comparable in resolution with PMMA, by using a scanning transmission electron microscope at 50 kV, demonstrating its high resolution capability in 1988 [507]. Acid diffusion plays a more serious role in limiting the resolution of equal line/space patterning and has remained as a longstanding important research subject in the history of chemical amplification [67–88,

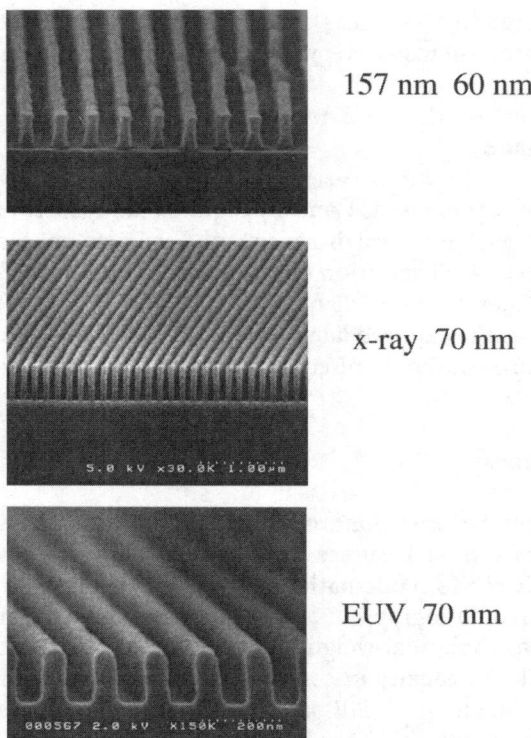

Fig. 178 Scanning electron micrographs of sub-100 nm features printed in chemical amplification resists; 60 nm line/space patterns by phase shifting 157 nm lithography [511], 70 nm line/space patterns by X-ray lithography [513], and 70 nm line/space patterns by EUV lithography [514]

508–510]. The diffusion coefficients reported ranged from 10^{-15} to 10^{-11} cm^2/s, which can be converted to three-dimensional diffusion lengths of the order of 85–850 nm during a 2 min PEB at 80–100 °C. The unrealistically large values result from inability to separate acid diffusion and deprotection chemistry. Although chemical amplification resists, first employed in manufacture of the 1000 nm device generation, have successfully printed features much smaller than 100 nm as demonstrated in Fig. 178 [511–514], perhaps much smaller than anyone expected, now the time has come to be truly concerned about the effect of acid diffusion on resolution as lithography moves to below 50 nm. In fact, it has been recently shown that the resolution limit of the tBOC- and ESCAP-type resists is about 50 nm (equal line/space) experimentally and theoretically [510].

Thermal and acid-catalyzed deprotection kinetics of PBOCST and PTBMA was monitored by UV and IR spectroscopy, respectively [515], and compared very favorably with models based on a stochastic kinetics simulator (CKS)

[516]. The chemistry and physics of the reaction-diffusion process in the *t*BOC resist containing di(4-*tert*-butylphenyl)iodonium nonaflate exposed to 254 and 193 nm radiation have been elucidated in detail using the CKS program [508]. The kinetics of the *t*BOC deprotection was very well explained by simulation using two diffusion paths; $D_{\text{PBOCST}}=3.5 \times 10^{-14}$ cm^2/s and $D_{\text{PHOST}}=8 \times 10^{-16}$ cm^2/s at 95 °C (Fig. 179). The diffusion coefficient determined in this work was 1×10^{-14} to 3.5×10^{-14} cm^2/s in PBOCST and 4×10^{-16} to 8×10^{-16} cm^2/s in PHOST at 85–95 °C [508], which are at the bottom of the range or lower than those previously reported using other techniques. The effective diffusion coefficient drops rapidly from 1.5×10^{-15} cm^2/s (pure PBOCST) to 1×10^{-15} cm^2/s as deprotection to PHOST proceeds. A one-dimensional diffusion coefficient of 10^{-15} cm^2/s gives a mean displacement of about 5 nm, the same as the range estimated from the catalytic chain length [70]. The movement of the deprotection boundary is substantial even when acid diffusion is completely shut down in simulation, due to the long catalytic chain length. The same analytical/modeling procedure has been successfully applied to a copolymer of BOCST and HOST (APEX-type) and to investigation of the impact of acid anion size and of added base on reaction-diffusion kinetics [509]. The simulation has shown that increased mobility of triflic acid degrades achievable resolution by roughly an order of magnitude compared to nonaflic acid and that only a proportional neutralization model provides an accurate quantitative description of the effect of base on deprotection kinetics. The simulation work has also indicated that the contribution of the deprotection chemistry to image spreading is a constant fraction of the image size (ca. 7%). Therefore, at line widths greater than 70 nm, the contribution of acid diffusion to image spreading is negligible. At feature dimensions near 40–50 nm, the contribution of acid diffusion to image spreading is equivalent to that of the deprotection chemistry and the impact of acid diffusion grows rapidly as feature size decreases further.

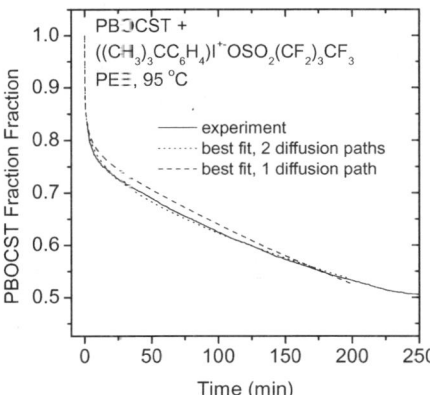

Fig. 179 Kinetics of *t*BOC resist deprotection at 95 °C; a single diffusion path vs two diffusion paths [508]

In addition to the image blur associated with the chemistry and physics of chemical amplification resists as discussed above, diffraction or scattering of the imaging radiation which generates acid at low concentrations in the nominally unexposed regions can contribute to image blur. It has been clearly demonstrated by using interferometric lithography that the aerial image contrast affects LER significantly [509a, 517]. Furthermore, statistical effects can limit the resolution of chemical amplification resists. The catalytic nature of imaging reduces the number of exposure events to a level where statistical dose fluctuations causes a loss in imaging precision for nanoscale features. The number of acid molecules required to image a nanoscale structure can become so small that statistical fluctuation in their spatial distribution and in the turnover of individual catalyst molecules can become evident and statistical dissolution effects at the line edge can further contribute to LER and limit resolution. An image spread function model has been applied to investigation of the resolution limit of the ESCAP deep UV resist [510]. A developable latent image formed under imaging conditions (exposure and PEB) from a very narrow, sharply defined line (1 nm wide exposed line) has been calculated. The final latent image formed after 2 min at 100 °C is greatly broadened compared to the initial distribution of acid (1 nm wide) and its form can be approximated as a truncated Lorentzian function with a full width at halfmaximum of 50 nm (Fig. 180), which is comparable to the imaging results of the ESCAP resist as shown in Fig. 181 [510]. The ESCAP-based resists are regarded as one of the most extensively developed classes of high performance chemical amplification resist materials and therefore these materials have found routine use in attempts to probe the resolution limits of chemical amplification resists. A qualitative sense of how imaging degrades as feature size is decreased can be gained by printing the ESCAP resist using interferometric lithography at 257 nm [518], 157 nm [321], and EUV (13.4 nm) [519] (Fig. 181). Imaging at the 45 nm line/space scale produces uniformly delineated resist images but the 30 nm

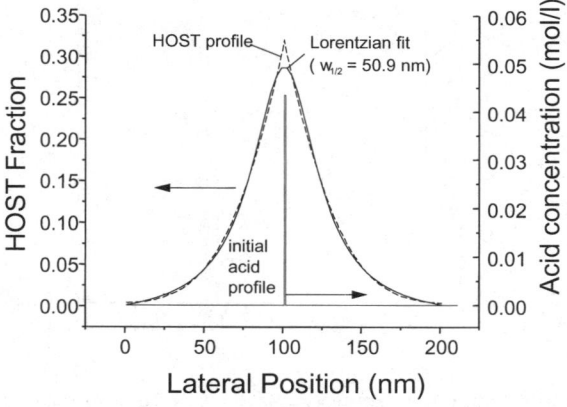

Fig. 180 Simulation of line spreading after 2 min bake of UVIIHS at 100 °C [510]

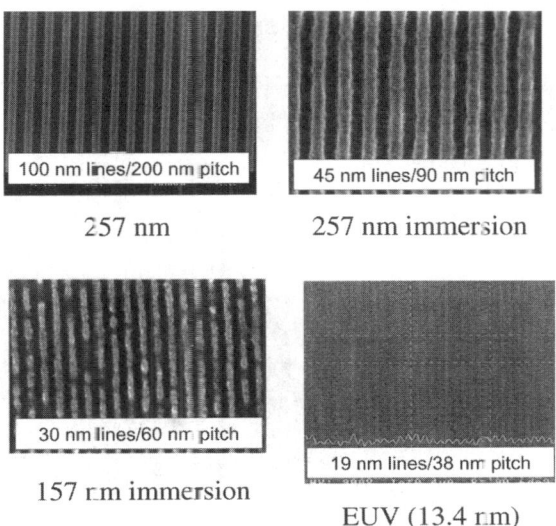

Fig. 181 SEM images of equal line/space patterns printed by interferometric lithography in ESCAP resists (257 nm, 257 nm immersion, 157 nm immersion, and EUV) [510]

quality is marginal. The resolution limit is not manifested as abrupt failure at a specific feature size but can be thought as a decline in image quality when a characteristic spatial scale has been reached [520].

10
Closing Remarks and Future Perspectives

The lithographic technology was adapted from printing initially. Now the chemical amplification concept for microlithography has been adopted by Polaroid to photography (Acid Amplified Imaging), replacing silver halide technology [521]. In addition to the semiconductor devices, chemical amplification resists have been implemented in manufacture of hard disk drives. It has been 20 years since the chemical amplification resist concept was invented, resulting in creation of new areas of science, foundation of new industries, and presentation/publication of numerous technical papers surrounding chemical amplification. The chemical amplification resist has enabled the information technology and high technology to advance at a remarkably rapid rate. Figure 182 presents the number of journal papers and united state patents issued as of December 2003 according to the literature search under "chemical amplification resists" by SciFinder. Since the very first paper on chemical amplification was presented in 1982, there was only a gradual increase in the number of publications in the 1980s. The research efforts on chemical amplification showed a significant increase in the early 1990s, resulting in more journal publications and

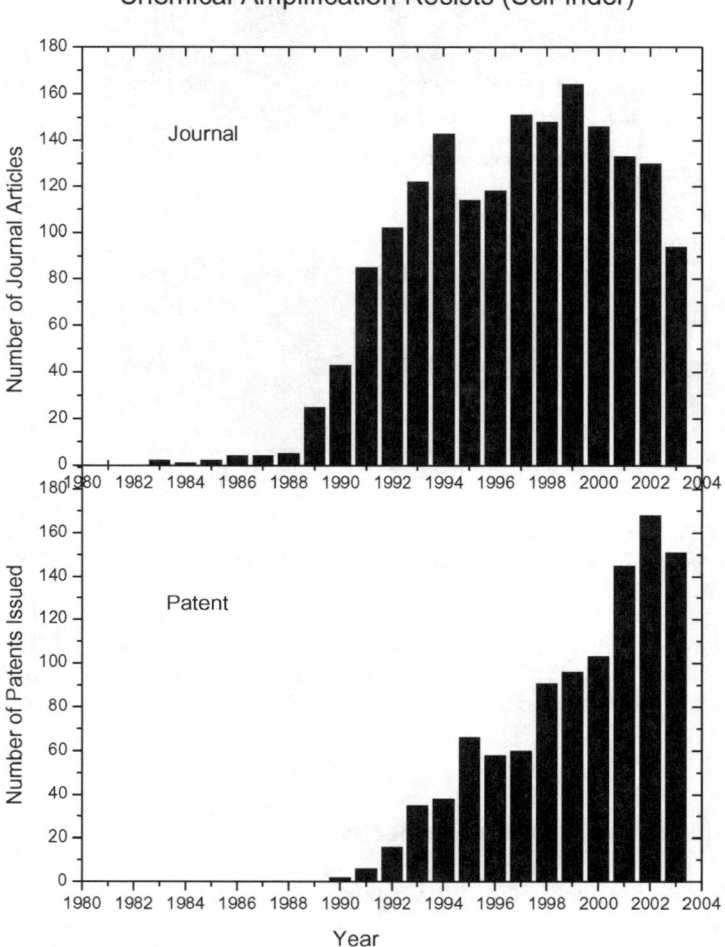

Fig. 182 Number of publications on chemical amplification resists (SciFinder)

patent applications. The journal publication has remained quite steady since mid-1990s but the issued patents continue to increase in number to this date. SciFinder deals with only chemistry publications while microlithography is a multi-disciplinary technology and a large number of researchers and engineers have been involved in chemical amplification resist in many different areas.

The industry roadmap specifies the need for chemically amplified resists that provide lithographic performance suitable to sustain their extension to 20 nm dimensional regime [522]. The ultimate resolution of ~0.3 nm has been demonstrated by moving atoms at will with a scanning tunneling microscope [523] but the process is too slow to be economically feasible (one atom/min or

0.017 pixel/s), while the current semiconductor technology at 80 nm resolution processes 100 wafers/h (300 Giga-pixels/s). Immersion lithography at 193 nm using water (refractive index of 1.44) [524] has recently emerged as a strong contender in competition with 157 nm lithography [525]. The chemical amplification resists have demonstrated excellent resolutions, much beyond anyone's expectation at the time of invention in the early 1980s. However, the resolution limit is approaching as mentioned in the earlier section. The potential problems waiting in the <50 nm regime may not be unique to chemical amplification resists but may be common to resist based on polymers. Could non-amplified resists resolve features smaller than 50 nm readily? What about the sensitivity requirement? All the NGL technologies demand extremely high resist sensitivity for throughput. Without an amplification mechanism, how could one achieve the required sensitivity higher than that currently available with chemical amplification resists? The next and ultimate question then would be what the effect of the shot noise on such a super-sensitive resist, which would receive only one photon/nm^2. In addition to the technical challenges in the area of lithography, the cost of making such devices may become prohibitively high. The prospects and issues tied to extension of semiconductor technology into the nanoscale regime have been examined in detail [526]. Factors that limit the use of lithographic exposure technology to support that miniaturization have been reviewed [505]. However, we have to remember that the microlithographic technology has survived seemingly impossible problems many times and thus it is interesting to observe what is going to emerge in the next few years.

Acknowledgement The author gratefully thanks R. Brainard, W. Hinsberg, F. Houle, F. Houlihan, D. Johnson, Y. Kikuchi, Q. Lin, D. Medeiros, H. Namatsu, S. Okazaki, and J. Thackeray for providing him with valuable figures.

References

1. Schaller B The origin, nature, and implications of "Moore's law": the benchmark of progress in semiconductor electronics. http://mason.gmu.edu/~rschalle/moorelaw.html
2. Süss O (1944) Liebigs Ann Chem 556:65
3. Pacansky J, Lyerla JR (1979) IBM J Res Dev 23:42
4. Dammel RR (1993) Diazonaphthoquinone-based resists. SPIE Press, Bellingham, WA
5. De Forest, W (1975) Photoresist materials and processes. McGraw Hill, New York, NY
6. Moreau WM (1988) Semiconductor lithography, principles, practices and materials. Plenum Press, New York, NY
7. Iwayanagi T, Ueno T, Nonogaki S, Ito H, Willson CG (1988) In: Bowden MJ, Turner SR (eds) Electronic and photonic applications of polymers. ACS Advances in Chemistry Series 218. American Chemical Society, Washington, D.C., p 107
8. Lai JH (1989) Polymers for electronic applications. CRC Press, Boca Raton, FL
9. Reiser A (1989) Photoreactive polymers, the science and technology of resists. Wiley, New York, NY
10. Willson CG (1994) In: Thompson LF, Willson CG, Bowden MJ (eds) Introduction to microlithography, 2nd edn. American Chemical Society, Washington, D.C., p 139

11. Reichmanis E, Thompson LF (1989) Chem Rev 89:1273
12. Reichmanis E, Houlihan FM, Nalamasu O, Neenan TX (1991) Chem Mater 3:394
13. Ito H (1993) In: Fouassier JP, Rabek JF (eds) Radiation curing in polymer science and technology. Elsevier, London, p 237
14. MacDonald SA, Willson CG, Fréchet JMJ (1994) Acc Chem Res 27:151
15. Ito H (1996) In: Salamone JC (ed) Polymeric materials encyclopedia. CRC Press, Boca Raton, FL, p 1146
16. Ito H (1997) In: Arshady R (ed) Desk reference of functional polymers. American Chemical Society, Washington, D.C., p 341
17. Ito H (1998) J Photopolym Sci Technol 11:379
18. Hinsberg WD, Wallraff GM, Allen RD (1998) In: Encyclopedia of chemical technology. Wiley, New York, NY, p 233
19. Ito H (1999) Proc SPIE 3678:2
20. Wallraff GM, Hinsberg WD (1999) Chem Rev 99:1801
21. Ito H (2000) IBM J Res Dev 44:119
22. Ito H (2003) J Polym Sci Part A Polym Chem 41:3863
23. Ito H, Willson CG, Fréchet JMJ (1982) Digest of Technical Papers of 1982 Symposium on VLSI Technology, p 86
24. a) Ito H, Willson CG (1982) Technical Papers of SPE Regional Technical Conference on Photopolymers, p 331; b) Ito H, Willson CG (1983) Polym Eng Sci 23:1012
25. Ito H, Willson CG (1984) In: Davidson T (ed) Polymers in electronics. ACS Symposium Series 242, American Chemical Society, Washington, D.C., p 11
26. Cameron J F, Fréchet JMJ (1990) J Org Chem 55:5919
27. Cameron JF, Fréchet JMJ (1991) J Photochem Photobiol 59:105
28. Cameron JF, Fréchet JMJ (1991) J Am Chem Soc 113:4303
29. Houlihan FM, Shugard A, Gooden R, Reichmanis E (1988) Macromolecules 21:2001
30. Pawlowski G, Dammel R, Przybilla H-J, Röschert H, Spiess W (1991) J Photopolym Sci Technol 4:389
31. Ito H, Breyta G, Hofer D, Fischer T, Prime B (1995) Proc SPIE 2438:53
32. Barclay GG, Medeiros DR, Sinta RF (1995) Chem Mater 7:1315
33. Crivello JV (1978) In: Pappas SP (ed) UV curing: science and technology. Technology Marketing Corporation, Stamford, CT, p 23, and references cited therein
34. Crivello JV, Lam JHW (1979) J Polym Sci Polym Lett Ed 17:759
35. Crivello JV, Lee JL (1981) Macromolecules 14:1141
36. Pappas SP (1985) J Imaging Technol 11:146
37. Dektar JL, Hacker NP (1987) J Chem Soc Chem Commun 1591
38. Dektar JL, Hacker NP (1988) J Org Chem 53:1833
39. Dektar JL, Hacker NP (1990) J Am Chem Soc 112:6004
40. Pohlers G, Scaiano JC, Sinta RF, Brainard R, Pai D (1997) Chem Mater 9:1353
41. a) Ohmori N, Nakazono Y, Hata M, Hoshino T, Tsuda M (1998) J Phys Chem B 102:927; b) Nakazono Y, Ohmori N, Hata M, Hoshino T, Tsuda M (1997) J Photopolym Sci Technol 10:485; c) Ohmori N, Hata M, Hoshino T, Tsuda M (1998) J Photopolym Sci Technol 11:395; d) Hirano Y, Ohmori N, Hata M, Hoshino T, Tsuda M (1999) J Photopolym Sci Technol 12:637; e) Hirano Y, Ohmori N, Okimoto N, Hata M, Hoshino T, Tsuda M (2000) J Photopolym Sci Technol 13:503
42. a) Tagawa S, Nagahara S, Iwamoto T, Wakita M, Kozawa T, Yamamoto Y, Werst D, Trifunac AD (2000) Proc SPIE 3999:204; b) Nagahara S, Sakurai Y, Wakita M, Yamamoto Y, Tagawa S, Komuro M, Yano E, Okazaki S (2000) Proc SPIE 3999:386
43. Maltabes JG, Holmes SJ, Morrow J, Barr RL, Hakey M, Reynolds G, Brunsvold WR, Willson CG, Clecak NJ, MacDonald SA, Ito H (1990) Proc SPIE 1262:2

44. Ito H, Flores E (1988) J Electrochem Soc 135:2322
45. Pappas SP, Gatechair LR, Jilek JH (1984) J Polym Sci Polym Chem Ed 22:77
46. Dektar JL, Hacker NP (1989) J Photochem Photobiol A 46:233
47. Crivello JV (1984) Adv Polym Sci 62:1
48. Buhr G, Dammel R, Lindley CR (1989) Proc ACS Div Polym Mater Sci Eng 61:269
49. Aoai T, Aotani Y, Umehara A, Kokubo T (1990) J Photopolym Sci Technol 3:389
50. Tsunooka M, Yanagi H, Kitayama M, Shirai M (1991) J Photopolym Sci Technol 4:239
51. Miller RD, Renaldo AF, Ito H (1988) J Org Chem 53:5571
52. a) Ahn K-D, Chung C-M, Koo D-I (1994) Chem Mater 6:1452 b) Chung C-M, Koo D-I, Ahn K-D (1994) J Photopolym Sci Technol 7:473; c) Ahn K-D, Koo J-S, Chung C-M (1996) J Polym Sci Polym Chem Ed 34:183; d) Chung C-M, Ahn K-D (1996) J Photopolym Sci Technol 9:553; e) Kim S-T, Kim J-B, Kim J-M, Chung C-M, Ahn K-D (1997) J Photopolym Sci Technol 10:489; f) Lee C-W, Shin J-H, Kang J-H, Kim J-M, Han D-K, Ahn K-D (1998) J Photopolym Sci Technol 11:405
53. Hanson JE, Reichmanis E, Houlihan FM, Neenan TX (1992) Chem Mater 4:837
54. a) Shrai M, Hayashi M, Tsunooka M (1992) Macromolecules 25:195; b) Shirai M, Sumino T, Tsunooka M (1994) In: Ito H, Tagawa S, Horie K (eds) Polymeric materials for microelectronic applications. ACS Symposium Series 579, American Chemical Society, Washington, D.C., p 183; c) Shirai M, Masuda M, Tsunooka M, Endo M, Matsuo T (1998) In: Ito H, Reichmanis E, Nalamasu O, Ueno T (eds) Micro- and nanopatterning polymers. ACS Symposium Series 706, American Chemical Society, Washington, D.C., p 306
55. Gonsalves KE, Hu Y, Wu H, Panepucci R, Merhari L (2001) In: Ito H, Khojasteh MM, Li W (eds) Forefront of lithographic materials research. Society of Plastics Engineers, Mid Hudson Section, Hopewell Junction, NY, 2001, p 51
56. Lamanna WL, Kessel CR, Savu PM, Cheburkov Y, Brinduse S, Kestner TA, Lillquist GJ, Parent MJ, Moorhouse KS, Zhang Y, Birznieks G, Kruger T, Pallazzotto MC (2002) Proc SPIE 4690:817
57. a) McKean DR, Schaedeli UP, MacDonald SA (1989) Proc ACS Div Polym Mater Sci Eng 60:45; b) McKean DR, Schaedeli UP, MacDoanld SA (1989) In: Reichmanis E, MacDonald SA, Iwayanagi T (eds) Polymers in microlithography. ACS Symposium Series 412, American Chemical Society, Washington, D.C., p 27; c) McKean DR, Schaedeli UP, MacDoanld SA (1989) J Polym Sci Part A Polym Chem Ed 27:3927
58. Takeyama N, Ueda Y, Kusumoto T, Ueki H, Hanabata M (1994) In: Thompson LF, Willson CG, Tagawa S (eds) Polymers for microelectronics. ACS Symposium Series 537, American Chemical Society, Washington, D.C., p 537
59. Thackeray J, Denison M, Fedynyshyn T, Kang D, Sinta R (1995) In: Reichmanis E, Ober CK, MacDonald SA, Iwayanagi T, Nishikubo T (eds) Microelectronics technology. ACS Symposium Series 614, American Chemical Society, Washington, D.C., p 110
60. a) Pohlers G, Virdee S, Scaiano JC, Sinta R (1996) Chem Mater 8:2654; b) Pohlers G, Scaiano JC, Sinta R (1997) Chem Mater 9:3222
61. a) Szmanda CR, Kavanagh R, Bohland J, Cameron J, Trefonas P, Blacksmith R (1999) Proc SPIE 3678:857; b) Szmanda CR, Brainard RL, Mackevich J, Awaji A, Tanaka T, Yamada Y, Bohland J, Tedesco S, Dal'Zotto B, Bruenger W, Torkler M, Fallman W, Loeschner H, Kaesmaier R, Nealey PM, Pawloski AR (1999) J Vac Sci Technol B17:3356
62. Cameron JF, Kang D, King M, Mori JMM, Virdee S, Zydowsky T, Sinta RF (1997) Proc. 10th International Conference on Photopolymers, p 120
63. Feke G, Grober R, Pohlers G, Moore K, Cameron JF (2001) In: Ito H, Khojasteh MM, Li W (ed) Forefront of lithographic materials research. Society of Plastics Engineers, Mid Hudson Section, Hopewell Junction, NY, 2001, p 307
64. Eckert AR, Moreau WM (1997) Proc SPIE 3049:879

65. a) Bukofsky SJ, Feke GD, Wu Q, Grober RD, Dentinger PM, Taylor JW (1998) Appl Phys Lett 73:408; b) Dentinger PM, Lu B, Taylor JW, Bukofsky SJ, Feke GD, Hessman D, Grober RD (1998) J Vac Sci Technol B16:3767; c) Feke GD, Hessman D, Grober RD, Lu B, Taylor JW (2000) J Vac Sci Technol, B18:136
66. a) Ito H, Sherwood M (1999) Proc SPIE 3678:160; b) Ito H, Sherwood M (1999) J Photopolym Sci Technol 12:493
67. Deguchi K, Ishiyama T, Horiuchi T, Yoshikawa A (1990) Jpn J Appl Phys 29:2207
68. a) Schlegel L, Ueno T, Hayashi N, Iwayanagi T (1991) J Vac Sci Technol B9:278; b) Schlegel L, Ueno T, Hayashi N, Iwayanagi T (1991) Jpn J Appl Phys 30:3132
69. a) Nakamura J, Ban H, Tanaka A (1991) J Photopolym Sci Technol 4:83; b) Nakamura J, Ban H, Deguchi K, Tanaka A (1991) Jpn J Appl Phys 10:6065
70. McKean DR, Allen RD, Kasai PH, Schaedeli UP, MacDoanld SA (1992) Proc SPIE 1672:94
71. Nakamura J, Ban H, Tanaka A (1992) Jpn J Appl Phys 31:4294
72. Nakamura J, Ban H, Deguchi K, Tanaka N (1992) J Photopolym Sci Technol 5:423
73. Nakamura J, Ban H, Tanaka A (1993) J Photopolym Sci Technol 6:31
74. Perkins FK, Dobisz EA, Marrian CR (1993) J Vac Sci Technol B11:2597
75. a) Asakawa K (1993) J Photopolym Sci Technol 6:505; b) Asakawa K, Ushirogouchi T, Nakase M (1995) Proc SPIE 2438:563
76. Fedynyshyn TH, Szmanda CR, Blacksmith RF, Houck WE (1993) Proc SPIE 1925:2
77. Cronin MF, Adams T, Fedynyshyn T, Georger J, Mori JM, Sinta R, Thackeray JW (1994) Proc SPIE 2195:214
78. Fedynyshyn TH, Thackeray JW, Georger JH, Denison MD (1994) J Vac Sci Technol B12:3888
79. Watanabe H, Sumitani H, Kumada T, Inoue M, Marumoto K, Matsui Y (1995) Jpn J Appl Phys 34:6780
80. Mack CA (1995) In: Reichmanis E, Ober CK, MacDonald SA, Iwayanagi T, Nishikubo T (eds) Microelectronics technology. ACS Symposium Series 614, American Chemical Society, Washington, D.C., p 56
81. Nakamura J, Ban H, Tanaka A (1995) In: Reichmanis E, Ober CK, MacDonald SA, Iwayanagi T, Nishikubo T (eds) Microelectronics technology. ACS Symposium Series 614, American Chemical Society, Washington, D.C., p 69
82. Itani T, Yoshino H, Hashimoto S, Yamada M, Samoto N, Kasama K (1996) J Vac Sci Technol B14:4226
83. Mueller KE, Koros WJ, Mack C, Willson CG (1997) Proc SPIE 3049:706
84. Sakamizu T, Arai T, Yamaguchi H, Shiraishi H (1997) Proc SPIE 3049:448
85. Zhang PL, Eckert AR, Willson CG, Webber SE, Byers J (1997) Proc SPIE 3049:898
86. Zhang PL, Webber S, Mendenhall J, Byers J, Chao K (1998) Proc SPIE 3333:794
87. Uchino S, Yamamoto J, Migitaka S, Kojima K, Hashimoto M, Murai F, Shiraishi H (1998) J Photopolym Sci Technol 11:555
88. Wallraff GM, Hinsberg WD, Houle FA, Morrison M, Larson CE, Sanchez M, Hoffnagle J, Brock PJ, Breyta G (1999) Proc SPIE 3678:138
89. Ito H (1988) Proc SPIE 920:33
90. a) Sundararajan N, Ogino K, Valiyaveettil S, Wang J, Yang S, Kameyama A, Ober CK, Allen RD, Byers J (1999) Proc SPIE 3678:78; b) Sundararajan N, Keimel CF, Bhargava N, Ober CK, Opitz J, Allen RD, Barclay G, Xu G (1999) J Photopolym Sci Technol 12:457
91. Lin Q, Angelopoulos M, Babich K, Medeiros D, Keimel C, Sundararajan N, Weibel G, Ober C (2001) In: Ito H, Khojasteh MK, Li W (eds) Forefront of lithographic materials research. Society of Plastics Engineers, Mid Hudson Section, Hopewell Junction, NY, p 347
92. Arai T, Sakamizu T, Katoh K, Hashimoto M, Shiraishi H (1997) J Photopolym Sci Technol 10:625

93. a) Ichimura K, Arimitsu K, Kazuaki K (1995) Chem Lett 551; b) Arimitsu K, Kudo K, Ohmori H, Ichimura K (1995) J Photopolym Sci Technol 8:43; c) Kudo K, Arimitsu K, Ohmori H, Ichimura K (1995) J Photopolym Sci Technol 8:45; d) Kudo K, Arimitsu K, Ichimura K (1996) Mol Cryst Liq Cryst 280:307; e) Ohmori H, Arimitsu K, Kudo K, Hayashi Y, Ichimura K (1996) J Photopolym Sci Technol 9:25; f) Arimitsu K, Kudo K, Hayashi Y, Ichimura K (1996) J Photopolym Sci Technol 9:29; g) Ichimura K, Arimitsu K, Noguchi S, Kudo K (1998) In: Ito H, Reichmanis E, Nalamasu O, Ueno T (eds) Micro- and nanopatterning polymers. ACS Symposium Series 706, American Chemical Society, Washington, D.C., p 161; h) Arimitsu K, Kudo K, Ichimura K (1998) J Am Chem Soc 120:37
94. Marshall JL, Minns RA, Stroud SG, Telfer SJ, Yang H (1997) Proc 11th International Conference on Photopolymers, p 245
95. a) Tsuda M, Ichikawa R, Oikawa S (1992) J Photopolym Sci Technol 5:31; b) Ichikawa R, Tsuda M, Oikawa S (1998) J Photopolym Sci Technol 6:23; c) Ichikawa R, Hata M, Okimoto N, Oikawa-Handa S, Tsuda M (1998) J Polym Sci Part A Polym Chem 36:1035
96. a) Houle F, Poliskie GM, Hinsberg WD, Pearson D, Sanchez MI, Ito H, Hoffnagle J (2000) Proc SPIE 3999:181; b) Hinsberg W, Houle F, Poliskie M, Pearson D, Sanchez M, Ito H, Hoffnagle J, Morrison M (2001) In: Ito H, Khojasteh MK, Li W (ed) Forefront of lithographic materials research. Society of Plastics Engineers, Mid Hudson Section, Hopewell Junction, NY, p 249; c) Hinsberg WD, Houle FA, Poliskie GM, Pearson D, Sanchez MI, Ito H (2002) J Phys Chem 42:9776
97. Fréchet JMJ, Eichler E, Ito H, Willson CG (1988) Polymer 24:995
98. Houlihan F, Bouchard F, Fréchet JMJ, Willson CG (1985) Can J Chem 63:153
99. Ito H, Knebelkamp A, Lundmark SB, Nguyen CV, Hinsberg WD (2000) J Polym Sci Part A Polym Chem 38:2415
100. Ito H, England WP, Lundmark SB (1992) Proc SPIE 1672:2
101. a) Turner SR, Arcus RA, Houle CG, Schleigh WR (1986) Polym Eng Sci 26:1096; b) Turner SR, Ahn KD, Willson CG (1987) In: Bowden MJ, Turner SR (eds) Polymers for high technology. ACS Symposium Series 346, American Chemical Society, Washington, D.C., p 200
102. Osuch CE, Brahim K, Hopf FR, McFarland A, Mooring CJ, Wu CJ (1986) Proc SPIE 631:68
103. Ahn K-D, Koo D-I, Kim S-J (1991) J Photopolym Sci Technol 4:433
104. Tarascon RG, Reichmanis E, Houlihan FM, Shugard A, Thompson LF (1989) Polym Eng Sci 29:850
105. Ito H, Willson CG, Fréchet JMJ, Farrall MJ, Eichler E (1983) Macromolecules 16:510
106. Houlihan FM, Shugard A, Gooden R, Reichmanis E (1988) Proc SPIE 920:67
107. Brunsvold W, Conley W, Montgomery W, Moreau W (1994) In: Thompson LF, Willson CG, Tagawa S (eds) Polymers for microelectronics. ACS Symposium Series 537, American Chemical Society, Washington, D.C., p 333
108. Gozdz AS, Shelburne JA, Lin FSD (1992) Proc ACS Div Polym Sci Eng 66:192
109. Houlihan FM, Reichmanis E, Tarascon RG, Taylor GN, Hellman MY, Thompson LF (1989) Macromolecules 22:2999
110. a) Ito H, England WP (1990) Polym Prep 31(1):427; b) Ito H, England WP, Ueda M (1992) Makromol Chem Macromol Symp 53:139
111. Novembre AE, Tai WW, Kometani JM, Hanson JE, Nalamasu O, Taylor GN, Reichmanis E, Thompson LF (1991) Proc SPIE 1466:89
112. Kometani JM, Galvin ME, Heffner SA, Houlihan FM, Nalamasu O, Chin E, Reichmanis E (1993) Macromolecules 26:2165
113. Shinoda T, Nishiwaki T, Inoue H (2000) J Polym Sci Part A Polym Chem 38:2760
114. Przybilla K-J, Röschert H, Pawlowski G (1992) Proc SPIE 1672:500

115. a) Ito H, Seehof N, Sato R (1997) Digest of Abstracts of 3rd International Symposium on 193 nm Lithography, p 43; b)) Ito H, Seehof N, Sato R (1997) Proc ACS Div Polym Mater Sci Eng 77:449; c) Ito H, Seehof N, Sato R, Nakayama T, Ueda M (1998) In: Ito H, Reichmanis E, Nalamasu O, Ueno T (eds) Micro- and nanopatterning polymers. ACS Symposium Series 706, American Chemical Society, Washington, D.C., p 208
116. Okamoto Y, Yeh TF, Skotheim TA (1993) J Polym Sci Part A Polym Chem 31:2573
117. a) Przybilla K-J, Dammel R, Röschert H, Spies W, Pawlowski G (1991) J Photopolym Sci Technol 4:421; b) Przybilla K-J, Röschert H, Spies W, Eckes C, Chatterjee S, Khanna D, Pawlowski G, Dammel R (1991) Proc SPIE 1466:174
118. Brunsvold W, Conley W, Crockatt D, Iwamoto N (1989) Proc SPIE 1086:357
119. Ito H, Willson CG, Fréchet JMJ (1987) Proc SPIE 771:24
120. Ito H, Pederson LA, Chiong KN, Sonchik S, Tsai C (1989) Proc SPIE 1086:11
121. Ito H, Ueda M (1988) Macromolecules 21:1475
122. Ito H, Ueda M, Ebina M (1989) In: Reichmanis E, MacDonald SA, Iwayanagi T (eds) Polymers in microlithography. ACS Symposium Series 412, American Chemical Society, Washington, D.C., p 57
123. a) Grant DH, Grassie N (1960) Polymer 1:125; b) Matsuzaki K, Okamoto T, Ishida A, Sobue H (1964) J Polym Sci Part A 2:1105; c) Lai JH (1984) Macromolecules 17:1010
124. Ohnishi Y, Niki H, Kobayashi Y, Hayase RH, Oyasato N, Sasaki O (1991) J Photopolym Sci Technol 4:337
125. Conlon DA, Crivello JV, Lee JL, O'Brien MJ (1989) Macromolecules 22:509
126. Hatada K, Kitayama T, Danjo S, Tsubokura Y, Yuki H, Morikawa K, Aritome H, Namba S (1983) Polym Bull 10:45
127. Ito H, Ueda M, England WP (1990) Macromolecules 23:2589
128. Ishizone T, Hirao A, Nakahama S (1989) Macromolecules 22:2895
129. Yamaoka T, Nishiki N, Koseki K, Koshiba M (1989) Polym Eng Sci 29:856
130. Murata M, Takahashi T, Koshiba M, Kawamura S, Yamaoka T (1990) Proc SPIE 1262:8
131. Hesp SAM, Hayashi N, Ueno T (1991) J Appl Polym Sci 42:877
132. Hayashi N, Schlegel L, Ueno T, Shiraishi H, Iwayanagi T (1991) Proc SPIE 1466:377
133. Jiang Y, Bassett DR (1994) In: Thompson LF, Willson CG, Tagawa S (eds) Polymers for microelectronics. ACS Symposium Series 537, American Chemical Society, Washington, D.C., p 40
134. Bowden M, Malik S, Ferreira L, Eisele J, Whewell A, Kokubo T, Kawabe Fujimori T, Tan S (2000) J Photopolym Sci Technol 13:507
135. Sovish RC (1959) J Org Chem 24:1345
136. Overberger CG, Salamone JC, Yaroslavsky S (1967) J Am Chem Soc 89:6231
137. Kato M (1969) J Polym Sci Part A-1 7:2175
138. Kato M (1969) J Polym Sci Part A-1 7:2405
139. Danusso F, Ferruti P, Crespi AMM (1965) Chim Ind (Milan) 47(1):55
140. Takai M, Asami R, Tsuzuki S (1981) Polym J 13:135
141. Arichi S, Sakamoto N, Himuro S, Miki M, Yoshida M (1985) Polymer 26:1175
142. Trumbo D (1996) Polym Bull 37:617
143. a) Barclay GG, Hawker CJ, Ito H, Orellana A, Malenfant PRL, Sinta R (1996) Proc SPIE 2724:249; b) Barclay GG, Hawker CJ, Ito H, Orellana A, Malenfant PRL, Sinta R (1998) Macromolecules 31:1024
144. Jones RL, Soles CL, Starr FW, Lin EK, Lenhart JL, Wu W-L, Goldfarb DL, Angelopoulos M (2002) Proc SPIE 4690:342
145. a) Hirao A, Yamaguchi K, Takenaka K, Suzuki K, Nakahama S, Yamazaki N (1982) Makromol Chem Rapid Commun 3:941; b) Hirao A, Takenaka K, Packirisamy S, Yamaguchi K, Nakahama S (1985) Makromol Chem 186:1157

146. Templeton MK, Szmanda CR, Zampini A (1987) Proc SPIE 771:136
147. Aoai T, Yamanaka T, Yagihara M (1997) J Photopolym Sci Technol 10:387
148. Ito H (2001) IBM J Res Dev 45:683
149. Przybilla K, Röschert H, Spiess W, Eckes C, Chatterjee S, Khanna D, Pawlowski G, Dammel R (1991) Proc SPIE 1456:174
150. Ito H, Schildknegt K, Mash EA (1991) Proc SPIE 1466:408
151. Sooriyakumaran R, Ito H, Mash EA (1991) Proc SPIE 1466:419
152. a) Pawlowski G, Sauer T, Dammel R, Gordon DJ, Hinsberg W, McKean D, Lindley CR, Merrem H-J, Vicari R, Willson CG (1990) Proc SPIE 1262:391; b) McKean DR, Sauer TP, Hinsberg WD, Willson CG, Vicari R, Gordon D (1990) Polym Preprints 31(2):599; c) McKean DR, Hinsberg WD, Sauer TP, Willson CG, Vicari R, Gordon DJ (1990) J Vac Sci Technol B8:1466
153. Willson CG, MacDonald SA, Ito H, Fréchet JMJ (1990) In: Tabata Y, Mita I, Nonogaki S, Horie K, Tagawa S (ed) Polymers for microelectronics. Kodansha, Tokyo, Japan, p 3
154. Kikuchi H, Kurata N, Hayashi K (1991) J Photopolym Sci Technol 4:357
155. Taylor GN, Stillwagon LE, Houlihan FM, Wolf TM Sogar DY, Hertler WR (1991) Chem Mater 3:1031
156. Choi S-J, Jung S-Y, Kim C-H, Park C-G, Han W-S, Koh Y-B, Lee M-Y (1996) Proc SPIE 2724:323
157. Kumada T, Kubota S, Koezuka H, Hanawa T, Kishimura S, Nagata H (1991) J Photopolym Sci Technol 4:469
158. Kawai Y, Tanaka A, Matsuda T (1992) Jpn J Appl Phys 31:4316
159. Ito H (1986) J Polym Sci Polym Chem Ed 24:2971
160. Schacht H-T, Munzel N, Falcigno P, Holzwarth H, Schneider J (1996) J Photopolym Sci Technol 9:573
161. a) Moon S, Naitoh K, Yamaoka T (1993) Chem Mater 5:1315; b) Moon S, Kamenosono K, Kondo S, Umehara A, Yamaoka T (1994) Chem Mater 6:1854; c) Taguchi T, Yamashita Y, Suzuki T, Yamaoka T (1995) J Vac Sci Technol B13(6):2972; d) Yamaoka T, Suzuki T, Taguchi T, Yamashita Y (1995) J Photopolym Sci Technol 8:665; e) Yamaoka T, Suzuki T, Takahara S, Taguchi T, Yamashita Y (1996) J Photopolym Sci Technol 9:723
162. Sakamizu T, Arai T, Shiraishi H (2000) J Photopolym Sci Technol 13:405
163. Ito H, Breyta G, Hofer D, Sooriyakumaran R, Petrillo K, Seeger D (1994) J Photopolym Sci Technol 7:433
164. Ito H, Dalby D, Pomeranz A, Sherwood M, Sato R, Sooriyakumaran R, Guy K, Breyta G (2000) Macromolecules 33:5080
165. Barclay GG, Mao Z, Xiong K, Trefonas P (2001) In: Ito H, Khojasteh MK, Li W (eds) Forefront of lithographic materials research. Society of Plastics Engineers, Mid-Hudson Section, Hopewell Junction, NY, p 153
166. a) Mehler C, Risse W (1991) Makromol Chem Rapid Commun 12:255; b) Risse W, Breunig S (1992) Makromol Chem 193:2915; c) Mehler C, Risse W (1992) Macromolecules 25:4226; d) Mathew JP, Reinmuth A, Melia J, Swords N, Risse W (1996) Macromolecules 29:2755
167. a) Goodall BL, Benedikt GM, McIntosh LH III, Barnes DA, Rhodes LF (1997) Proc ACS Div Polym Mater Sci Eng 75:56; b) Goodall BL, Benedikt GM, Jayaraman S, McIntosh LH III, Barnes DA, Rhodes LF, Shick RA (1998) Poly Preprints 39(1):216; c) Hennis AD, Polley JD, Long GS, Sen A, Yandulov D, Lipian J, Benedikt GM, Rhodes LF (2001) Organometallics 20:2802; d) Lipian J, Mimna RA, Fondran JC, Yandulov D, Shick RA, Goodall BL, Rhodes LF, Huffman JC (2002) Macromolecules 35:8969
168. Okoroanyanwu U, Shimokawa T, Byers J, Willson CG (1998) Chem Mater 10:3319
169. a) Gaylord NG, Mandal BM, Martan M (1976) J Polym Sci Polym Lett Ed 14:555; b) Gaylord NG, Desphande AB, Mandal BM, Martan M (1977) J Macromol Sci Chem A11:1053

170. Kennedy JP, Makowski HS (1967) J Macromol Sci Chem A1(3):345
171. a) Andersen AW, Merkling NG (1954) US Patent 2721189; b) Truett WL, Johnson DR, Robinson IM, Montague BA (1960) J Am Chem Soc 82:2337; c) Tsujino T, Saegusa T, Furukawa J (1965) Makromol Chem 85:71; d) Michelotti FW, Keaveney WP (1965) J Polym Sci A 3:895; e) Rhinehart RE, Smith HP (1965) J Polym Sci B 1049
172. McKean DR, MacDonald SA, Clecak NJ, Willson CG (1988) Proc SPIE 920:60
173. a) O'Brien MJ, Crivello JV (1988) Proc SPIE 920:42; b) O'Brien MJ (1989) Polym Eng Sci 29:846
174. a) Lingnau J, Dammel R, Theis J (1989) Polym Eng Sci 29:874; b) Lingnau J, Dammel R, Lindley CR, Pawlowski G, Scheunemann U, Theis J (1990) In: Tabata Y, Mita I, Nonogaki S, Horie K, Tagawa S (eds) Polymers for microelectronics. VHS Publishers, Deerfield Beach, FL, p 445; c) Eckes C, Pawlowski G, Przybilla K, Meier W, Madore M, Dammel R (1991) Proc SPIE 1466:394
175. Poot A, Delzenne G, Pollet R, Laridon U (1971) J Photogr Sci 19:88
176. Naitoh K, Kanai K, Yamaoka T, Umehara A (1991) J Photopolym Sci Technol 4:411
177. a) Schwalm R, Binder H, Dunbay B, Krause A (1990) In: Tabata Y, Mita I, Nonogaki S, Horie K, Tagawa S (eds) Polymers for microelectronics. VCH Publishers, Deerfield Beach, FL, p 425; b) Schwalm R (1989) Proc ACS Div Polym Mater Sci Eng 61:278
178. Houlihan FM, Chin E, Nalamasu O, Kometani JM (1994) In: Thompson LF, Willson CG, Tagawa S (eds) Polymers for microelectronics. ACS Symposium Series 537, American Chemical Society, Washington, D.C., p 23
179. Allen RD, Quan PL, Wallraff GM, Larson CE, Hinsberg WD, Conley WE, Muller KP (1993) Proc SPIE 1925:246
180. Allen RD, Wallraff GM, Hinsberg WD, Simpson LL (1991) J Vac Sci Technol B9:3357
181. Kunz RR, Allen RD, Hinsberg WD, Wallraff GM (1993) Proc SPIE 1925:167
182. Pawlowski G, Przybilla K-J, Spiess W, Wengenroth H, Röschert H (1992) J Photopolym Sci Technol 5:55
183. Ahn K-D, Kang J-H, Kim S-J, Park B-S, Park C-E, Park C-G (1992) J Photopolym Sci Technol 5:67
184. Gutsche CD (1983) Acc Chem Res 16:161
185. a) Fujita J, Ohnishi Y, Ochiai Y, Matsui S (1996) Appl Phys Lett 68:1297; b) Ohnishi Y, Wamme N, Fujita J (1998) In: Ito H, Reichmanis E, Nalamasu O, Ueno T (eds) Micro- and nanopatterning polymers. ACS Symposium Series 706, American Chemical Society, Washington, D.C., p 249
186. a) Ito H, Nakayama T, Ueda M, Sherwood M, Miler D (1999) Proc ACS Div Polym Mater Sci Eng 81:51; b) Ito H, Nakayama T, Ueda M (2000) US Patent 6,093,517
187. a) Aoai T, Yamanaka T, Kokubo T (1994) Proc SPIE 2195:111; b) Aoai T, Yamanaka T, Yagihara M (1997) J Photopolym Sci Technol 10:387
188. Kihara N, Saito S, Naito T, Ushirogouchi T, Asakawa K, Nakase M (1997) J Photopolym Sci Technol 10:417
189. Ito H, Ueda M (1990) Macromolecules 23:2885
190. Ito H, Ueda M (1992) Makromol Chem Macromol Symp 54/55:551
191. Pinnow MJ, Noyes BF III, Tran HV, Tattersall PI, Cho S, Klopp JM, Bensel N, Fréchet JMJ, Sanders DP, Grubbs RH, Willson CG (2002) Proc ACS Div Polym Mater Sci Eng 87:403
192. a) Tomalia DA, Baker H, Dewald J, Hall M, Kallos G, Martin S, Roeck J, Ryder J, Smith P (1985) Polym J 17:117; b) Tomalia DA, Baker H, Dewald J, Hall M, Kallos G, Martin S, Roeck J, Ryder J, Smith P (1986) Macromolecules 19:2466
193. a) Trimble AR, Tully DC, Fréchet JMJ, Medeiros DR, Anhelopoulos M (2000) Proc SPIE 3999:1198; b) Tully DC, Trimble AR, Fréchet JMJ (2000) Proc SPIE 3999:1202

194. a) Williamson M, Neureuther A (2000) Proc SPIE 3999:1189; b) Williamson M, Neureuther A (2000) J Vac Sci Technol B18(6):3345
195. Fujigaya T, Ueda M (2000) J Photopolym Sci Technol 13:339(2000) (not chemically amplified)
196. a) Ishikawa W, Inada H, Nakano H, Shirota Y (1991) Chem Lett 1731; b) Ueta E, Nakano H, Shirota Y (1994) Chem Lett 2397; c) Shirota Y, Kobata T, Noma N (1994) Chem Lett 1145; d) Shirota Y (2000) J Mater Chem 10:1; e) Utsumi H, Nagahama D, Nakano H, Shirota Y (2000) J Mater Chem 10:2436
197. a) Yoshiiwa M, Kageyama H, Wakaya F, Takai M, Gamo K, Shirota Y (1996) J Photopolym Sci Technol 9:57; b) Yoshiiwa M, Kageyama H, Shirota Y, Wakaya F, Gamo K, Takai M (1996) Appl Phys Lett 69:2605; c) Kadota T, Kageyama H, Wakaya F, Gamo K, Shirota Y (1998) J Photopolym Sci Technol 11:147; d) Kadota T, Kageyama H, Wakaya F, Gamo K, Shirota Y (1999) J Photopolym Sci Technol 12:375; e) Kadota T, Kageyama H, Wakaya F, Gamo K, Shirota Y (2000) J Photopolym Sci Technol 13:203
198. a) Takeshi K, Nakayama T, Ueda M (1998) Chem Lett 865; b) Ueda M, Takahashi D, Nakayama T, Haba O (1998) Chem Mater 10:2230; c) Haba O, Takahashi D, Haga K, Sakai Y, Nakayama T, Ueda M (1998) In: Ito H, Reichmanis E, Nalamasu O, Ueno T (eds) Micro- and nanopatterning polymers. ACS Symposium Series 706, American Chemical Society, Washington, D.C., p 237; d) Nakayama T, Ueda M (1999) J Mater Chem 9:697; e) Nakayama T, Takahashi D, Takeshi K, Ueda M (1999) J Photopolym Sci Technol 12:347; f) Haba Haga K, Ueda M (1999) Chem Mater 11:427; g) Fujigaya T, Shibasaki Y, Ueda M (2001) J Photopolym Sci Technol 14:275; h) Young-Gil K, Kim JB, Fujigaya T, Shibasaki Y, Ueda M (2002) J Mater Chem 12(1):53
199. Woods RL, Lyons CF, Mueller R, Conway J (1988) Proc KTI Microelectronics Seminar, p 341
200. MacDonald SA, Clecak NJ, Wendt HR, Willson CG, Snyder CD, Knors CJ, Ceyoe NB, Maltabes JG, Morrow JR, McGuire AE, Holmes SJ (1991) Proc SPIE 1466:2
201. Hinsberg WD, MacDonald SA, Clecak NJ, Snyder CD (1992) Proc SPIE 1672:24
202. Nalamasu O, Cheng M, Timko AG, Pol V, Reichmanis E, Thompson LF (1991) J Photopolym Sci Technol 4:299
203. Kumada T, Tanaka Y, Ueyama A, Kubota S, Koezuka H, Hanawa T, Morimoto H (1991) Proc SPIE 1925:31
204. Oikawa A, Santoh N, Miyata S, Hatakenaka Y, Tanaka H, Nakagawa K (1993) Proc SPIE 1925:92
205. Röschert H, Przybilla K-J, Spiess W, Wengenroth H, Pawlowski G (1992) Proc SPIE 1672:33
206. Funhoff DJH, Binder H, Schwalm R (1992) Proc SPIE 1672:46
207. Ushirogouchi T, Kihara H, Saito S, Naito T, Asakawa K, Tada T, Nakase M (1994) Proc SPIE 2195:205
208. Houlihan FM, Chin E, Kometani JM, Hartley R, Nalamasu O (1994) Proc 10th Int Conf on Photopolymers 48
209. a) Huang W-S, Kwong R, Katnani A, Khojasteh M (1994) Proc SPIE 2195:37; b) Huang W-S, Katnani AD, Yang D, Brunsvold B, Bantu R, Khojasteh M, Sooriyakumaran S, Kwong R, Lee KY, Hefferon G (1995) J Photopolym Sci Technol 8:525
210. a) Ito H, England WP, Clecak NJ, Breyta G, Lee H, Yoon DY, Sooriyakumaran R, Hinsberg WD (1993) Proc SPIE 1925:65; b) Ito H, England WP, Sooriyakumaran R, Clecak NJ, Breyta G, Hinsberg WD, Lee H, Yoon DY (1993) J Photopolym Sci Technol 6:547; c) Ito H, Breyta G, Hofer DC, Sooriyakumaran R (1995) In: Reichmanis E, Ober CK, MacDonald SA, Iwayanagi T, Nishikubo T (eds) Microelectronics technology. ACS Symposium Series 614, American Chemical Society, Washington, D.C., p 21

211. Nakamura J, Ban H, Kawai Y, Tanaka A (1995) J Photopolym Sci Technol 8:555
212. Funato S, Kinoshita Y, Kudo T, Masuda S, Okazaki H, Padmanaban M, Przybilla KJ, Suehiro N, Pawlowski G(1995) J Photopolym Sci Technol 8:543
213. Hinsberg W, MacDonald S, Clecak N, Snyder C, Ito H (1993) Proc SPIE 1925:43
214. Hinsberg WD, MacDonald SA, Snyder CD, Ito H, Allen RD (1993) In: Thompson LF, Willson CG, Tagawa S (eds) Polymers for microelectronics. ACS Symposium Series 537, American Chemical Society, Washington, D.C., p 101
215. a) Conley W, Breyta G, Brunsvold B, DiPietro R, Hofer D, Holmes S, Ito H, Nunes R, Fichtl G, Hagerty P, Thackeray J (1996) Proc SPIE 2724:34; b) Conley W et al. (1997) Proc SPIE 3049:282
216. a) Levenson MD (1993) Phys Today 28; b) Levenson MD (1995) Solid State Technol 38:57; c) Dusa M, van Praagh J, Finders J, Ridley A (2001) Solid State Technol 44(9):52
217. Levenson MD, Viswanathan NS, Simpson RA (1982) IEEE Trans Electron Devices 29:1828
218. a) Kim JS, Choi CI, Kim MS, Nbok CK, Kim HS, Baik KH (1998) Jpn J Appl Phys 37:6863; b) Kim J-S, Lee G, Jung M-H, Baik K-H (2000) J Photopolym Sci Technol 13:471
219. Thackeray JW (2001) In: Ito H, Khojasteh M, Li W (eds) Forefront of lithographic materials research. Society of Plastics Engineers, Mid Hudson Section, Hopewell Junction, NY, p 133
220. Fedynyshyn TH, Kunz RR, Doran SP, Goodman RB, Lind ML, Curtin JE (2000) Proc SPIE 3999:335
221. a) Brainard RL, Henderson C, Cobb J, Rao V, Mackevich JF, Okoroanyanwu U, Gunn S, Chambers J, Connolly S (1998) J Vac Sci Technol B 17(6):3384; b) Brainard RL, Cobb J, Cutler CA (2003) J Photopolym Sci Technol 16:401
222. Medeiros DR (2002) J Photopolym Sci Technol 15:411
223. Srinivasan R, Mayne-Banton V (1985) Appl Phys Lett 41:576
224. Kawamura Y, Toyoda T, Namba S (1982) J Appl Phys 53:6489
225. Craighead HG, White JC, Howard RE, Jackel LD, Behringer RE, Sweeney JE, Epworth RW (1983) J Vac Sci Technol, B1, 1186(1983)
226. a) Takechi S, Kaimoto Y, Nozaki K, Abe N (1992) J Photopolym Sci Technol 5:439; b) Kaimoto Y, Nozaki K, Takechi S, Abe N (1992) Proc SPIE 1672:66
227. Gokan H, Esho S, Ohnishi Y (1983) J Electrochem Soc 130:143
228. Kunz R, Palmateer S, Forte A, Allen R, Wallraff G, DiPietro R, Hofer D (1996) Proc SPIE 2724:365
229. Wallow T, Brock P, DiPietro R, Allen R, Opitz J, Sooriyakumaran R, Hofer D, Mewherter A-M, Cui Y, Yan W, Worth G, Moreau W, Meute J, Byers J, Rich GK, McCallum M, Jayaraman S, Vicari R, Cagles J, Sun S, Hullihen K (1999) Proc SPIE 3678:26
230. a) Imoto M, Otsu T, Tsuda K, Ito T (1964) J Polym Sci A2:1407; b) Otsu T, Ito T, Imoto M (1965) J Polym Sci B3:113; c) Otsu T, Ito T, Imoto M (1967) J Polym Sci C16:2121; d) Otsu T, Ito T, Fukumizu T, Imoto M (1966) Bull Chem Soc Jpn 39:2257
231. Takahashi M, Takechi S, Nozaki K, Kaimoto Y, Abe N (1994) J Photopolym Sci Technol 7:31
232. a) Nozaki K, Kaimoto Y, Takahashi M, Takechi S, Abe N (1994) Chem Mater 6:1492; b) Takahashi M, Takechi S, Kaimoto Y, Hanyu I, Abe N, Nozaki K (1995) Proc SPIE 2438:422
233. a) Nozaki K, Watanabe K, Yano E, Kotachi A, Takechi S, Hanyu I (1996) J Photopolym Sci Technol 9:509; b) Nozaki K, Yano E (1997) J Photopolym Sci Technol 10:545
234. a) Allen RD, Wallraff GM, DiPietro RA, Kunz RR (1994) J Photopolym Sci Technol 7:507; b) Allen RD, Wang IY, Wallraff GM, DiPietro RA, Hofer DC, Kunz RR (1995) J Photopolym Sci Technol 8:623
235. Allen RD, Wallraff GM, DiPietro RA, Hofer DC, Kunz RR (1995) Proc SPIE 2438:474

236. Aoai T, Sato K, Kodama K, Kawabe Y, Nakao H, Yagihara M (1999) J Photopolym Sci Technol 12:477
237. Nakano K, Maeda K, Iwasa S, Ohfuji T, Hasegawa E (1995) Proc SPIE 2438:433
238. Nakano K, Iwasa S, Maeda K, Hasegawa E (2001) J Photopolym Sci Technol 14:357
239. a) Maeda K, Nakano K, Iwasa S, Hasegawa E (1997) Proc SPIE 3049:55; b) Nakano K, Maeda K, Iwasa S, Hasegawa E (1997) J Photopolym Sci Technol 10:561
240. Nakano K, Iwasa S, Maeda K, Hasegawa E (1998) Proc SPIE 3333:43
241. a) Allen RD, Sooriyakumaran R, Opitz J, Wallraff GM, Breyta G, DiPietro RA, Hofer DC, Kunz RR, Okoroanyanwu U, Willson CG (1996) J Photopolym Sci 9:465; b) Allen RD, Sooriyakumaran R, Opitz J, Wallraff GM, DiPietro RA, Breyta G, Hofer DC, Kunz RR, Jayaraman S, Shick R, Goodall B, Okoroanyanwu U, Willson CG (1996) Proc SPIE 2724:334
242. a) Ushirogouchi T, Asakawa K, Shida N, Okino T, Saito S, Funaki Y, Takaragi A, Tsutsumi K, Nakano T (2000) Proc SPIE 3999:1147; b) Shida N, Ushirogouchi T, Asakawa K, Okino T, Saito S, Funaki Y, Takaragi A, Tsutsumi K, Inoue K, Nakano T (2000) J Photopolym Sci 13:601
243. a) Ishii Y (1997) J Molecular Catalysis A 117:124; b) Ishii Y, Kato S, Iwahama T, Sakaguchi S (1996) Tetrahedron Lett 37:4993; c) Kato S, Iwahama T, Sakaguchi S, Ishii Y (1998) J Org Chem 63:222; d) Sakaguchi S, Eikawa M, Ishii Y (1997) Tetrahedron Lett 38:7075
244. Nakase M, Naito T, Asakawa K, Hongu A, Shida N, Ushirogouchi T (1995) Proc SPIE 2438:445
245. Wallow TI, Houlihan FM, Nalamasu O, Chandross EA, Neenan T, Reichmanis E (1996) Proc SPIE 2724:355
246. Houlihan FM, Wallow T, Timko A, Neria E, Hutton R, Cirelli R, Kometani JM, Nalamasu O, Reichmanis E (1997) J Photopolym Sci Technol 10:511
247. Dabbagh G, Houlihan FM, Rushkin I, Hutton RS, Nalamasu O, Reichmanis E, Gabor AH, Medina AN (1999) Proc SPIE 3678:86
248. Mirau PA, Heffner SA, Rushkin I, Houlihan F (2000) Proc SPIE 3999:104
249. Houlihan FM, Kometani JM, Timko AG, Hutton RS, Cirelli RA, Reichmanis E, Nalamasu O, Gabor A, Medina A, Biafore J, Slater S (1998) Proc SPIE 3333:73
250. Choi S-J, Kang Y, Jung D-W, Park C-G, Moon J-T, Lee M-Y (1997) J Photopolym Sci Technol 10:521
251. a) Okoroanyanwu U, Shimokawa T, Byers J, Medeiros D, Willson CG, Niu QJ, Fréchet JMJ, Allen R (1997) Proc SPIE 3049:92; b) Patterson K, Okoroanyanwu U, Shimokawa T, Cho S, Byers J, Willson CG (1998) Proc SPIE 3333:425; c) Okoroanyanwu U, Byers J, Shimokawa T, Willson CG (1998) Chem Mater 10:3328; d) Byers J, Patterson K, Cho S, McCallum M, Willson CG (1998) J Photopolym Sci Technol 11:465
252. Allen RD, Opitz J, Ito H, Wallow TI, Casmier DV, DiPietro RA, Brock P, Breyta G, Sooriyakumaran R, Larson CE, Hofer DC, Varanasi PR, Mewherter AM, Jayaraman S, Vicari R, Rhodes LF (1999) Proc SPIE 3678:66
253. Ito H, Miller D, Sveum N, Sherwood M (2000) J Polym Sci Part A Polym Chem 38:3521
254. Pasquale AJ, Allen RD, Long TE (2001) Macromolecules 34:8064
255. Giese B (1983) Angew Chem Int Ed Engl 22:753 and references cited therein
256. a) Jones SA, Tirrell DA (1986) Macromolecules 19:2080; b) Jones SA, Tirrell DA (1987) J Polym Sci Part A Polym Chem 25:3177
257. Ito H, Miller D, Sherwood M (2000) J Photopolym Sci Technol 13:559
258. Ito H, Allen RD, Opitz J, Wallow TI, Truong HD, Hofer DC, Varanasi PR, Jordhamo GM, Jayaraman S, Vicari RP (2000) Proc SPIE 3999:2
259. Rahman MD, Bar J-B, Cook M, Durham DL, Kudo T, Kim W-K, Padmanaban M, Dammel RR (2000) Proc SPIE 3999:220

260. a) Choi S-J, Kang Y, Jung D-W, Park C-G, Moon J-T (1997) Proc SPIE 3049:104; b) Choi S-J, Kang Y, Jung D-W, Park C-G, Moon J-T, Lee M-Y (1997) J Photopolym Sci Technol 10:521
261. Hinsberg WD, Lee SW, Ito H, Horne DE, Kanazawa KK (2001) Proc SPIE 4345:1
262. a) Kim J-B, Lee B-W, Kang J-S, Kim S-J, Park J-H, Seo D-C, Baik K-H, Jung J-C, Roh C-H (1999) Proc SPIE 3678:36; b) Kim J-B, Lee B-W, Kang JS, Seo D-C, Roh C-H (1999) Polymer 40(26):7423; c) Kim JB, Lee B-W, Yun H-J, Kwon Y-G (2000) Chem Lett 414
263. Kim J-B, Kwon Y-G, Yun H-J, Jung M-H (2001) J Photopolym Sci Technol 14:401
264. a) Kim J-B, Jung M-H, Cheong J-H, Kim J-Y, Bok C-K, Koh C-W, Baik K-H (1997) J Photopolym Sci Technol 10:493; b) Bok C-K, Koh C-W, Jung M-H, Baik K-H, Kim J-B, Cheong J-H (1997) Proc SPIE 3049; c) Kim J-B, Choi J-H, Kwon Y-G, Jung M-H, Chang K-H (1998) Polymer 40:1087; d) Kim J-B, Kwon Y-G, Choi J-H, Jung M-H (1999) Proc SPIE 3678:536
265. Rahman MD, McKenzie D, Bae J-B, Kudo T, Kim W-K, Padmanaban M, Dammel RR (2001) Proc SPIE 4345:159
266. Sartori G, Ciampelli F, Cameli N (1963) Chim E Ind 45:1479
267. Kaminsky W, Bark A, Däke I (1990) Stud Surf Sci Catal 56:425
268. a) Sen A, Lai TW (1982) Organometallics 1:415; b) Sen A, Lai TW, Thomas RR (1988) J Organomet Chem 358:567
269. a) Mehler C, Risse W (1991) Makromol Chem Rapid Commun 12:255; b) Seehof N, Mehler C, Breunig S, Risse W (1992) J Mol Catal 76:219
270. a) Hennis AD, Polley JD, Long GS, Sen A, Yandulov D, Lipian J, Benedikt GM, Rhodes LF (2001) Organometallics 20:2802; b) Lipian J, Mimna RA, Fondran JC, Yandulov D, Shick RA, Goodall BL, Rhodes LF, Huffman JC (2002) Macromolecules 35:8969; c) Benedikt GM, Elce E, Goodall BL, Kalamarides HA, McIntosh LH III, Rhodes LF, Selvy KT, Andes C, Oyler K, Sen A (2002) Macromolecules 35:8978
271. Varanasi PR, Maniscalco J, Mewherter AM, Lawson MC, Jordhamo G, Allen R, Opitz J, Ito H, Wallow T, Hofer D, Langsdorf L, Jayaraman S, Vicari R (1999) Proc SPIE 3678:51
272. Varanasi PR, Allen RD, Ito H, Wallow T, Truong H, Chen R, Brunsvold B, Jordhamo G, Kwong R, Kajita T, Nishimura Y, Slezak M, Peterson W, Koshiba M (2001) J Photopolym Sci Technol 14:385
273. Li W, Varanasi PR, Lawson MC, Kwong RW, Chen K-J, Ito H, Truong H, Allen RD, Yamamoto M, Kobayashi E, Slezak M (2003) Proc SPIE 5039:61
274. a) Choi S-J, Kim H-W, Woo S-G, Moon J-T (2000) Proc SPIE 3999:54; b) Kim H-W, Choi S-J, Jung D-W, Lee S, Lee S-H, Kang Y, Woo S-G, Moon J-T, Kavanagh R, Barclay G, Orsula G, Mattia J, Caporale S, Adams T, Tanaka T, Kang D (2001) J Photopolym Sci Technol 14:363
275. a) Niu QJ, Fréchet JMJ, Okoroanyanwu U, Byers JD, Willson CG (1997) Proc SPIE 3049:113; b) Niu QJ, Fréchet JMJ (1998) Angew Chem Int Ed Engl 37:667
276. a) Meagley RP, Park LY, Fréchet JMJ (1998) Proc SPIE 3333:83; b) Meagley RP, Pasini D, Park LY, Fréchet JMJ (1999) Chem Commun 1587; c) Pasini D, Low E, Meagley RP, Fréchet JMJ, Willson CG, Byers JD (1999) Proc SPIE 3678:94; d) Klopp JM, Pasini D, Fréchet JMJ, Byers JD (2000) Proc SPIE 3999:23
277. Hattori T, Tuchiya Y, Yamanaka R, Hattori K, Shiraishi H (1997) J Photopolym Sci Technol 10:535
278. Gabor A H, Ober CK (1995) In: Reichmanis E, Ober C K, MacDonald S A, Iwayanagi T, Nishikubo T (eds) Microelectronics technology. ACS Symposium Series 614, American Chemical Society, Washington, D.C., p 281
279. Sundararajan N, Yang S, Ogino K, Valiyaveettil S, Wang J, Zhou X, Ober CK, Obendorf SK, Allen RD (2000) Chem Mater 12:41

280. a) Gallagher-Wetmore P, Wallraff GM, Allen RD (1995) Proc SPIE 2438:694; b) Gabor AH, Allen RD, Gallagher-Wetmore F, Ober CK (1996) Proc SPIE 2724:410; c) Gallagher-Wetmore P, Ober CK, Gabor AH, Allen RD (1996) Proc SPIE 2725 289; d) Allen RD, Wallraff GM (1997) US Patent 5,665,527
281. Jung J-C, Kong K-K, Lee G, Shin K-S, Baik K-H (2001) In: Ito H, Khojasteh M, Li W (eds) Forefront of lithographic materials research. Society of Plastics Engineers, Mid Hudson Section, NY, p 89
282. a) Lucas K, Slezak M, Ercken M, Van Roey F (2001) Proc SPIE 4345:725; b) Van Driessche V, Lucas K, Van Roey F, Grozev G, Tzviatkov P (2002) Proc SPIE 4690:631
283. a) Toyoshima T, Ishibashi T, Minamide A, Sugino K, Katayama K, Shoya T, Arimoto I, Yasuda N, Adachi H, Matsui Y (1998) Technical Digest – International Electron Devices Meeting, p 333; b) Ishibashi T, Toyoshima T, Yasuda N, Kanda T, Tanaka H, Kinoshita Y, Watase N, Eakin R (2001) Jpn J Appl Phys 40(1):419; c) Toyoshima T, Ishibashi T, Yasuda N, Tarutani S, Kanda T, Takahashi K, Takano Y, Tanaka H (2002) J Photopolym Sci Technol 15:377
284. Jung J-C, Lee S-K, Lee G, Koh C-W, Kong K-K, Hwang Y-S, Kim J-S, Shin K-S (2001) J Photopolym Sci Technol 14:419
285. Satyanarayana S, Cohan C (2003) Proc SPIE 5039:257
286. Kunz RR, Bloomstein TM, Hardy DE, Goodman RB, Downs DK, Curtin JE (1999) Proc SPIE 3678:13
287. a) Bloomstein TM, Horn MW, Rothchild M, Kunz RR, Palmacci ST, Goodman RB (1997) J Vac Sci Technol B15:2112; b) Bloomstein TM, Rothchild M, Kunz RR, Hardy DE, Goodman RB, Palmacci ST (1998) J Vac Sci Technol B16:3154
288. a) Ito H, Wallraff GM, Brock P, Fender N, Truong H, Breyta G, Miller DC, Sherwood MH, Allen RD (2001) Proc SPIE 4345:273; b) Ito H, Wallraff GM, Fender N, Brock P, Larson CE, Truong HD, Breyta G, Miller DC, Sherwood MH, Allen RD (2001) J Photopolym Sci Technol 14:583; c) Ito H, Wallraff GM, Fender N, Brock PJ, Hinsberg WD, Mahorowala A, Larson CE, Truong HD, Breyta G, Allen RD (2001) J Vac Sci Technol B19(6):2678
289. a) Patterson K, Yamachika M, Hung R, Brodsky C, Yamada S, Somervelle M, Osborn B, Hall D, Dukovic G, Byers J, Conley W, Willson CG (2000) Proc SPIE 3999:365; b) Chiba T, Hung RJ, Yamada S, Trinque B, Yamachika M, Brodsky C, Paterson K, Heyden AV, Jamison A, Lin S-H, Somervelle M, Byers J, Conley W, Willson CG (2000) J Photopolym Sci Technol 13:657; c) Hung RJ, Tran HV, Trinque BC, Chiba T, Yamada S, Sanders DP, Connor EF, Grubbs RH, Klopp J, Fréchet JM, Thomas BH, Shafer GJ, DesMarteau DD, Conley W, Willson CG (2001) Proc SPIE 4345:385; d) Tran HV, Hung RJ, Chiba T, Yamada S, Mrozek T, Hsieh Y-T, Chambers CR, Osborn BP, Trinque BC, Pinnow MJ, Sanders DP, Connor EC, Grubbs RH, Conley W, MacDonald SA, Willson CG (2001) J Photopolym Sci Technol 14:669
290. a) Trinque BC, Osborn BP, Chambers CR, Hsieh Y-T, Corry S, Chiba T, Hung RJ, Tran HV, Zimmerman P, Miller D, Conley W, Willson C G (2002) Proc SPIE 4690:58; b) Willson CG, Trinque BC, Osborn BP, Chambers CR, Hsieh Y-T, Chiba T, Zimmerman P, Miller D, Conley W (2002) J Photopolym Sci Technol 15:583
291. a) Crawford MK, Feiring AE, Feldman J, French RH, Periyasamy M, Schadt FL III, Smalley RJ, Zumsteg FC, Kunz RR, Rao V, Liao L, Holl SM (2001) Proc SPIE 4345:428; b) Crawford MK, Farnham WB, Feiring AE, Feldman J, French RH, Leffew KW, Petrov VA, Schadt FL III, Zumsteg FC (2002) J Photopolym Sci Technol 15:677
292. Itani T, Toriumi M, Naito T, Ishikawa S, Miyoshi S, Yamazaki T, Watanabe M (2001) J Vac Sci Technol B19(6):2705
293. a) Toriumi M, Shida N, Watanabe H, Yamazaki T, Ishikawa S, Itani T (2002) Proc SPIE 4690:191; b) Toriumi M, Yamazaki T, Furukawa T, Irie S, Ishikawa S, Itani T (2002) J Vac Sci Technol B20:2909

294. a) Schmaljohann D, Bae YC, Weibel GL, Hamad AH, Ober CK (2000) Proc SPIE 3999:330; b) Schmaljohann D, Bae YC, Weibel GL, Hamad AH, Ober CK (2000) J Photopolym Sci Technol 13:451; c) Schmaljohann D, Hamad AH, Pham VQ, Yu T, Bae YC, Weibel GL, Ober CK (2001) In: Ito H, Khojasteh M, Li W (eds) Forefront of lithographic materials research. Society of Plastics Engineers, Mid-Hudson Section, NY, p 81
295. a) Hamad AH, Schmaljohann D, Bae YC, Yu T, Pham VQ, Dai J, Ober CK (2001) In: Ito H, Khojasteh M, Li W (eds) Forefront of lithographic materials research. Society of Plastics Engineers, Mid-Hudson Section, NY, p 65; b) Hamad AH, Bae YC, Liu X-Q, Ober CK, Houlihan FM, Dabbagh G, Novembre AE (2002) Proc SPIE 4690:477
296. Bae YC, Weibel GL, Hamad AH, Schmaljohann D, Ober CK (2001) In: Ito H, Khojasteh M, Li W (eds) Forefront of lithographic materials research. Society of Plastics Engineers, Mid Hudson Section, NY, p 75
297. a) Urry WH, Niu JHY, Lundsted LG (1968) J Org Chem 33:2302; b) Snow AW, Spraue LG, Soulen RL, Grate JW, Wohltjen H (1991) J Appl Polym Sci 43:1659; c) Coleman MM, Yang X, Painter PC, Kim YH (1994) J Polym Sci Part A Polym Chem 32:1817
298. Bae YC, Douki K, Yu T, Dai J, Schmaljohann D, Kang SH, Kim KH, Koerner H, Conley W, Miller D, Balasubramanian R, Holl S, Ober CK (2001) J Photopolym Sci Technol 14:613
299. Willson CG, Ito H, Miller DC, Tessier TG (1983) Polym Eng Sci 23:1000
300. Matsuzawa NN, Mori S, Yano E, Okazaki S, Ishitani A, Dixon DA (2000) Proc SPIE 3999 375
301. Ito H, Miller DC, Willson CG (1982) Macromolecules 15:915
302. Ito H, Giese B, Engelbrecht R (1984) Macromolecules 17:2204
303. Ito H, Schwalm R (1987) In: Hogen-Esch TE, Smid J (eds) Recent advances in anionic polymerization. Elsevier, New York, NY, p 421
304. Ito H, Renaldo AF, Ueda M (1989) Macromolecules 22:45
305. a) Ito H, Truong HD, Okazaki M, Miller DC, Fender N, Breyta G, Brock PJ, Wallraff GM, Larson CE, Allen RD (2002) Proc SPIE 4690:18; b) Ito H, Truong HD, Okazaki M, Miller DC, Fender N, Brock PJ, Wallraff GM, Larson CE, Allen RD (2002) J Photopolym Sci Technol 15:591
306. Dammel RR, Sakamuri R, Lee S-H, Rahman MD, Kudo T, Romano A, Rhodes L, Lipian J, Hacker C, Barnes DA (2002) Proc SPIE 4690:101
307. a) Vohra V R, Douki K, Kwark Y-J, Liu X-Q, Ober C K, Bae YC, Conley W, Miller D, Zimmerman P (2002) Proc SPIE 4690:84; b) Ober C K, Douki K, Vohra VR, Kwark Y-J, Liu X-Q, Conley W, Miller D, Zimmerman P (2002) J Photopolym Sci Technol 15:603
308. Fender N, Brock PJ, Chau W, Bangsaruntip S, Mahorowala A, Wallraff GM, Hinsberg WD, Larson CE, Ito H, Breyta G, Burnham K, Truong H, Lawson P, Allen RD (2001) Proc SPIE 4345:417
309. Kunz RR, Sinta R, Sworin M, Mowers WA, Fedynyshyn TH, Liberman V, Curtin JE (2001) Proc SPIE 4345:285
310. a) Fedynyshyn TH, Kunz RR, Sinta RF, Sworin M, Mowers WA, Goodman RB, Doran SP (2001) Proc SPIE 4345:296; b) Fedynyshyn TH, Kunz RR, Sinta RF, Sworin M, Mowers WA, Goodman RB, Dora SP (2001) Proc SPIE 4345:396; c) Fedynyshyn TH, Kunz RR, Sinta RF, Sworin M, Mowers WA, Goodman RB, Cabral A (2002) J Photopolym Sci Technol 15:655
311. Cho S, Klauck-Jacobs A, Yamada S, Xu CB, Leonard J, Zampini A (2002) Proc SPIE 4690:522
312. a) Conley W, Miller D, Chambers C, Osborn B, Hung RJ, Tran HV, Trinque BC, Pinnow M, Chiba T, McDonald S, Zimmerman P, Dammel R, Romano A, Willson CG (2002) Proc SPIE 4690:69; b) Conley W, Miller D, Chambers C, Trinque BC, Osborn B, Chiba T, Zimmerman P, Dammel R, Romano A, Willson CG (2002) J Photopolym Sci Technol 15:613

313. Willson CG, Miller RD, McKean DR, Clecak N, Thompkins T, Hofer D, Michl J, Downing J (1983) Polym Eng Sci 23:1004
314. a) Kodama S, Kaneko I, Takebe Y, Okada S, Kawaguchi Y, Shida N, Ishikawa S, Toriumi M, Itani T (2002) Proc SPIE, 4690:76; b) Shida N, Watanabe H, Yamazaki T, Ishikawa S, Toriumi M, Itani T (2002) Proc SPIE 4690:497
315. a) Ando S, Fujigaya T, Ueda M (2002) Jpn J Appl Phys 41:L105; b) Ando S, Fujigaya T, Ueda M (2002) J Photopolym Sci Technol 15:559
316. Fujigaya T, Ando S, Shibasaki Y, Kishimura S, Endo M, Sasago M, Ueda M (2002) J Photopolym Sci Technol 15:643
317. a) Shirai M, Kataoka A, Shinozuka T, Tsunooka M, Kishimura S, Endo M, Sasago M (2002) In: Ito H, Khojasteh M, Li W (eds) Forefront of lithographic materials research. Society of Plastics Engineers, Mid Hudson Section, Hopewell Junction, NY, p 209; b) Shirai M, Shinozuka T, Okamura H, Tsunooka M, Kishimura S, Endo M, Sasago M (2001) J Photopolym Sci Technol 14:621; c) Shinozuka T, Kawakami T, Okamura H, Tsunooka M, Shirai M (2002) J Photopolym Sci Technol 15:629
318. a) Dammel RR, Sakamuri R, Romano A, Vicari R, Hacker C, Conley W, Miller D (2001) Proc SPIE 4345:350; b) Dammel RR, Sakamuri R, Kudo T, Romano A, Rhodes L, Vicari R, Hacker C, Conley W, Miller D (2001) J Photopolym Sci Technol 14:603
319. Matsuzawa NN, Ishitani A, Dixon DA, Uda T (2001) Proc SPIE 4345:396
320. Yamazaki T, Itani T (2002) Jpn J Appl Phys 41:4065
321. Switkes M, Rothschild M (2001) J Vac Sci Technol B19(6):2353
322. a) Novembre AE, Ocola LE, Houlihan F, Knurek C, Blakey M (1998) J Photopolym Sci Technol 11:541; b) Ocola LE, Blakey MI, Orphanos PA, Li W-Y, Novembre AE (2001) In: Ito H, Khojasteh M, Li W (eds) Forefront of lithographic materials research. Society of Plastics Engineers, Mid Hudson Section, Hopewell Junction, NY, p 31; c) Ocola LE, Blakey MI, Orphanos PA, Li W-Y, Kasica RJ, Novembre AE (2001) J Photopolym Sci Technol 14:547
323. Brainard RL, Guevremont JM, Reeves SD, Zhou X, Nguyen TB, Mackevich JF, Taylor GN, Anderson EH (2001) J Photopolym Sci Technol 14:531
324. a) Utsumi T (1999) Jpn J Appl Phys 38(12B):7046; b) Utsumi T (1999) J Vac Sci Technol B17(6):2897; c) Endo A, Higuchi A, Kasahara H, Nozue H, Shimazu N, Fukui T, Yasumitsu N, Miyatake T, Anazawa N (2002) J Photopolym Sci Technol 15:403
325. Haller I, Hatzakis M, Srinivasan R (1968) IBM J Res Develop 251
326. Lin BJ (1975) J Vac Sci Technol 12:1317
327. Ito H (1997) In: Arshady R (ed) Desk reference of functional polymers. American Chemical Society, Washington, D C, p 311
328. Ito H, England WP, Ueda M (1990) J Photopolym Sci Technol 3:219
329. Vogl O (1960) J Polym Sci Polym Chem Ed 46:261
330. Aso C, Tagami S, Kunitake T (1969) J Polym Sci A-1 7:497
331. a) Tsuda M, Hata M, Nishida R, Oikawa S (1993) J Photopolym Sci Technol 6:491(1993); b) Tsuda M, Hata M, Nishida R, Oikawa S (1997) J Polym Sci Part A Polym Chem 35:77
332. Hatada K, Kitayama T, Danjo S, Yuki H, Aritome H, Namba S, Nate K, Yokono H (1982) Polym Bull 8:469
333. Ito H, Schwalm R (1989) J Electrochem Soc 136:241
334. Ito H, Ueda M, Schwalm R (1988) J Vac Sci Technol B6(6):2259
335. Ito H, Ueda M, Renaldo A (1989) J Electrochem Soc 136:245
336. Ito H (1992) J Photopolym Sci Technol 5:123
337. Ito H, Flores E, Renaldo AF (1988) J Electrochem Soc 135:2328
338. a) Ito H, Ueda M (1990) Macromolecules 23:2589; b) Ito H, Ueda M, Ito T (1990) J Photopolym Sci Technol 3:335

339. Ueda M, Ito T, Ito H (1990) Macromolecules 23:2895
340. Nagasaki Y (1998) In: Ito H, Reichmanis E, Nalamasu O, Ueno T (eds) Micro- and nanopatterning polymers. ACS Symposium Series 706, American Chemical Society, Washington, D.C., p 276
341. Houlihan FM, Bouchard F, Fréchet JMJ, Willson CG (1986) Macromolecules 19:13
342. Fréchet JMJ, Eichler E, Stanciulescu M, Iizawa T, Bouchard F, Houlihan FM, Willson CG (1987) In: Bowden MJ, Turner SR (eds) Polymers for high technology. ACS Symposium Series 346, American Chemical Society, Washington, D.C., p 138
343. Fréchet JMJ, Willson CG, Iizawa T, Nishikubo T, Igarashi K, Fahey J (1989) In: Reichmanis E, MacDonald SA, Iwayanagi T (eds) Polymers in microlithography. ACS Symposium Series 412, American Chemical Society, Washington, D.C., p 100
344. Eschbaumer C, Heusinger N, Hohle C, Sebald M (2002) J Photopolym Sci Technol 15:673
345. Inaki Y, Matsumura N, Takemoto K (1994) In: Thompson LF, Willson CG, Tagawa S (eds) Polymers for microelectronics, ACS Symposium Series 537, American Chemical Society Washington, D.C., p 142
346. a) Darling GD, Vekselman AM (1995) Chem Mater 7:850; b) Vekselman AM, Darling GD (1996) Proc SPIE 2724:296; c) Darling GD, Vekselman AM (1997) Proc 11th International Conference on Photopolymers, p 197; d) Darling GD, Vekselman AM, Yamada S (1998) Proc SPIE 3333:438
347. Stöver H, Matuszczak S, Chin R, Shimizu K, Willson CG, Fréchet JMJ (1989) Proc ACS Div Polym Mater Sci Eng 61:412
348. Ito H (1993) In: Reichmanis E, Frank CW, O'Donnell JH (eds) Irradiation of polymeric materials. ACS Symposium Series 527, American Chemical Society, Washington, D.C., p 197
349. Uchino S, Iwayanagi T, Ueno T, Hayashi N (1991) Proc SPIE 1466:429
350. Ito H, Sooriyakumaran R, Mash EA (1991) J Photopolym Sci Technol 4:319
351. Ito H, Maekawa Y (1994) In: Ito H, Tagawa S, Horie K (eds) Polymeric materials for microelectronic applications. ACS Symposium Series 579, American Chemical Society, Washington, D.C., p 70
352. Cho S, Heyden AV, Byers J, Willson CG (2000) Proc SPIE 3999:62
353. Ito H, Maekawa Y, Sooriyakumaran R, Mash EA (1994) In: Thompson LF, Willson CG, Tagawa S (eds) Polymers for microelectronics. ACS Symposium Series 537, American Chemical Society, Washington, D.C., p 64
354. Feeley WE, Imhof JC, Stein CM, Fisher TA, Legenza MW (1986) Polym Eng Sci 26:1101
355. Liu H-Y, de Grandpre MP, Feeley WE (1988) J Vac Sci Technol B6:379
356. Thackeray JW, Orsula GW, Pavelcheck EK, Canistro D (1989) Proc SPIE 1086:34
357. Thackeray JW, Orsula GW, Bohland JF, McCullough AW (1989) J Vac Sci Technol B7(6):1620
358. Berry AK, Graziano KA, Bogan LE Jr, Thackeray JW (1989) In: Reichmanis E, MacDonald SA, Iwayanagi T (eds) Polymers in microlithography. ACS Symposium Series 412, American Chemical Society, Washington, D.C., p 87
359. Lingnau J, Dammel R, Theis J (1989) Solid State Technol 32(9):105
360. Lingnau J, Dammel R, Theis J (1989) Solid State Technol 32(10):107
361. Bohland JF, Calabrese GS, Cronin MF, Canistro D, Fedynyshyn TH, Ferrari J, Lamola AA, Orsula GW, Pavelcheck EK, Sinta R, Thackeray JW, Berry AK Bogan LE Jr, de Grandpre MP, Feeley WE, Grazinano KA, Olsen R, Thompson S, Winkle MR (1990) J Photopolym Sci Technol 3:355
362. Thackeray JW, Orsula GW, Rajaratnam MM, Sinta R, Herr D, Pavelcheck E (1991) Proc SPIE 1466:39

363. Allen MT, Calabrese GS, Lamola AA, Orsula GW, Rajaratnam MM, Sinta R, Thackeray JW (1991) J Photopolym Sci Technol 4:379
364. Thackeray JW, Adams T, Cronin MF, Denison M, Fedynyshyn TH, Georger J, Mori M, Orsula GW, Sinta R (1994) J Photopolym Sci Technol 7:619
365. Brunner TA, Fonseca C (2001) Proc SPIE 4345:30
366. Reck B, Allen RD, Twieg RJ, Willson CG, Matuszczk S, Stover HDH, Li NH, Fréchet JMJ (1989) Polym Eng Sci 29:960
367. Fréchet JMJ, Matuszczak S, Stover HDH, Willson CG, Reck B (1989) In: Reichmanis E, MacDonald SA, Iwayanagi T (eds) Polymers in microlithography. ACS Symposium Series 412, American Chemical Society, Washington, D.C., p 74
368. Fréchet JMJ, Kryczka B, Matuszczak S, Reck B, Stanciulescu M, Willson CG (1990) J Photopolym Sci Technol 3:235
369. Stöver HDH, Matuszczak S, Willson CG, Fréchet JMJ (1991) Macromolecules 24:1741
370. Fréchet JMJ, Matuszczak S, Reck B, Stöver HDH, Willson CG (1991) Macromolecules 24:1746
371. Fréchet JMJ, Lee SM (1993) Proc SPIE 1925:102
372. Schaedeli U, Muenzel N, Holzwarth H (1993) Proc SPIE 1925:109
373. Huang W-S, Lee KY, Chen RK-J, Schepis D (1996) Proc SPIE 2724:315
374. a) Haba O, Takahashi D, Haga K, Sakai Y, Nakayama T, Ueda M (1998) In: Ito H, Reichmanis E, Nalamasu O, Ueno T (eds) Micro- and nanopatterning polymers. ACS Symposium Series 706, American Chemical Society, Washington, D.C., p 237; b) Ueda M, Takahashi D, Nakayama T, Haba O (1998) Chem Mater 10:2230; c) Haba O, Haga K, Ueda M, Morikawa O, Konishi H (1999) Chem Mater 11:427
375. Fahey JT, Fréchet JMJ (1991) Proc SPIE 1466:67
376. a) Ueno T, Uchino S, Hattori KT, Onozuka T, Shirai S, Moriuchi N, Hashimoto M, Koibuchi S (1994) Proc SPIE 2195:173; b) Migitaka S, Uchino S, Ueno T, Yamamoto J, Kojima K, Hashimoto M, Shiraishi H (1996) Proc SPIE 2724:613
377. Uchino S, Yamamoto J, Migitaka S, Kojima K, Hashimoto M, Murai F, Shiraishi H (1998) J Photopolym Sci Technol 11:555
378. Uchino S, Ueno T, Migitaka S, Yamamoto J, Tanaka T, Murai F, Shiraishi H, Hashimoto M (1996) Proc SPIE 2724:438
379. a) Ueno T, Shiraishi H, Hayashi N, Tadano K, Fukuma E, Iwayanagi T (1990) Proc SPIE 1262:26; b) Shiraishi H, Fukuma E, Hayashi N, Ueno T, Tadano K, Iwayanagi T (1990) J Photopolym Sci Technol 3:385; c) Shiraishi H, Fukuma E, Hayashi N, Tadano K, Ueno T (1990) Chem Mater 3:621
380. McKean DR, Clecak NJ, Pederson LA (1990) Proc SPIE 1262:110
381. a) Iwasa S, Nakano K, Maeda K, Hasegawa E (1998) Proc SPIE 3333:417; b) Iwasa S, Maeda K, Hasegawa E (1999) J Photopolym Sci Technol 12:487
382. Katsuyama A, Matsuo T, Endo M (1997) Digest of Abstracts of the 3rd International Symposium on 193 nm Lithography, p 51
383. Naito T, Takahashi M, Ohfuji T, Sasago M (1998) Proc SPIE 3333:503
384. a) Hattori T, Tsuchiya Y, Yokoyama Y, Shiraishi H (1998) Chem Mater 10:1789; b) Hattori T, Tsuchiya Y, Yokoyama Y, Oizumi H, Morisawa T, Yamaguchi A, Shiraishi H (1999) Proc SPIE 3678:411; c) Hattori T, Tsuchiya Y, Yokoyama Y, Oizumi H, Morisawa T, Yamaguchi A, Shiraishi H (1999) J Photopolym Sci Technol 12:537
385. a) Yokoyama Y, Hattori T, Kimura K, Tanaka T, Shiraishi H (2000) J Photopolym Sci Technol 13:579; b) Yokoyama Y, Hattori T, Kimura K, Tanaka T, Shiraishi H (2001) J Photopolym Sci Technol 14:393
386. Hattori T, Yokoyama Y, Kimura K, Yamanaka R, Tanaka T, Fukuda H (2003) J Photopolym Sci Technol 16:489

387. Aoai T, Lee J-S, Watanabe H, Kondo S, Miyagawa N, Takahara S, Yamaoka T (1999) J Photopolym Sci Technol 12:303
388. Lee S, Toshiaki A, Kondo S, Miyagawa N, Takahara S, Yamaoka T (2002) J Polym Sci Part A Polym Chem 40:1858
389. LaBianca NC, Gelorme JD (1994) Proc 10th International Conference on Photopolymers, p 239
390. http://aveclafauxfreeserverscom/SU-8html
391. a) Stewart KJ, Hatzakis M, Shaw JM, Seeger DE, Neumann E (1989) J Vac Sci Technol B7:1734; b) Stewart KJ, Hatzakis M, Shaw JM (1989) Polym Eng Sci 29:907
392. Dubois JC, Eranian A, Datmanti E (1978) Proc Electrochem Soc 78(5):303
393. Crivello JV (1984) In: Davidson T (ed) Polymers in electronics. ACS Symposium Series 242, American Chemical Society, Washington, D.C., p 3
394. Conley WE, Moreau W, Perreault S, Spinillo G, Wood R (1990) Proc SPIE 1262:49
395. Allen RD, Conley W, Gelorme JD (1992) Proc SPIE 1672:513
396. a) Yamaoka T, Kamenosono K, Moon S, Naitoh K, Kondo S, Umehara A (1994) J Photopolym Sci Technol 7:533; b) Watanabe H, Maeshima K, Aoai T, Kondo S, Naitoh T, Ohfuji T, Sasago M, Miyagawa N, Takahara S, Yamaoka T (1998) J Photopolym Sci Technol 11:537; c) Noppakunddilograt S, Miyagawa N, Takahara S, Yamaoka T (1999) J Photopolym Sci Technol 12:773; d) Noppakunddilograt S, Miyagawa N, Takahara S, Yamaoka T (2000) J Photopolym Sci Technol 13:719; e) Lee J-S, Suzuki H, Odoi K, Miyagawa N, Takahara S, Yamaoka T (2002) Proc SPIE 4690:541
397. Shadman F (2000) Proc SPIE 3999:1237
398. a) Hult A, Skolling O, Cothe S, Mellstrom U (1986) Preprints of ACS Div of Polym Mater Sci Eng 55:594; b) Hult A, Skolling O, Göthe S, Mellström U (1987) In: Bowden MJ, Turner SR (eds) Polymers for high technology. ACS Symposium Series 346, American Chemical Society, Washington, D.C., p 162
399. Lin Q, Steinhäusler T, Simpson L, Wilder M, Medeiros DR, Willson CG, Havard J, Fréchet JMJ (1997) Chem Mater 9:1725
400. Havard JM, Shim S-Y, Fréchet JMJ, Lin Q, Medeiros DR, Willson CG, Byers JD (1999) Chem Mater 11:719
401. Havard JM, Pasini D, Fréchet JMJ, Medeiros D, Patterson K, Yamada S, Willson CG (1998) Proc SPIE 3333:111
402. Havard JM, Vladimirov N, Fréchet JMJ, Yamada S, Willson CG, Byers JD (1999) Macromolecules 32:86
403. Havard JM, Yoshida M, Vladimirov N, Fréchet JMJ, Medeiros DR, Patterson K, Yamada S, Willson CG, Byers JD (1999) J Polym Sci Part A Polym Chem 37:1225
404. Yamada S, Medeiros DR, Patterson K, Jen W-LK, Rager T, Lin Q, Lenci C, Byers JD, Havard JM, Pasini D, Fréchet JMJ, Willson CG (1998) Proc SPIE 3333:245
405. a) Havard JM, Pasini D, Fréchet JMJ, Medeiros D, Yamada S, Willson CG (1998) In: Ito H, Reichmanis E, Nalamasu O, Ueno T (eds) Micro- and nanopatterning polymers. ACS Symposium Series 706, American Chemical Society, Washington, D.C., p 262; b) Havard JM, Fréchet JMJ, Pasini D, Mar B, Yamada S, Medeiros D, Willson CG (1997) Proc SPIE 3049:437
406. a) Yamada S, Owens J, Rager T, Nielsen M, Byers JD, Willson CG (2000) Proc SPIE 3999:569; b) Owens J, Yamada S, Willson CG (2001) In: Ito H, Khojasteh M, Li W (eds) Forefront of lithographic materials research. Society of Plastics Engineers, Mid Hudson Section, Hopewell Junction, p 229
407. Allen RD, Wallraff GM (1997) US Patent 5,665,527 (09/09/1997)
408. Ober CK, Gabor AH (1996) J Photopolym Sci Technol 9:1
409. Sundararajan N, Yang S, Ogino K, Valiyaveettil S, Wang J, Zhou X, Ober CK, Obendorf SK, Allen RD (2000) Chem Mater 12:41

410. a) DeSimone JM, Guan Z, Elsbernd CS (1992) Science 257:945; b) McAdams CL, Flowers D, Hoggan EN, Carbonell RG, DeSimone JM (2001) Proc SPIE 4345:327
411. Tanaka T, Morigami M, Atoda N (1993) J Electrochem Soc 140:L115
412. a) Namatsu H, Yamazaki K, Kurihara K (2000) J Vac Sci Technol B18(2):780; b) Namatsu H (2002) J Photopolym Sci Technol 15:381
413. Goldfarb DL, de Pablo JJ, Nealey PF, Simons JP, Moreau WM, Angelopoulos M (2000) J Vac Sci Technol B18(6):3313
414. Smith JN, Highs HG, Keller JV, Goodner WR, Wood TE (1979) Semiconductor International 2(10):41
415. Taylor GN, Wolf TM (1980) J Electrochem Soc 127:2665
416. Shaw JM, Hatzakis M, Paraszczak J, Liutkus J, Babich E (1983) Polym Eng Sci 23:1054
417. Sakata M, Kosuge M, Ito T, Yamashita Y (1991) J Photopolym Sci Technol 4:75
418. Sakata M, Ito T, Kosuge M, Yamashita Y (1992) J Photopolym Sci Technol 5:181
419. Kawai Y, Tanaka A, Ban H, Nakamura J, Matsuda T (1992) J Photopolym Sci Technol 5:431
420. Brunsvold W, Stewart K, Jaganathan P, Sooriyakumaran R, Parrill J, Muller KP, Sachdev H (1993) Proc SPIE 1925:377
421. a) Lin Q, Katnani A, Brunner T, DeWan C, Fairchok C, LaTulipe D, Simons J, Petrillo K, Babich K, Sooriyakumaran Wallraff G, Hofer D (1998) Proc SPIE 3333:278; b) Wallraff GM, Larson CE, Sooriyakumaran R, Opitz J, Fenzel-Alexander D, DiPietro R, Hofer D, Breyta G, Sherwood M, Meute J, Lin Q, Simons J, Babich K, Petrillo K, Angelopoulos M (1998) J Photopolym Sci Technol 11:673
422. a) Wallraff GM, Larson CE, Sooriyakumaran R, Opitz J, Fenzel-Alexander D, DiPietro R, Hofer DC, Sherwood M, Meute J, LaTulipe D, Simons J, Seeger D, Petrillo K, Angelopoulos M, Lin Q, Fairchok C (1997) Proc of 11th International Conference on Photopolymers, p 102; b) Sooriyakumaran R, Wallraff GM, Larson CE, Fenzel-Alexander D, DiPietro RA, Opitz J, Hofer DC, LaTulipe DC Jr, Simons JP, Petrillo KE, Babich K, Angelopoulos M, Lin Q, Katnani AD (1998) Proc SPIE 3333:219; c) Lin Q, Petrillo K, Babich K, LaTulipe D, Medeiros D, Mahorowala A, Simons J, Angelopoulos M, Wallraff G, Larson C, Fenzel-Alexander D, Sooriyakumaran R, Breyta G, Brock P, DiPietro R, Hofer D (1999) Proc SPIE 3678:241
423. a) Bassindale A, Taylor P (1989) In: Patai S, Rappoport Z (eds) The chemistry of organic silicon compounds, p 893; b) Lambert J, Zhao Y (1996) J Am Chem Soc 118:7867
424. Schaedeli U, Tinguely E, Blakeney AJ, Falcigno P, Kunz R (1996) Proc SPIE 2724:344
425. White D, Steinhäusler T, Blakeney AJ, Gabor AH, Beauchemin BT Jr, Deady WR, Jarmalowicz J, Kunz RR, Dean KR, Rich GK, Stark D (1997) Proc 11th International Conference on Photopolymers, p 84
426. Bowden MJ, Gabor AH, Dimov O, Medina AN, Foster P, Steinhäusler T, Biafore JJ, Spaziano G, Slater SG, Blakeney AJ, Neisser MO, Houlihan FM, Cirelli RA, Dabbagh G, Hutton RS, Rushkin IL, Sweeney JR, Timko AG, Nalamasu O, Reichmanis E (1999) J Photopolym Sci Technol 12:423
427. Goethals AM, Jaenen P, Pollers I, Van Roey F, Ronse K, Heskamp B, Davies G (1999) J Photopolym Sci Technol 12:445
428. Kang Y-J, Lee H, Kim E-R, Choi S-J, Park C-G (1997) J Photopolym Sci Technol 10:585
429. a) Sooriyakumaran R, Fenzel-Alexander D, Brock PJ, Larson CE, DiPietro RA, Wallraff GM, Hofer DC, Dawson DJ, Mahorowala AP, Angelopoulos M (2000) Proc SPIE 3999:1171; b) Kwong R, Varanasi PR, Lawson M, Hughes T, Jordhamo G, Khojasteh M, Mahorowala A, Sooriyakumaran Brock P, Larson C, Fenzel-Alexander D, Truong H, Allen R (2001) Proc SPIE 4345:50
430. Hatakeyama J, Kaneko I, Nagura S, Ishihara T (1997) Digest of Abstracts of 3rd International Conference on 193 nm Lithography, p 53

431. Morisawa T, Matsuzawa N, Mori S, Kaimoto Y, Endo M, Kuhara K, Ohfuji T, Sasago M (1997) J Photopolym Sci Technol 10:589
432. Morisawa T, Matsuzawa NN, Mori S, Kaimoto Y, Endo M, Ohfuji T, Kuhara K, Sasago M (1998) J Photopolym Sci Technol 11:667
433. Namba Y, Takahashi H (1998) J Photopolym Sci Technol 11:663
434. Fujigaya T, Shibasaki Y, Ueda M, Kishimura S, Endo M, Sasago M (2001) In: Ito H, Khojasteh M, Li W (eds) Forefront of lithographic materials research. Society of Plastics Engineers, Mid Hudson Section, Hopewell Junction, NY, p 219
435. Sooriyakumaran R, Fenzel-Alexander D, Fender N, Wallraff GM, Allen RD (2001) Proc SPIE 4345:319
436. Hung RJ, Yamachika M, Chiba T, Iwasawa H, Hayashi A, Yamahara N, Shimokawa T (2002) J Photopolym Sci Technol 15:693
437. Shinozuka T, Kawakami T, Okamura H, Tsunooka M, Shirai M (2002) J Photopolym Sci Technol 15:629
438. Palmateer SC, Cann SG, Curtin JE, Doran SP, Eriksen LM, Forte AR, Kunz RR, Lyszczarz TM, Stern MB, Nelson C (1998) Proc SPIE 3333:634
439. Kunz RR, Downs DK (1999) J Vac Sci Technol 17(6):3330
440. Steinman A (1988) Proc SPIE 920:13
441. Meyer WH, Curtis BJ, Brunner HR (1983) Microelectronic Eng 1:29
442. a) Shirai M, Miwa T, Tsunooka M (1993) J Photopolym Sci Technol 6:27; b) Shirai M, Miwa T, Tsunooka M (1995) Chem Mater 7:642
443. a) MacDonald SA, Ito H, Hiraoka H, Willson CG (1985) Technical Papers of SPE Regional Conference on Photopolymers, Society of Plastics Engineers, Brookfield, CT, p 177; b) Willson CG, MacDonald SA, Ito H, Fréchet JMJ (1990) In: Tabata Y, Mita I, Nonogaki S, Horie K, Tagawa S (eds) Polymers for microelectronics – science and technology. VHC Publishers, Deerfield Beach, FL, p 3; c) MacDonald SA, Schlosser H, Ito H, Clecak NJ, Willson CG (1991) Chem Mater 3:435
444. MacDonald SS, Schlosser H, Clecak NJ, Willson CG (1992) Chem Mater 4:1364
445. Postnikov SV, Somervell MH, Henderson CL, Katz S, Willson CG, Byers J, Qin A, Lin Q (1998) Proc SPIE 3333:997
446. Koh C-W, Baik K-H (2000) J Photopolym Sci Technol 13:539
447. a) Kaimoto Y, Mori S, Matsuzawa N, Kuhara K (1997) Digest of Abstracts of the 3rd International Symposium on 193 nm Lithography, p 93; b) Kaimoto Y, Mori S, Matsuzawa N, Kuhara K, Sasago M (1998) J Photopolym Sci Technol 11:633
448. a) Somervell MH, Fryer DS, Osborn B, Patterson K, Cho S, Byers J, Willson CG (2000) Proc SPIE 3999:270; b) Somervell M, Fryer D, Osborn B, Patterson K, Byers J, Willson CG (2000) J Vac Sci Tech B18(5):2251; c) Jamieson A, Somervell M, Tran HV, Hung R, MacDoanld SA, Willson CG (2001) Proc SPIE 4345:406
449. Coopmans F, Roland B (1986) Proc SPIE 631:34
450. Goethals AM, Baik KH, Van den hove L, Tedesco S (1991) Proc SPIE 1466:604
451. a) Pierrat C, Tedesco S, Vinet F, Lerme M, Dal'Zotto B (1989) J Vac Sci Technol B7(6):1782; b) Pierrat C, Bono H, Vinet F, Mourier T, Chevallier M, Guibert JC (1990) Proc SPIE 1262:244
452. a) Hartney MA, Kunz RR, Ehrlich DJ, Shaver DC (1990) Proc SPIE 1262:119; b) Hartney MA, Johnson DW, Spencer AC (1991) Proc SPIE 1466:238; c) Johnson DW, Hartney MA (1992) Jpn J Appl Phys 31:4321
453. Palmateer SC, Kunz RR, Horn MW, Forte AR, Rothschild M (1995) Proc SPIE 2438:455
454. Schellekens JPW, Visser RJ (1989) Proc SPIE 1086:220
455. a) Hutchinson JM, Das S, Zhang G, Pawloski AR (1997) Digest of Abstracts of the 3rd International Symposium on 193 nm Lithography, p 35; b) Hutchinson J, Rao V, Zhang

G, Pawloski A, Fonseca C, Chambers J, Holl S, Das S, Henderson C, Wheeler D (1998) Proc SPIE 3333:165
456. Oizumi H, Yamashita Y, Ogawa T, Soga T, Yamanaka R (1995) Jpn J Appl Phys 34: 6734
457. Yang B-JL, Yang M, Chiong KN (1989) J Vac Sci Technol B7(6):1729
458. Maeda M, Ohfuji T, Aizaki N, Hasegawa E (1995) Proc SPIE 2438:465
459. Matsuo T, Endo M (1996) J Photopolym Sci Technol 9:523
460. Satou I, Kuhara K, Endo M, Morimoto H (1999) Proc SPIE 3678:251
461. a) Satou I, Kuhara K, Endo M, Morimoto H (1999) J Vac Sci Technol B17:3326; b) Satou I, Kuhara K, Endo M, Morimoto H (1999) Jpn J Appl Phys 38:7008
462. Watanabe H, Satou I, Itani T (2000) J Photopolym Sci Technol 13:545
463. McColgin W, Daly RC, Jech J Jr, Brust TB (1988) Proc SPIE 920:260
464. Shaw JM, Hatzakis M, Babich ED, Paraszczak JR, Whitman DF, Stewart KJ (1989) J Vac Sci Technol B7(6):1709
465. Watanabe M, Watanabe H, Satou I, Itani T (2001) J Photopolym Sci Technol 14:643
466. a) Sebald M, Sezi R, Leuschner R, Ahne H, Birkle S (1990) Microelectronic Eng 11:531; b) Sezi R, Sebald M, Leuschner R, Ahne H, Birkle S, Borndörfer H (1990) Proc SPIE 1262:84; c) Sebald M, Leuschner R, Sezi R, Ahne H, Birkle S (1990) Proc SPIE 1262:528
467. Sebald M, Berthold J, Beyer M, Leuschner R, Nölscher C, Scheler U, Sezi R, Ahne H, Birkle S (1991) Proc SPIE 1466:227
468. a) Hien S, Czech G, Domke W-D, Raske H, Sebald M, Stiebert I (1998) Proc SPIE 3333:154; b) Hien S, Czech G, Domke W-D, Richter E, Sebald M, Stiebert I (1999) J Photopolym Sci Technol 12:673
469. a) Sugita K, Ikagawa M, Harada K, Kushida M, Saito K (2000) Jpn J Appl Phys 39:669; b) Sugita K, Ikagawa M, Ming LC, Yamashita M, Harada K, Kushida M, Sato K (2000) J Photopolym Sci Technol 13:535; c) Sugita K, Ikagawa M, Ming LC, Yamashita M, Harada K, Kushida M, Saito K (2001) In: Ito H, Khojasteh M, Li W (eds) Forefront of lithographic materials research. Society of Plastics Engineers, Mid Hudson Section, Hopewell Junction, NY, p 203
470. Venkatesan T, Taylor GN, Wagner A, Wilkens B, Barr D (1981) J Vac Sci Technol 19:1379
471. Taylor GN, Stillwagon LE, Venkatesan T (1984) J Electrochem Soc 131:1664
472. Stillwagon LE, Silverman PJ, Taylor GN (1985) Technical Papers of SPE Regional Technical Conference on Photopolymers, Society of Plastics Engineers, Ridgefield, CT, p 87
473. Nalamasu O, Baiocchi FA, Taylor GN (1989) In: Reichmanis E, MacDonald SA, Iwayanagi T (eds) Polymers in microlithography. ACS Symposium Series 412, American Chemical Society, Washington, D.C., p 189
474. a) Hult A, MacDonald SA, Willson CG (1985) Macromolecules 18:1804; b) Hult A, Ito H, MacDonald SA, Willson CG (1985) US Patent 4,551,418 (11/05/85)
475. a) Brodsky CJ, Trinque BC, Johnson HF, Willson CG (2001) Proc SPIE 4343:415; b) Brodsky CJ, Johnson HF, Trinque BC, Willson CG (2001) In: Ito H, Khojasteh M, Li W (eds) Forefront of lithographic materials research. Society of Plastics Engineers, Mid Hudson Section, Hopewell Junction, NY, p 187
476. Shirai M, Endo M, Tsunooka M, Endo M (1999) J Photopolym Sci Technol 12:669
477. Shirai M, Nakanishi J, Tsunooka M, Matsuo T, Endo M (1998) J Photopolym Sci Technol 11:641
478. Matsuo T, Endo M, Mori S, Kuhara K, Sasago M, Shirai M, Tsunooka M (1998) J Photopolym Sci Technol 11:645
479. Shirai M, Nakaseko H, Tsunooka M (2000) J Photopolym Sci Technol 13:531
480. Paniez PJ, Brun S, Derrough S (1997) Proc SPIE 3049:168
481. Keddie JL, Jones RAL, Cory RA (1994) Europhys Lett 27:59

482. Dutcher JR, Forrest JA, Dalnoki-Veress K (1998) Abstr Pap Am Cem Soc 215:276
483. DeMaggio GB, Frieze WE, Gidley DW, Zhu M, Hristov HA, Yee AF (1997) Phys Rev Lett 78:1524
484. Forrest JA, Svanberg C, Revesz K, Rodahl M, Torell LM, Kasemo B (1998) Phys Rev E 58:R1226
485. Satomi N, Takahara A, Kajiyama T (1999) Macromolecules 32:4474
486. Wallace WE, Vanzanten JH, Wu WL (1995) Phys Rev E 52:R3329
487. a) Soles CL, Lin EK, Wu W-L, Lin Q, Angelopoulos M (2001) In: Ito H, Khojasteh M, Li W (eds) Forefront of lithographic materials research. Society of Plastics Engineers, Mid Hudson Section, Hopewell Junction, NY, p 263; b) Soles CL, Lin EK, Lenhart JL, Jones RL, Wu W-L, Goldfarb DL, Angelopoulos M (2001) J Vac Sci Technol B19(6):2690
488. Fryer DS, Nealey PF, de Pablo JJ (2000) Macromolecules 33:6439
489. Reichmanis E, Galvin ME, Uhrich KE, Mirau P, Heffner SA (1994) In: Ito H, Tagawa S, Horie K (eds) Polymeric materials for microelectronic applications. ACS Symposium Series 579, American Chemical Society, Washington, D.C., p 52
490. Ito H, Fenzel-Alexander D, Breyta G (1997) J Photopolym Sci Technol 10:397
491. Ito H, Hinsberg WD, Rhodes LF, Chang C (2003) Proc SPIE 5039:70
492. a) Ito H, Miller DC (2004) J Polym Sci Part A Polym Chem 42:1468; b) Ito H, Okazaki M, Miller DC (2004) J Polym Sci Part A Polym Chem 42:1478; c) Ito H, Okazaki M, Miller DC (2004) J Polym Sci Part A Polym Chem 42:1506
493. Wallraff GM, Bangsaruntip S, Fender N, Hinsberg W, Houle F, Brock P, Hoffnagle J, Sanchez M, Larson CE, Breyta G, Chau W (2001) In: Ito H, Khojasteh M, Li W (eds) Forefront of lithographic materials research. Society of Plastics Engineers, Mid Hudson Section, Hopewell Junction, NY, p 375
494. Medeiros DR, Mahorowala AP, Huang W-S, Kwong R W, Lang RN, Moreau WM, Angelopoulos M (2001) In: Ito H, Khojasteh M, Li W (eds) Forefront of lithographic materials research. Society of Plastics Engineers, Mid Hudson Section, p 17
495. Uhrich KE, Reichmanis E, Baiocchi FA (1994) Chem Mater 6:295
496. a) Lenhart JL, Jones RL, Lin EK, Soles CL, Wu W, Fischer DA, Sambasivan S (2002) J Vac Sci Technol B20:2920; b) Jablonski EL, Lenhart JL, Sambasivan S, Fischer DA, Jones RL, Lin EK, Wu W-L, Goldfarb D, Temple K, Angelopoulos K, Ito H (2003) AIP Conf Proc 683, Characterization and Metrology for ULSI Technology, p 439; c) Jablonski EL, Prabhu VM, Sambasivan S, Fischer DA, Lin EK, Goldfarb DM, Temple K, Angelopoulos K, Ito H (2003) J Vac Sci Technol B21(6):3162
497. Toriumi M, Masuhara H (1993) In: Reichmanis E, Frank CW, O'Donnell JH (eds) Irradiation of polymeric materials, ACS Symposium Series 527, American Chemical Society, Washington, D.C., p 167
498. a) Jones RL, Kumar SK, Ho DL, Briber RM, Russell TP (2001) Macromolecules 34:559; b) Jones RL, Kumar SK, Ho DL, Briber RM, Russell TP (1999) Nature 400:146
499. a) Lin EK, Soles CL, Goldfarb DL, Trinque BC, Burns SD, Jones RL, Lenhart JL, Angelopoulos M, Willson CG, Satija SK, Wu W-L (2002) Proc SPIE 4690:313; b) Lin EK, Soles CL, Goldfarb DL, Trinque BC, Burns SD, Jones RL, Lenhart JL, Angelopoulos M, Willson CG, Satija SK, Wu W-L (2002) Science 297:372
500. Rao V, Hinsberg WD, Frank CW, Pease RFW (1993) Proc SPIE 1925:538
501. Buttry D, Ward M (1992) Chem Rev 92:1355
502. Shirai M, Shinozuka T, Tsunooka M, Ishikawa S, Itani T (2003) Proc SPIE 5039:113
503. Hinsberg W, Willson CG, Kanazawa K (1986) J Electrochem Soc 133:1448
504. Hilfiker JN, Singh B, Synowicki RA, Bungay CL (2000) Proc SPIE 3998:390
505. Ito T, Okazaki S (2000) Nature 406:1027
506. Hayashida T et al. (1987) Proc SPIE 772:66

507. Umbach CP, Broers AN, Willscn CG, Koch R, Laibowitz RB (1988) J Vac Sci Technol B6(1):319
508. Houle FA, Hinsberg WD, Morrison M, Sanchez MI, Wallraff G, Larson C, Hoffnagle J (2000) J Vac Sci Technol B18(4):1874
509. a) Hinsberg W, Houle F, Wallraff G, Sanchez M, Morrison M, Hoffnagle J, Ito H, Nguyen C, Larson CE, Brock PJ, Breyta G (1999) J Photopolym Sci Technol 12:649; b) Hinsberg W, Houle F, Sanchez M, Morrison M, Wallraff G, Larson C, Hoffnagle J, Brock P, Breyta G (2000) Proc SPIE 3999:148; c) Sanchez MI, Hinsberg WD, Houle FA, Morrison M, Wallraff GM, Larson C, Hoffnagle JA, Brock PJ, Breyta G (2001) In: Ito, H, Khojasteh M, Li W (eds) Forefront of lithographic materials research. Society of Plastics Engineers, Mid Hudson Section, Hopewell Junction, NY, p 283
510. Hinsberg W, Houle F, Sanchez M, Hoffnagle J, Wallraff G, Medeiros D, Gallatin G, Cobb J (2003) Proc SPIE 5039:1
511. Houlihan F, Romano A, Rentkiewicz D, Sakamuri R, Dammel RR, Conley W, Rich G, Miller D, Rhodes L, McDaniels J, Chang C (2003) J Photopolym Sci Technol 16:581
512. Hojo T, Sato M, Komano H (2003) J Photopolym Sci Technol 16:455
513. Kikuchi Y, Fukuda T, Yanazawa H (2003) J Photopolym Sci Technol 16:369
514. Oizumi H et al. (2000) Digest of International Microprocesses and Nanotechnology Conference, p 48
515. Wallraff G, Hutchinscn J, Hinsberg W, Houle F, Seidel P, Johnson R, Oldham W (1994) J Vac Sci Technol B12(6):3857
516. The Chemical Kinetics Simulator (CKS) program package is available for a no-cost license from IBM at http://www.almaden.ibm.com/st/msim/
517. a) Sanchez MI, Hinsberg WD, Houle FA, Hoffnagle JA, Ito H, Nguyen C (1999) Proc SPIE 3678:104; b) Sanchez MI, Hinsberg WD, Houle FA, Hoffnagle JA, Ito H, Nguyen C (1999) Microlithography World 8(2):19
518. Hoffnagle JA, Hinsberg WD, Sanchez M, Houle FA (1999) J Vac Sci Technol B17(6):3306
519. Solak H, He D, Li W, Cerrina F, Sohn B, Yang X, Nealey P (1999) Proc SPIE 3676:278
520. Hinsberg W, Hoffnagle J, Houle F, Ito H, Sanchez M, Sherwood M, Wallraff G (2000) Microlithography World 9(2):16
521. Marshall JL, Telfer SJ, Young MA, Lindholm EP, Minns RA, Takiff L (2002) Science 297:1516
522. SEMATECH (2001) International Technical Roadmap for Semiconductors, SEMATECH, Inc, Austin, TX
523. Eigler DM, Schweizer EK (1990) Nature 344:524
524. SEMATECH (2003) 2nd Immersion Lithography Workshop, July 11, 2003, San Jose, CA; CD available from International SEMATECH, Austin, TX
525. SEMATECH (2003) 4th International Symposium on 157 nm Lithography, August 25–28, 2003, Yokohama, Japan; CD available from Selete/International SEMATECH
526. IEEE (2001) Proc IEEE, Special Issue on Limits of Semiconductor Technology 89(3)

Editor: Karel Dušek
Received: February 2004

Author Index Volumes 101–172

Author Index Volumes 1–100 see Volume 100

de, Abajo, J. and *de la Campa, J. G.*: Processable Aromatic Polyimides.Vol. 140, pp. 23–60.
Abetz, V. see Förster, S.: Vol. 166, pp. 173–210.
Adolf, D. B. see Ediger, M. D.: Vol. 116, pp. 73–110.
Aharoni, S. M. and *Edwards, S. F.*: Rigid Polymer Networks.Vol. 118, pp. 1–231.
Albertsson, A.-C., Varma, I. K.: Aliphatic Polyesters: Synthesis, Properties and Applications. Vol. 157, pp. 99–138.
Albertsson, A.-C. see Edlund, U.: Vol. 157, pp. 53–98.
Albertsson, A.-C. see Söderqvist Lindblad, M.: Vol. 157, pp. 139–161.
Albertsson, A.-C. see Stridsberg, K. M.: Vol. 157, pp. 27–51.
Albertsson, A.-C. see Al-Malaika, S.: Vol. 169, pp. 177–199.
Al-Malaika, S.: Perspectives in Stabilisation of Polyolefins. Vol. 169, pp. 121–150.
Améduri, B., Boutevin, B. and *Gramain, P.*: Synthesis of Block Copolymers by Radical Polymerization and Telomerization. Vol. 127, pp. 87–142.
Améduri, B. and *Boutevin, B.*: Synthesis and Properties of Fluorinated Telechelic Monodispersed Compounds. Vol. 102, pp. 133–170.
Amselem, S. see Domb, A. J.: Vol. 107, pp. 93–142.
Andrady, A. L.: Wavelenght Sensitivity in Polymer Photodegradation. Vol. 128, pp. 47–94.
Andreis, M. and *Koenig, J. L.*: Application of Nitrogen–15 NMR to Polymers.Vol. 124, pp. 191–238.
Angiolini, L. see Carlini, C.: Vol. 123, pp. 127–214.
Anjum, N. see Gupta, B.: Vol. 162, pp. 37–63.
Anseth, K. S., Newman, S. M. and *Bowman, C. N.*: Polymeric Dental Composites: Properties and Reaction Behavior of Multimethacrylate Dental Restorations. Vol. 122, pp. 177–218.
Antonietti, M. see Cölfen, H.: Vol. 150, pp. 67–187.
Armitage, B. A. see O'Brien, D. F.: Vol. 126, pp. 53–58.
Arndt, M. see Kaminski, W.: Vol. 127, pp. 143–187.
Arnold Jr., F. E. and *Arnold, F. E.*: Rigid-Rod Polymers and Molecular Composites.Vol. 117, pp. 257–296.
Arora, M. see Kumar, M. N. V. R.: Vol. 160, pp. 45–118.
Arshady, R.: Polymer Synthesis via Activated Esters: A New Dimension of Creativity in Macromolecular Chemistry. Vol. 111, pp. 1–42.

Bahar, I., Erman, B. and *Monnerie, L.*: Effect of Molecular Structure on Local Chain Dynamics: Analytical Approaches and Computational Methods. Vol. 116, pp. 145–206.
Ballauff, M. see Dingenouts, N.: Vol. 144, pp. 1–48.
Ballauff, M. see Holm, C.: Vol. 166, pp. 1–27.
Ballauff, M. see Rühe, J.: Vol. 165, pp. 79–150.

Baltá-Calleja, F. J., González Arche, A., Ezquerra, T. A., Santa Cruz, C., Batallón, F., Frick, B. and López Cabarcos, E.: Structure and Properties of Ferroelectric Copolymers of Poly(vinylidene) Fluoride. Vol. 108, pp. 1–48.
Barnes, M. D. see Otaigbe, J.U.: Vol. 154, pp. 1–86.
Barshtein, G. R. and Sabsai, O. Y.: Compositions with Mineralorganic Fillers.Vol. 101, pp. 1–28.
Baschnagel, J., Binder, K., Doruker, P., Gusev, A. A., Hahn, O., Kremer, K., Mattice, W. L., Müller-Plathe, F., Murat, M., Paul, W., Santos, S., Sutter, U. W., Tries, V.: Bridging the Gap Between Atomistic and Coarse-Grained Models of Polymers: Status and Perspectives. Vol. 152, pp. 41–156.
Batallán, F. see Baltá-Calleja, F. J.: Vol. 108, pp. 1–48.
Batog, A. E., Pet'ko, I.P., Penczek, P.: Aliphatic-Cycloaliphatic Epoxy Compounds and Polymers. Vol. 144, pp. 49–114.
Barton, J. see Hunkeler, D.: Vol. 112, pp. 115–134.
Bell, C. L. and Peppas, N. A.: Biomedical Membranes from Hydrogels and Interpolymer Complexes. Vol. 122, pp. 125–176.
Bellon-Maurel, A. see Calmon-Decriaud, A.: Vol. 135, pp. 207–226.
Bennett, D. E. see O'Brien, D. F.: Vol. 126, pp. 53–84.
Berry, G. C.: Static and Dynamic Light Scattering on Moderately Concentraded Solutions: Isotropic Solutions of Flexible and Rodlike Chains and Nematic Solutions of Rodlike Chains. Vol. 114, pp. 233–290.
Bershtein, V. A. and Ryzhov, V. A.: Far Infrared Spectroscopy of Polymers. Vol. 114, pp. 43–122.
Bhargava R., Wang S.-Q., Koenig J. L: FTIR Microspectroscopy of Polymeric Systems. Vol. 163, pp. 137–191.
Biesalski, M.: see Rühe, J.: Vol. 165, pp. 79–150.
Bigg, D. M.: Thermal Conductivity of Heterophase Polymer Compositions.Vol. 119, pp. 1–30.
Binder, K.: Phase Transitions in Polymer Blends and Block Copolymer Melts: Some Recent Developments. Vol. 112, pp. 115–134.
Binder, K.: Phase Transitions of Polymer Blends and Block Copolymer Melts in Thin Films. Vol. 138, pp. 1–90.
Binder, K. see Baschnagel, J.: Vol. 152, pp. 41–156.
Bird, R. B. see Curtiss, C. F.: Vol. 125, pp. 1–102.
Biswas, M. and Mukherjee, A.: Synthesis and Evaluation of Metal-Containing Polymers. Vol. 115, pp. 89–124.
Biswas, M. and Sinha Ray, S.: Recent Progress in Synthesis and Evaluation of Polymer-Montmorillonite Nanocomposites. Vol. 155, pp. 167–221.
Bogdal, D., Penczek, P., Pielichowski, J., Prociak, A.: Microwave Assisted Synthesis, Crosslinking, and Processing of Polymeric Materials. Vol. 163, pp. 193–263.
Bohrisch, J., Eisenbach, C.D., Jaeger, W., Mori H., Müller A.H.E., Rehahn, M., Schaller, C., Traser, S., Wittmeyer, P.: New Polyelectrolyte Architectures. Vol. 165, pp. 1–41.
Bolze, J. see Dingenouts, N.: Vol. 144, pp. 1–48.
Bosshard, C.: see Gubler, U.: Vol. 158, pp. 123–190.
Boutevin, B. and Robin, J. J.: Synthesis and Properties of Fluorinated Diols. Vol. 102. pp. 105–132.
Boutevin, B. see Amédouri, B.: Vol. 102, pp. 133–170.
Boutevin, B. see Améduri, B.: Vol. 127, pp. 87–142.
Bowman, C. N. see Anseth, K. S.: Vol. 122, pp. 177–218.
Boyd, R. H.: Prediction of Polymer Crystal Structures and Properties. Vol. 116, pp. 1–26.
Briber, R. M. see Hedrick, J. L.: Vol. 141, pp. 1–44.

Bronnikov, S. V., Vettegren, V. I. and *Frenkel, S. Y.*: Kinetics of Deformation and Relaxation in Highly Oriented Polymers. Vol. 125, pp. 103–146.
Brown, H. R. see Creton, C.: Vol. 156, pp. 53–135.
Bruza, K. J. see Kirchhoff, R. A.: Vol. 117, pp. 1–66.
Budkowski, A.: Interfacial Phenomena in Thin Polymer Films: Phase Coexistence and Segregation. Vol. 148, pp. 1–112.
Burban, J. H. see Cussler, E. L.: Vol. 110, pp. 67–80.
Burchard, W.: Solution Properties of Branched Macromolecules. Vol. 143, pp. 113–194.

Calmon-Decriaud, A., Bellon-Maurel, V., Silvestre, F.: Standard Methods for Testing the Aerobic Biodegradation of Polymeric Materials. Vol 135, pp. 207–226.
Cameron, N. R. and *Sherrington, D. C.*: High Internal Phase Emulsions (HIPEs)-Structure, Properties and Use in Polymer Preparation. Vol. 126, pp. 163–214.
de la Campa, J. G. see de Abajo, J.: Vol. 140, pp. 23–60.
Candau, F. see Hunkeler, D.: Vol. 112, pp. 115–134.
Canelas, D. A. and *DeSimone, J. M.*: Polymerizations in Liquid and Supercritical Carbon Dioxide. Vol. 133, pp. 103–140.
Canva, M., Stegeman, G. I.: Quadratic Parametric Interactions in Organic Waveguides. Vol. 158, pp. 87–121.
Capek, I.: Kinetics of the Free-Radical Emulsion Polymerization of Vinyl Chloride. Vol. 120, pp. 135–206.
Capek, I.: Radical Polymerization of Polyoxyethylene Macromonomers in Disperse Systems. Vol. 145, pp. 1–56.
Capek, I.: Radical Polymerization of Polyoxyethylene Macromonomers in Disperse Systems. Vol. 146, pp. 1–56.
Capek, I. and *Chern, C.-S.*: Radical Polymerization in Direct Mini-Emulsion Systems. Vol. 155, pp. 101–166.
Cappella, B. see Munz, M.: Vol. 164, pp. 87–210.
Carlesso, G. see Prokop, A.: Vol. 160, pp. 119–174.
Carlini, C. and *Angiolini, L.*: Polymers as Free Radical Photoinitiators. Vol. 123, pp. 127–214.
Carter, K. R. see Hedrick, J. L.: Vol. 141, pp. 1–44.
Casas-Vazquez, J. see Jou, D.: Vol. 120, pp. 207–266.
Chandrasekhar, V.: Polymer Solid Electrolytes: Synthesis and Structure. Vol 135, pp. 139–206.
Chang, J. Y. see Han, M. J.: Vol. 153, pp. 1–36.
Chang, T.: Recent Advances in Liquid Chromatography Analysis of Synthetic Polymers. Vol. 163, pp. 1–60.
Charleux, B., Faust R.: Synthesis of Branched Polymers by Cationic Polymerization. Vol. 142, pp. 1–70.
Chen, P. see Jaffe, M.: Vol. 117, pp. 297–328.
Chern, C.-S. see Capek, I.: Vol. 155, pp. 101–166.
Chevolot, Y. see Mathieu, H. J.: Vol. 162, pp. 1–35.
Choe, E.-W. see Jaffe, M.: Vol. 117, pp. 297–328.
Chow, T. S.: Glassy State Relaxation and Deformation in Polymers. Vol. 103, pp. 149–190.
Chujo, Y. see Uemura, T.: Vol. 167, pp. 81–106.
Chung, S.-J. see Lin, T.-C.: Vol. 161, pp. 157–193
Chung, T.-S. see Jaffe, M.: Vol. 117, pp. 297–328.
Cölfen, H. and *Antonietti, M.*: Field-Flow Fractionation Techniques for Polymer and Colloid Analysis. Vol. 150, pp. 67–187.
Comanita, B. see Roovers, J.: Vol. 142, pp. 179–228.
Connell, J. W. see Hergenrother, P. M.: Vol. 117, pp. 67–110.

Creton, C., Kramer, E. J., Brown, H. R., Hui, C.-Y.: Adhesion and Fracture of Interfaces Between Immiscible Polymers: From the Molecular to the Continuum Scale. Vol. 156, pp. 53–135.
Criado-Sancho, M. see *Jou, D.*: Vol. 120, pp. 207–266.
Curro, J. G. see *Schweizer, K. S.*: Vol. 116, pp. 319–378.
Curtiss, C. F. and Bird, R. B.: Statistical Mechanics of Transport Phenomena: Polymeric Liquid Mixtures. Vol. 125, pp. 1–102.
Cussler, E. L., Wang, K. L. and Burban, J. H.: Hydrogels as Separation Agents. Vol. 110, pp. 67–80.

Dalton, L. Nonlinear Optical Polymeric Materials: From Chromophore Design to Commercial Applications. Vol. 158, pp. 1–86.
Dautzenberg, H. see *Holm, C.*: Vol. 166, pp.113–171.
Davidson, J. M. see *Prokop, A.*: Vol. 160, pp.119–174.
Desai, S. M., Singh, R. P.: Surface Modification of Polyethylene. Vol. 169, pp. 231–293.
DeSimone, J. M. see *Canelas D. A.*: Vol. 133, pp. 103–140.
DiMari, S. see *Prokop, A.*: Vol. 136, pp. 1–52.
Dimonie, M. V. see *Hunkeler, D.*: Vol. 112, pp. 115–134.
Dingenouts, N., Bolze, J., Pötschke, D., Ballauf, M.: Analysis of Polymer Latexes by Small-Angle X-Ray Scattering. Vol. 144, pp. 1–48.
Dodd, L. R. and Theodorou, D. N.: Atomistic Monte Carlo Simulation and Continuum Mean Field Theory of the Structure and Equation of State Properties of Alkane and Polymer Melts. Vol. 116, pp. 249–282.
Doelker, E.: Cellulose Derivatives. Vol. 107, pp. 199–266.
Dolden, J. G.: Calculation of a Mesogenic Index with Emphasis Upon LC-Polyimides.Vol. 141, pp. 189–245.
Domb, A. J., Amselem, S., Shah, J. and Maniar, M.: Polyanhydrides: Synthesis and Characterization. Vol. 107, pp. 93–142.
Domb, A. J. see *Kumar, M. N. V. R.*: Vol. 160, pp. 45118.
Doruker, P. see *Baschnagel, J.*: Vol. 152, pp. 41–156.
Dubois, P. see *Mecerreyes, D.*: Vol. 147, pp. 1–60.
Dubrovskii, S. A. see *Kazanskii, K. S.*: Vol. 104, pp. 97–134.
Dunkin, I. R. see *Steinke, J.*: Vol. 123, pp. 81–126.
Dunson, D. L. see *McGrath, J. E.*: Vol. 140, pp. 61–106.
Dziezok, P. see *Rühe, J.*: Vol. 165, pp. 79–150.

Eastmond, G. C.: Poly(ε-caprolactone) Blends. Vol. 149, pp. 59–223.
Economy, J. and Goranov, K.: Thermotropic Liquid Crystalline Polymers for High Performance Applications. Vol. 117, pp. 221–256.
Ediger, M. D. and Adolf, D. B.: Brownian Dynamics Simulations of Local Polymer Dynamics. Vol. 116, pp. 73–110.
Edlund, U. Albertsson, A.-C.: Degradable Polymer Microspheres for Controlled Drug Delivery. Vol. 157, pp. 53–98.
Edwards, S. F. see *Aharoni, S. M.*: Vol. 118, pp. 1–231.
Eisenbach, C. D. see *Bohrisch, J.*: Vol. 165, pp. 1–41.
Endo, T. see *Yagci, Y.*: Vol. 127, pp. 59–86.
Engelhardt, H. and Grosche, O.: Capillary Electrophoresis in Polymer Analysis. Vol.150, pp. 189–217.
Engelhardt, H. and Martin, H.: Characterization of Synthetic Polyelectrolytes by Capillary Electrophoretic Methods. Vol. 165, pp. 211–247.
Eriksson, P. see *Jacobson, K.*: Vol. 169, pp. 151–176.

Erman, B. see Bahar, I.: Vol. 116, pp. 145–206.
Eschner, M. see Spange, S.: Vol. 165, pp. 43–78.
Estel, K. see Spange, S.: Vol. 165, pp. 43–78.
Ewen, B. and *Richter, D.*: Neutron Spin Echo Investigations on the Segmental Dynamics of Polymers in Melts, Networks and Solutions. Vol. 134, pp. 1–130.
Ezquerra, T. A. see Baltá-Calleja, F. J.: Vol. 108, pp. 1–48.

Fatkullin, N. see Kimmich, R.: Vol. 170, pp. 1–113.
Faust, R. see Charleux, B.: Vol. 142, pp. 1–70.
Faust, R. see Kwon, Y.: Vol. 167, pp. 107–135.
Fekete, E. see Pukánszky, B.: Vol. 139, pp. 109–154.
Fendler, J. H.: Membrane-Mimetic Approach to Advanced Materials. Vol. 113, pp. 1–209.
Fetters, L. J. see Xu, Z.: Vol. 120, pp. 1–50.
Förster, S., Abetz, V., Müller, A. H. E.: Polyelectrolyte Block Copolymer Micelles. Vol. 166, pp. 173–210.
Förster, S. and *Schmidt, M.*: Polyelectrolytes in Solution. Vol. 120, pp. 51–134.
Freire, J. J.: Conformational Properties of Branched Polymers: Theory and Simulations. Vol. 143, pp. 35–112.
Frenkel, S. Y. see Bronnikov, S.V.: Vol. 125, pp. 103–146.
Frick, B. see Baltá-Calleja, F. J.: Vol. 108, pp. 1–48.
Fridman, M. L.: see Terent'eva, J. P.: Vol. 101, pp. 29–64.
Fukui, K. see Otaigbe, J. U.: Vol. 154, pp. 1–86.
Funke, W.: Microgels-Intramolecularly Crosslinked Macromolecules with a Globular Structure. Vol. 136, pp. 137–232.
Furusho, Y. see Takata, T.: Vol. 171, pp. 1–75.

Galina, H.: Mean-Field Kinetic Modeling of Polymerization: The Smoluchowski Coagulation Equation. Vol. 137, pp. 135–172.
Ganesh, K. see Kishore, K.: Vol. 121, pp. 81–122.
Gaw, K. O. and *Kakimoto, M.*: Polyimide-Epoxy Composites. Vol. 140, pp. 107–136.
Geckeler, K. E. see Rivas, B.: Vol. 102, pp. 171–188.
Geckeler, K. E.: Soluble Polymer Supports for Liquid-Phase Synthesis. Vol. 121, pp. 31–80.
Gedde, U. W., Mattozzi, A.: Polyethylene Morphology. Vol. 169, pp. 29–73.
Gehrke, S. H.: Synthesis, Equilibrium Swelling, Kinetics Permeability and Applications of Environmentally Responsive Gels. Vol. 110, pp. 81–144.
de Gennes, P.-G.: Flexible Polymers in Nanopores. Vol. 138, pp. 91–106.
Georgiou, S.: Laser Cleaning Methodologies of Polymer Substrates. Vol. 168, pp. 1–49.
Geuss, M. see Munz, M.: Vol. 164, pp. 37–210
Giannelis, E. P., Krishnamoorti, R., Manias, E.: Polymer-Silicate Nanocomposites: Model Systems for Confined Polymers and Polymer Brushes. Vol. 138, pp. 107–148.
Godovsky, D. Y.: Device Applications of Polymer-Nanocomposites. Vol. 153, pp. 163–205.
Godovsky, D. Y.: Electron Behavior and Magnetic Properties Polymer-Nanocomposites. Vol. 119, pp. 79–122.
González Arche, A. see Baltá-Calleja, F. J.: Vol. 108, pp. 1–48.
Goranov, K. see Economy, J.: Vol. 117, pp. 221–256.
Gramain, P. see Améduri, B.: Vol. 127, pp. 87–142.
Grest, G. S.: Normal and Shear Forces Between Polymer Brushes. Vol. 138, pp. 149–184.
Grigorescu, G., Kulicke, W.-M.: Prediction of Viscoelastic Properties and Shear Stability of Polymers in Solution. Vol. 152, p. 1–40.
Gröhn, F. see Rühe, J.: Vol. 165, pp. 79–150.
Grosberg, A. and *Nechaev, S.*: Polymer Topology. Vol. 106, pp. 1–30.

Grosche, O. see *Engelhardt, H.:* Vol. 150, pp. 189–217.
Grubbs, R., Risse, W. and *Novac, B.:* The Development of Well-defined Catalysts for Ring-Opening Olefin Metathesis. Vol. 102, pp. 47–72.
Gubler, U., Bosshard, C.: Molecular Design for Third-Order Nonlinear Optics. Vol. 158, pp. 123–190.
van Gunsteren, W. F. see *Gusev, A. A.:* Vol. 116, pp. 207–248.
Gupta, B., Anjum, N.: Plasma and Radiation-Induced Graft Modification of Polymers for Biomedical Applications. Vol. 162, pp. 37–63.
Gusev, A. A., Müller-Plathe, F., van Gunsteren, W. F. and *Suter, U. W.:* Dynamics of Small Molecules in Bulk Polymers. Vol. 116, pp. 207–248.
Gusev, A. A. see *Baschnagel, J.:* Vol. 152, pp. 41–156.
Guillot, J. see *Hunkeler, D.:* Vol. 112, pp. 115–134.
Guyot, A. and *Tauer, K.:* Reactive Surfactants in Emulsion Polymerization. Vol. 111, pp. 43–66.

Hadjichristidis, N., Pispas, S., Pitsikalis, M., Iatrou, H., Vlahos, C.: Asymmetric Star Polymers Synthesis and Properties. Vol. 142, pp. 71–128.
Hadjichristidis, N. see *Xu, Z.:* Vol. 120, pp. 1–50.
Hadjichristidis, N. see *Pitsikalis, M.:* Vol. 135, pp. 1–138.
Hahn, O. see *Baschnagel, J.:* Vol. 152, pp. 41–156.
Hakkarainen, M.: Aliphatic Polyesters: Abiotic and Biotic Degradation and Degradation Products. Vol. 157, pp. 1–26.
Hakkarainen, M., Albertsson, A.-C.: Environmental Degradation of Polyethylene. Vol. 169, pp. 177–199.
Hall, H. K. see *Penelle, J.:* Vol. 102, pp. 73–104.
Hamley, I. W.: Crystallization in Block Copolymers. Vol. 148, pp. 113–138.
Hammouda, B.: SANS from Homogeneous Polymer Mixtures: A Unified Overview. Vol. 106, pp. 87–134.
Han, M. J. and *Chang, J. Y.:* Polynucleotide Analogues. Vol. 153, pp. 1–36.
Harada, A.: Design and Construction of Supramolecular Architectures Consisting of Cyclodextrins and Polymers. Vol. 133, pp. 141–192.
Haralson, M. A. see *Prokop, A.:* Vol. 136, pp. 1–52.
Hassan, C. M. and *Peppas, N. A.:* Structure and Applications of Poly(vinyl alcohol) Hydrogels Produced by Conventional Crosslinking or by Freezing/Thawing Methods. Vol. 153, pp. 37–65.
Hawker, C. J.: Dentritic and Hyperbranched Macromolecules Precisely Controlled Macromolecular Architectures. Vol. 147, pp. 113–160.
Hawker, C. J. see *Hedrick, J. L.:* Vol. 141, pp. 1–44.
He, G. S. see *Lin, T.-C.:* Vol. 161, pp. 157–193.
Hedrick, J. L., Carter, K. R., Labadie, J. W., Miller, R. D., Volksen, W., Hawker, C. J., Yoon, D. Y., Russell, T. P., McGrath, J. E., Briber, R. M.: Nanoporous Polyimides. Vol. 141, pp. 1–44.
Hedrick, J. L., Labadie, J. W., Volksen, W. and *Hilborn, J. G.:* Nanoscopically Engineered Polyimides. Vol. 147, pp. 61–112.
Hedrick, J. L. see *Hergenrother, P. M.:* Vol. 117, pp. 67–110.
Hedrick, J. L. see *Kiefer, J.:* Vol. 147, pp. 161–247.
Hedrick, J. L. see *McGrath, J. E.:* Vol. 140, pp. 61–106.
Heinrich, G. and *Klüppel, M.:* Recent Advances in the Theory of Filler Networking in Elastomers. Vol. 160, pp. 1–44.
Heller, J.: Poly (Ortho Esters). Vol. 107, pp. 41–92.
Helm, C. A.: see *Möhwald, H.:* Vol. 165, pp. 151–175.

Hemielec, A. A. see Hunkeler, D.: Vol. 112, pp. 115–134.
Hergenrother, P. M., Connell, J. W., Labadie, J. W. and *Hedrick, J. L.*: Poly(arylene ether)s Containing Heterocyclic Units. Vol. 117, pp. 67–110.
Hernández-Barajas, J. see Wandrey, C.: Vol. 145, pp. 123–182.
Hervet, H. see Léger, L.: Vol. 138, pp 185–226.
Hilborn, J. G. see Hedrick, J. L.: Vol. 147, pp. 61–112.
Hilborn, J. G. see Kiefer, J.: Vol. 147, pp. 161–247.
Hiramatsu, N. see Matsushige, M.: Vol. 125, pp. 147–186.
Hirasa, O. see Suzuki, M.: Vol. 110, pp. 241–262.
Hirotsu, S.: Coexistence of Phases and the Nature of First-Order Transition in Poly-N-isopropylacrylamide Gels. Vol. 110, pp. 1–26.
Höcker, H. see Klee, D.: Vol. 149, pp. 1–57.
Holm, C., Hofmann, T., Joanny, J. F., Kremer, K., Netz, R. R., Reineker, P., Seidel, C., Vilgis, T. A., Winkler, R. G.: Polyelectrolyte Theory. Vol. 166, pp. 67–111.
Holm, C., Rehahn, M., Oppermann, W., Ballauff, M.: Stiff-Chain Polyelectrolytes. Vol. 166, pp. 1–27.
Hornsby, P.: Rheology, Compounding and Processing of Filled Thermoplastics. Vol. 139, pp. 155–216.
Houbenov, N. see Rühe, J.: Vol. 165, pp. 79–150.
Huber, K. see Volk, N.: Vol. 166, pp. 29–65.
Hugenberg, N. see Rühe, J.: Vol. 165, pp. 79–150.
Hui, C.-Y. see Creton, C.: Vol. 156, pp. 53–135.
Hult, A., Johansson, M., Malmström, E.: Hyperbranched Polymers. Vol. 143, pp. 1–34.
Hunkeler, D., Candau, F., Pichot, C., Hemielec, A. E., Xie, T. Y., Barton, J., Vaskova, V., Guillot, J., Dimonie, M. V., Reichert, K. H.: Heterophase Polymerization: A Physical and Kinetic Comparision and Categorization. Vol. 112, pp. 115–134.
Hunkeler, D. see Macko, T.: Vol. 163, pp. 61–136.
Hunkeler, D. see Prokop, A.: Vol. 136, pp. 1–52; 53–74.
Hunkeler, D. see Wandrey, C.: Vol. 145, pp. 123–182.

Iatrou, H. see Hadjichristidis, N.: Vol. 142, pp. 71–128.
Ichikawa, T. see Yoshida, H.: Vol. 105, pp. 3–36.
Ihara, E. see Yasuda, H.: Vol. 133, pp. 53–102.
Ikada, Y. see Uyama,Y.: Vol. 137, pp. 1–40.
Ikehara, T. see Jinnuai, H.: Vol. 170, pp. 115–167.
Ilavsky, M.: Effect on Phase Transition on Swelling and Mechanical Behavior of Synthetic Hydrogels. Vol. 109, pp. 173–206.
Imai, Y.: Rapid Synthesis of Polyimides from Nylon-Salt Monomers. Vol. 140, pp. 1–23.
Inomata, H. see Saito, S.: Vol. 106, pp. 207–232.
Inoue, S. see Sugimoto, H.: Vol. 146, pp. 39–120.
Irie, M.: Stimuli-Responsive Poly(N-isopropylacrylamide), Photo- and Chemical-Induced Phase Transitions. Vol. 110, pp. 49–66.
Ise, N. see Matsuoka, H.: Vol. 114, pp. 187–232.
Ito, H.: Chemical Amplification Resists for Microlithography. Vol. 172, pp. 37–245.
Ito, K., Kawaguchi, S.: Poly(macronomers), Homo- and Copolymerization. Vol. 142, pp. 129–178.
Ito, Y. see Suginome, M.: Vol. 171, pp. 77–136.
Ivanov, A. E. see Zubov, V. P.: Vol. 104, pp. 135–176.

Jacob, S. and *Kennedy, J.*: Synthesis, Characterization and Properties of OCTA-ARM Polyisobutylene-Based Star Polymers. Vol. 146, pp. 1–38.

Jacobson, K., Eriksson, P., Reitberger, T., Stenberg, B.: Chemiluminescence as a Tool for Polyolefin. Vol. 169, pp. 151–176.
Jaeger, W. see Bohrisch, J.: Vol. 165, pp. 1–41.
Jaffe, M., Chen, P., Choe, E.-W., Chung, T.-S. and *Makhija, S.*: High Performance Polymer Blends. Vol. 117, pp. 297–328.
Jancar, J.: Structure-Property Relationships in Thermoplastic Matrices. Vol. 139, pp. 1–66.
Jen, A. K-Y. see Kajzar, F.: Vol. 161, pp. 1–85.
Jerome, R. see Mecerreyes, D.: Vol. 147, pp. 1–60.
Jiang, M., Li, M., Xiang, M. and *Zhou, H.*: Interpolymer Complexation and Miscibility and Enhancement by Hydrogen Bonding. Vol. 146, pp. 121–194.
Jin, J. see Shim, H.-K.: Vol. 158, pp. 191–241.
Jinnai, H., Nishikawa, Y., Ikehara, T. and *Nishi, T.*: Emerging Technologies for the 3D Analysis of Polymer Structures. Vol. 170, pp. 115–167.
Jo, W. H. and *Yang, J. S.*: Molecular Simulation Approaches for Multiphase Polymer Systems. Vol. 156, pp. 1–52.
Joanny, J.-F. see Holm, C.: Vol. 166, pp. 67–111.
Joanny, J.-F. see Thünemann, A. F.: Vol. 166, pp. 113–171.
Johannsmann, D. see Rühe, J.: Vol. 165, pp. 79–150.
Johansson, M. see Hult, A.: Vol. 143, pp. 1–34.
Joos-Müller, B. see Funke, W.: Vol. 136, pp. 137–232.
Jou, D., Casas-Vazquez, J. and *Criado-Sancho, M.*: Thermodynamics of Polymer Solutions under Flow: Phase Separation and Polymer Degradation. Vol. 120, pp. 207–266.

Kaetsu, I.: Radiation Synthesis of Polymeric Materials for Biomedical and Biochemical Applications. Vol. 105, pp. 81–98.
Kaji, K. see Kanaya, T.: Vol. 154, pp. 87–141.
Kajzar, F., Lee, K.-S., Jen, A. K.-Y.: Polymeric Materials and their Orientation Techniques for Second-Order Nonlinear Optics. Vol. 161, pp. 1–85.
Kakimoto, M. see Gaw, K. O.: Vol. 140, pp. 107–136.
Kaminski, W. and *Arndt, M.*: Metallocenes for Polymer Catalysis. Vol. 127, pp. 143–187.
Kammer, H. W., Kressler, H. and *Kummerloewe, C.*: Phase Behavior of Polymer Blends – Effects of Thermodynamics and Rheology. Vol. 106, pp. 31–86.
Kanaya, T. and *Kaji, K.*: Dynamcis in the Glassy State and Near the Glass Transition of Amorphous Polymers as Studied by Neutron Scattering. Vol. 154, pp. 87–141.
Kandyrin, L. B. and *Kuleznev, V. N.*: The Dependence of Viscosity on the Composition of Concentrated Dispersions and the Free Volume Concept of Disperse Systems. Vol. 103, pp. 103–148.
Kaneko, M. see Ramaraj, R.: Vol. 123, pp. 215–242.
Kang, E. T., Neoh, K. G. and *Tan, K. L.*: X-Ray Photoelectron Spectroscopic Studies of Electroactive Polymers. Vol. 106, pp. 135–190.
Karlsson, S. see Söderqvist Lindblad, M.: Vol. 157, pp. 139–161.
Karlsson, S.: Recycled Polyolefins. Material Properties and Means for Quality Determination. Vol. 169, pp. 201–229.
Kato, K. see Uyama,Y.: Vol. 137, pp. 1–40.
Kautek, W. see Krüger, J.: Vol. 168, pp. 247–290.
Kawaguchi, S. see Ito, K.: Vol. 142, p 129–178.
Kawata, S. see Sun, H.-B.: Vol. 170, pp. 169–273.
Kazanskii, K. S. and *Dubrovskii, S. A.*: Chemistry and Physics of Agricultural Hydrogels. Vol. 104, pp. 97–134.
Kennedy, J. P. see Jacob, S.: Vol. 146, pp. 1–38.
Kennedy, J. P. see Majoros, I.: Vol. 112, pp. 1–113.

Khokhlov, A., Starodybtzev, S. and *Vasilevskaya, V.*: Conformational Transitions of Polymer Gels: Theory and Experiment. Vol. 109, pp. 121–172.
Kiefer, J., Hedrick J. L. and *Hiborn, J. G.*: Macroporous Thermosets by Chemically Induced Phase Separation. Vol. 147, pp. 161–247.
Kihara, N. see Takata, T.: Vol. 171, pp. 1–75.
Kilian, H. G. and *Pieper, T.*: Packing of Chain Segments. A Method for Describing X-Ray Patterns of Crystalline, Liquid Crystalline and Non-Crystalline Polymers. Vol. 108, pp. 49–90.
Kim, J. see Quirk, R. P.: Vol. 153, pp. 67–162.
Kim, K.-S. see Lin, T.-C.: Vol. 161, pp. 157–193.
Kimmich, R., Fatkullin, N.: Polymer Chain Dynamics and NMR. Vol. 170, pp. 1–113.
Kippelen, B. and *Peyghambarian, N.*: Photorefractive Polymers and their Applications. Vol. 161, pp. 87–156.
Kirchhoff, R. A. and *Bruza, K. J.*: Polymers from Benzocyclobutenes. Vol. 117, pp. 1–66.
Kishore, K. and *Ganesh, K.*: Polymers Containing Disulfide, Tetrasulfide, Diselenide and Ditelluride Linkages in the Main Chain. Vol. 121, pp. 81–122.
Kitamaru, R.: Phase Structure of Polyethylene and Other Crystalline Polymers by Solid-State 13C/MNR. Vol. 137, pp 41–102.
Klee, D. and *Höcker, H.*: Polymers for Biomedical Applications: Improvement of the Interface Compatibility. Vol. 149, pp. 1–57.
Klier, J. see Scranton, A. B.: Vol. 122, pp. 1–54.
v. Klitzing, R. and *Tieke, B.*: Polyelectrolyte Membranes. Vol. 165, pp. 177–210.
Klüppel, M.: The Role of Disorder in Filler Reinforcement of Elastomers on Various Length Scales. Vol. 164, pp. 1–86
Klüppel, M. see Heinrich, G.: Vol. 160, pp 1–44.
Knuuttila, H., Lehtinen, A., Nummila-Pakarinen, A.: Advanced Polyethylene Technologies – Controlled Material Properties. Vol. 169, pp. 13–27.
Kobayashi, S., Shoda, S. and *Uyama, H.*: Enzymatic Polymerization and Oligomerization. Vol. 121, pp. 1–30.
Köhler, W. and *Schäfer, R.*: Polymer Analysis by Thermal-Diffusion Forced Rayleigh Scattering. Vol. 151, pp. 1–59.
Koenig, J. L. see Bhargava, R.: Vol. 163, pp. 137–191.
Koenig, J. L. see Andreis, M.: Vol. 124, pp. 191–238.
Koike, T.: Viscoelastic Behavior of Epoxy Resins Before Crosslinking. Vol. 148, pp. 139–188.
Kokko, E. see Löfgren, B.: Vol. 169, pp. 1–12.
Kokufuta, E.: Novel Applications for Stimulus-Sensitive Polymer Gels in the Preparation of Functional Immobilized Biocatalysts. Vol. 110, pp. 157–178.
Konno, M. see Saito, S.: Vol. 109, pp. 207–232.
Konradi, R. see Rühe, J.: Vol. 165, pp. 79–150.
Kopecek, J. see Putnam, D.: Vol. 122, pp. 55–124.
Koßmehl, G. see Schopf, G.: Vol. 129, pp. 1–145.
Kozlov, E. see Prokop, A.: Vol. 160, pp. 119–174.
Kramer, E. J. see Creton, C.: Vol. 156, pp. 53–135.
Kremer, K. see Baschnagel, J.: Vol. 152, pp. 41–156.
Kremer, K. see Holm, C.: Vol. 166, pp. 67–111.
Kressler, J. see Kammer, H. W.: Vol. 106, pp. 31–86.
Kricheldorf, H. R.: Liquid-Cristalline Polyimides. Vol. 141, pp. 83–188.
Krishnamoorti, R. see Giannelis, E. P.: Vol. 138, pp. 107–148.
Krüger, J. and *Kautek, W.*: Ultrashort Pulse Laser Interaction with Dielectrics and Polymers, Vol. 168, pp. 247–290.

Kuchanov, S. I.: Modern Aspects of Quantitative Theory of Free-Radical Copolymerization. Vol. 103, pp. 1–102.
Kuchanov, S. I.: Principles of Quantitive Description of Chemical Structure of Synthetic Polymers. Vol. 152, p. 157–202.
Kudaibergennow, S. E.: Recent Advances in Studying of Synthetic Polyampholytes in Solutions. Vol. 144, pp. 115–198.
Kuleznev, V. N. see *Kandyrin, L. B.*: Vol. 103, pp. 103–148.
Kulichkhin, S. G. see *Malkin, A. Y.*: Vol. 101, pp. 217–258.
Kulicke, W.-M. see *Grigorescu, G.*: Vol. 152, p. 1–40.
Kumar, M. N. V. R., Kumar, N., Domb, A. J. and *Arora, M.*: Pharmaceutical Polymeric Controlled Drug Delivery Systems. Vol. 160, pp. 45–118.
Kumar, N. see *Kumar M. N. V. R.*: Vol. 160, pp. 45–118.
Kummerloewe, C. see *Kammer, H. W.*: Vol. 106, pp. 31–86.
Kuznetsova, N. P. see *Samsonov, G. V.*: Vol. 104, pp. 1–50.
Kwon, Y. and *Faust, R.*: Synthesis of Polyisobutylene-Based Block Copolymers with Precisely Controlled Architecture by Living Cationic Polymerization. Vol. 167, pp. 107–135.

Labadie, J. W. see *Hergenrother, P. M.*: Vol. 117, pp. 67–110.
Labadie, J. W. see *Hedrick, J. L.*: Vol. 141, pp. 1–44.
Labadie, J. W. see *Hedrick, J. L.*: Vol. 147, pp. 61–112.
Lamparski, H. G. see *O'Brien, D. F.*: Vol. 126, pp. 53–84.
Laschewsky, A.: Molecular Concepts, Self-Organisation and Properties of Polysoaps. Vol. 124, pp. 1–86.
Laso, M. see *Leontidis, E.*: Vol. 116, pp. 283–318.
Lazár, M. and *Rychlý, R.*: Oxidation of Hydrocarbon Polymers. Vol. 102, pp. 189–222.
Lechowicz, J. see *Galina, H.*: Vol. 137, pp. 135–172.
Léger, L., Raphaël, E., Hervet, H.: Surface-Anchored Polymer Chains: Their Role in Adhesion and Friction. Vol. 138, pp. 185–226.
Lenz, R. W.: Biodegradable Polymers. Vol. 107, pp. 1–40.
Leontidis, E., de Pablo, J. J., Laso, M. and *Suter, U. W.*: A Critical Evaluation of Novel Algorithms for the Off-Lattice Monte Carlo Simulation of Condensed Polymer Phases. Vol. 116, pp. 283–318.
Lee, B. see *Quirk, R. P.*: Vol. 153, pp. 67–162.
Lee, K.-S. see *Kajzar, F.*: Vol. 161, pp. 1–85.
Lee, Y. see *Quirk, R. P*: Vol. 153, pp. 67–162.
Lehtinen, A. see *Knuuttila, H.*: Vol. 169, pp. 13–27.
Leónard, D. see *Mathieu, H. J.*: Vol. 162, pp. 1–35.
Lesec, J. see *Viovy, J.-L.*: Vol. 114, pp. 1–42.
Li, M. see *Jiang, M.*: Vol. 146, pp. 121–194.
Liang, G. L. see *Sumpter, B. G.*: Vol. 116, pp. 27–72.
Lienert, K.-W.: Poly(ester-imide)s for Industrial Use. Vol. 141, pp. 45–82.
Lin, J. and *Sherrington, D. C.*: Recent Developments in the Synthesis, Thermostability and Liquid Crystal Properties of Aromatic Polyamides. Vol. 111, pp. 177–220.
Lin, T.-C., Chung, S.-J., Kim, K.-S., Wang, X., He, G. S., Swiatkiewicz, J., Pudavar, H. E. and *Prasad, P. N.*: Organics and Polymers with High Two-Photon Activities and their Applications. Vol. 161, pp. 157–193.
Lippert, T.: Laser Application of Polymers. Vol. 168, pp. 51–246.
Liu, Y. see *Söderqvist Lindblad, M.*: Vol. 157, pp. 139–161
López Cabarcos, E. see *Baltá-Calleja, F. J.*: Vol. 108, pp. 1–48.

Löfgren, B., Kokko, E., Seppälä, J.: Specific Structures Enabled by Metallocene Catalysis in Polyethenes. Vol. 169, pp. 1–12.
Löwen, H. see Thünemann, A. F.: Vol. 166, pp. 113–171.

Macko, T. and Hunkeler, D.: Liquid Chromatography under Critical and Limiting Conditions: A Survey of Experimental Systems for Synthetic Polymers. Vol. 163, pp. 61–136.
Majoros, I., Nagy, A. and Kennedy, J. P.: Conventional and Living Carbocationic Polymerizations United. I. A Comprehensive Model and New Diagnostic Method to Probe the Mechanism of Homopolymerizations. Vol. 112, pp. 1–113.
Makhija, S. see Jaffe, M.: Vol. 117, pp. 297–328.
Malmström, E. see Hult, A.: Vol. 143, pp. 1–34.
Malkin, A. Y. and Kulichkhin, S. G.: Rheokinetics of Curing. Vol. 101, pp. 217–258.
Maniar, M. see Domb, A. J.: Vol. 107, pp. 93–142.
Manias, E. see Giannelis, E. P.: Vol. 138, pp. 107–148.
Martin, H. see Engelhardt, H.: Vol. 165, pp. 211–247.
Marty, J. D. and Mauzac, M.: Molecular Imprinting: State of the Art and Perspectives. Vol. 172, pp. 1–35.
Mashima, K., Nakayama, Y. and Nakamura, A.: Recent Trends in Polymerization of a-Olefins Catalyzed by Organometallic Complexes of Early Transition Metals.Vol. 133, pp. 1–52.
Mathew, D. see Reghunadhan Nair, C.P.: Vol. 155, pp. 1–99.
Mathieu, H. J., Chevolot, Y, Ruiz-Taylor, L. and Leónard, D.: Engineering and Characterization of Polymer Surfaces for Biomedical Applications. Vol. 162, pp. 1–35.
Matsumoto, A.: Free-Radical Crosslinking Polymerization and Copolymerization of Multivinyl Compounds. Vol. 123, pp. 41–80.
Matsumoto, A. see Otsu, T.: Vol. 136, pp. 75–138.
Matsuoka, H. and Ise, N.: Small-Angle and Ultra-Small Angle Scattering Study of the Ordered Structure in Polyelectrolyte Solutions and Colloidal Dispersions. Vol. 114, pp. 187–232.
Matsushige, K., Hiramatsu, N. and Okabe, H.: Ultrasonic Spectroscopy for Polymeric Materials. Vol. 125, pp. 147–186.
Mattice, W. L. see Rehahn, M.: Vol. 131/132, pp. 1–475.
Mattice, W. L. see Baschnagel, J.: Vol. 152, pp. 41–156.
Mattozzi, A. see Gedde, U. W.: Vol. 169, pp. 29–73.
Mauzac, M. see Marty, J. D.: Vol. 172, pp. 1–35.
Mays, W. see Xu, Z.: Vol. 120, pp. 1–50.
Mays, J. W. see Pitsikalis, M.: Vol. 135, pp. 1–138.
McGrath, J. E. see Hedrick, J. L.: Vol. 141, pp. 1–44.
McGrath, J. E., Dunson, D. L., Hedrick, J. L.: Synthesis and Characterization of Segmented Polyimide-Polyorganosiloxane Copolymers. Vol. 140, pp. 61–106.
McLeish, T. C. B., Milner, S. T.: Entangled Dynamics and Melt Flow of Branched Polymers. Vol. 143, pp. 195–256.
Mecerreyes, D., Dubois, P. and Jerome, R.: Novel Macromolecular Architectures Based on Aliphatic Polyesters: Relevance of the Coordination-Insertion Ring-Opening Polymerization. Vol. 147, pp. 1–60.
Mecham, S. J. see McGrath, J. E.: Vol. 140, pp. 61–106.
Menzel, H. see Möhwald, H.: Vol. 165, pp. 151–175.
Meyer, T. see Spange, S.: Vol. 165, pp. 43–78.
Mikos, A. G. see Thomson, R. C.: Vol. 122, pp. 245–274.
Milner, S. T. see McLeish, T. C. B.: Vol. 143, pp. 195–256.
Mison, P. and Sillion, B.: Thermosetting Oligomers Containing Maleimides and Nadiimides End-Groups. Vol. 140, pp. 137–180.

Miyasaka, K.: PVA-Iodine Complexes: Formation, Structure and Properties. Vol. 108, pp. 91–130.
Miller, R. D. see Hedrick, J. L.: Vol. 141, pp. 1–44.
Minko, S. see Rühe, J.: Vol. 165, pp. 79–150.
Möhwald, H., Menzel, H., Helm, C. A., Stamm, M.: Lipid and Polyampholyte Monolayers to Study Polyelectrolyte Interactions and Structure at Interfaces. Vol. 165, pp. 151–175.
Monnerie, L. see Bahar, I.: Vol. 116, pp. 145–206.
Mori, H. see Bohrisch, J.: Vol. 165, pp. 1–41.
Morishima, Y.: Photoinduced Electron Transfer in Amphiphilic Polyelectrolyte Systems. Vol. 104, pp. 51–96.
Morton M. see Quirk, R. P: Vol. 153, pp. 67–162.
Motornov, M. see Rühe, J.: Vol. 165, pp. 79–150.
Mours, M. see Winter, H. H.: Vol. 134, pp. 165–234.
Müllen, K. see Scherf, U.: Vol. 123, pp. 1–40.
Müller, A. H. E. see Bohrisch, J.: Vol. 165, pp. 1–41.
Müller, A. H. E. see Förster, S.: Vol. 166, pp. 173–210.
Müller, M. see Thünemann, A. F.: Vol. 166, pp. 113–171.
Müller-Plathe, F. see Gusev, A. A.: Vol. 116, pp. 207–248.
Müller-Plathe, F. see Baschnagel, J.: Vol. 152, p. 41–156.
Mukerherjee, A. see Biswas, M.: Vol. 115, pp. 89–124.
Munz, M., Cappella, B., Sturm, H., Geuss, M., Schulz, E.: Materials Contrasts and Nanolithography Techniques in Scanning Force Microscopy (SFM) and their Application to Polymers and Polymer Composites. Vol. 164, pp. 87–210
Murat, M. see Baschnagel, J.: Vol. 152, p. 41–156.
Mylnikov, V.: Photoconducting Polymers. Vol. 115, pp. 1–88.

Nagy, A. see Majoros, I.: Vol. 112, pp. 1–11.
Naka, K. see Uemura, T.: Vol. 167, pp. 81–106.
Nakamura, A. see Mashima, K.: Vol. 133, pp. 1–52.
Nakayama, Y. see Mashima, K.: Vol. 133, pp. 1–52.
Narasinham, B., Peppas, N. A.: The Physics of Polymer Dissolution: Modeling Approaches and Experimental Behavior. Vol. 128, pp. 157–208.
Nechaev, S. see Grosberg, A.: Vol. 106, pp. 1–30.
Neoh, K. G. see Kang, E. T.: Vol. 106, pp. 135–190.
Netz, R.R. see Holm, C.: Vol. 166, pp. 67–111.
Netz, R.R. see Rühe, J.: Vol. 165, pp. 79–150.
Newman, S. M. see Anseth, K. S.: Vol. 122, pp. 177–218.
Nijenhuis, K. te: Thermoreversible Networks. Vol. 130, pp. 1–252.
Ninan, K. N. see Reghunadhan Nair, C.P.: Vol. 155, pp. 1–99.
Nishi, T. see Jinnai, H.: Vol. 170, pp. 115–167.
Nishikawa, Y. see Jinnai, H.: Vol. 170, pp. 115–167.
Noid, D. W. see Otaigbe, J. U.: Vol. 154, pp. 1–86.
Noid, D. W. see Sumpter, B. G.: Vol. 116, pp. 27–72.
Novac, B. see Grubbs, R.: Vol. 102, pp. 47–72.
Novikov, V. V. see Privalko, V. P.: Vol. 119, pp. 31–78.
Nummila-Pakarinen, A. see Knuuttila, H.: Vol. 169, pp. 13–27.

O'Brien, D. F., Armitage, B. A., Bennett, D. E. and *Lamparski, H. G.*: Polymerization and Domain Formation in Lipid Assemblies. Vol. 126, pp. 53–84.
Ogasawara, M.: Application of Pulse Radiolysis to the Study of Polymers and Polymerizations. Vol.105, pp. 37–80.

Okabe, H. see Matsushige, K.: Vol. 125, pp. 147–186.
Okada, M.: Ring-Opening Polymerization of Bicyclic and Spiro Compounds. Reactivities and Polymerization Mechanisms. Vol. 102, pp. 1–46.
Okano, T.: Molecular Design of Temperature-Responsive Polymers as Intelligent Materials. Vol. 110, pp. 179–198.
Okay, O. see Funke, W.: Vol. 136, pp. 137–232.
Onuki, A.: Theory of Phase Transition in Polymer Gels. Vol. 109, pp. 63–120.
Oppermann, W. see Holm, C.: Vol. 166, pp. 1–27.
Oppermann, W. see Volk, N.: Vol. 166, pp. 29–65.
Osad'ko, I. S.: Selective Spectroscopy of Chromophore Doped Polymers and Glasses. Vol. 114, pp. 123–186.
Osakada, K., Takeuchi, D.: Coordination Polymerization of Dienes, Allenes, and Methylenecycloalkanes. Vol. 171, pp. 137–194.
Otaigbe, J. U., Barnes, M. D., Fukui, K., Sumpter, B. G., Noid, D. W.: Generation, Characterization, and Modeling of Polymer Micro- and Nano-Particles. Vol. 154, pp. 1–86.
Otsu, T. and Matsumoto, A.: Controlled Synthesis of Polymers Using the Iniferter Technique: Developments in Living Radical Polymerization. Vol. 136, pp. 75–138.

de Pablo, J. J. see Leontidis, E.: Vol. 116, pp. 283–318.
Padias, A. B. see Penelle, J.: Vol. 102, pp. 73–104.
Pascault, J.-P. see Williams, R. J. J.: Vol. 128, pp. 95–156.
Pasch, H.: Analysis of Complex Polymers by Interaction Chromatography. Vol. 128, pp. 1–46.
Pasch, H.: Hyphenated Techniques in Liquid Chromatography of Polymers. Vol. 150, pp. 1–66.
Paul, W. see Baschnagel, J.: Vol. 152, p. 41–156.
Penczek, P. see Batog, A. E.: Vol. 144, pp. 49–114.
Penczek, P. see Bogdal, D.: Vol. 163, pp. 193–263.
Penelle, J., Hall, H. K., Padias, A. B. and Tanaka, H.: Captodative Olefins in Polymer Chemistry. Vol. 102, pp. 73–104.
Peppas, N. A. see Bell, C. L.: Vol. 122, pp. 125–176.
Peppas, N. A. see Hassan, C. M.: Vol. 153, pp. 37–65
Peppas, N. A. see Narasimhan, B.: Vol. 128, pp. 157–208.
Pet'ko, I. P. see Batog, A. E.: Vol. 144, pp. 49–114.
Pheyghambarian, N. see Kippelen, B.: Vol. 161, pp. 87–156.
Pichot, C. see Hunkeler, D.: Vol. 112, pp. 115–134.
Pielichowski, J. see Bogdal, D.: Vol. 163, pp. 193–263.
Pieper, T. see Kilian, H. G.: Vol. 108, pp. 49–90.
Pispas, S. see Pitsikalis, M.: Vol. 135, pp. 1–138.
Pispas, S. see Hadjichristidis, N.: Vol. 142, pp. 71–128.
Pitsikalis, M., Pispas, S., Mays, J. W., Hadjichristidis, N.: Nonlinear Block Copolymer Architectures. Vol. 135, pp. 1–138.
Pitsikalis, M. see Hadjichristidis, N.: Vol. 142, pp. 71–128.
Pleul, D. see Spange, S.: Vol. 165, pp. 43–78.
Plummer, C. J. G.: Microdeformation and Fracture in Bulk Polyolefins. Vol. 169, pp. 75–119.
Pötschke, D. see Dingenouts, N.: Vol. 144, pp. 1–48.
Pokrovskii, V. N.: The Mesoscopic Theory of the Slow Relaxation of Linear Macromolecules. Vol. 154, pp. 143–219.
Pospíšil, J.: Functionalized Oligomers and Polymers as Stabilizers for Conventional Polymers. Vol. 101, pp. 65–168.

Pospíśil, J.: Aromatic and Heterocyclic Amines in Polymer Stabilization. Vol. 124, pp. 87–190.
Powers, A. C. see Prokop, A.: Vol. 136, pp. 53–74.
Prasad, P. N. see Lin, T.-C.: Vol. 161, pp. 157–193.
Priddy, D. B.: Recent Advances in Styrene Polymerization. Vol. 111, pp. 67–114.
Priddy, D. B.: Thermal Discoloration Chemistry of Styrene-co-Acrylonitrile. Vol. 121, pp. 123–154.
Privalko, V. P. and *Novikov, V. V.*: Model Treatments of the Heat Conductivity of Heterogeneous Polymers. Vol. 119, pp 31–78.
Prociak, A. see Bogdal, D.: Vol. 163, pp. 193–263
Prokop, A., Hunkeler, D., DiMari, S., Haralson, M. A., Wang, T. G.: Water Soluble Polymers for Immunoisolation I: Complex Coacervation and Cytotoxicity. Vol. 136, pp. 1–52.
Prokop, A., Hunkeler, D., Powers, A. C., Whitesell, R. R., Wang, T. G.: Water Soluble Polymers for Immunoisolation II: Evaluation of Multicomponent Microencapsulation Systems. Vol. 136, pp. 53–74.
Prokop, A., Kozlov, E., Carlesso, G. and *Davidsen, J. M.*: Hydrogel-Based Colloidal Polymeric System for Protein and Drug Delivery: Physical and Chemical Characterization, Permeability Control and Applications. Vol. 160, pp. 119–174.
Pruitt, L. A.: The Effects of Radiation on the Structural and Mechanical Properties of Medical Polymers. Vol. 162, pp. 65–95.
Pudavar, H. E. see Lin, T.-C.: Vol. 161, pp. 157–193.
Pukánszky, B. and *Fekete, E.*: Adhesion and Surface Modification. Vol. 139, pp. 109–154.
Putnam, D. and *Kopecek, J.*: Polymer Conjugates with Anticancer Acitivity. Vol. 122, pp. 55–124.

Quirk, R. P., Yoo, T., Lee, Y., M., Kim, J. and *Lee, B.*: Applications of 1,1-Diphenylethylene Chemistry in Anionic Synthesis of Polymers with Controlled Structures. Vol. 153, pp. 67–162.

Ramaraj, R. and *Kaneko, M.*: Metal Complex in Polymer Membrane as a Model for Photosynthetic Oxygen Evolving Center. Vol. 123, pp. 215–242.
Rangarajan, B. see Scranton, A. B.: Vol. 122, pp. 1–54.
Ranucci, E. see Söderqvist Lindblad, M.: Vol. 157, pp. 139–161.
Raphaël, E. see Léger, L.: Vol. 138, pp. 185–226.
Reddinger, J. L. and *Reynolds, J. R.*: Molecular Engineering of p-Conjugated Polymers. Vol. 145, pp. 57–122.
Reghunadhan Nair, C. P., Mathew, D. and *Ninan, K. N.*: Cyanate Ester Resins, Recent Developments. Vol. 155, pp. 1–99.
Reichert, K. H. see Hunkeler, D.: Vol. 112, pp. 115–134.
Rehahn, M., Mattice, W. L., Suter, U. W.: Rotational Isomeric State Models in Macromolecular Systems. Vol. 131/132, pp. 1–475.
Rehahn, M. see Bohrisch, J.: Vol. 165, pp. 1–41.
Rehahn, M. see Holm, C.: Vol. 166, pp. 1–27.
Reineker, P. see Holm, C.: Vol. 166, pp. 67–111.
Reitberger, T. see Jacobson, K.: Vol. 169, pp. 151–176.
Reynolds, J. R. see Reddinger, J. L.: Vol. 145, pp. 57–122.
Richter, D. see Ewen, B.: Vol. 134, pp. 1–130.
Risse, W. see Grubbs, R.: Vol. 102, pp. 47–72.
Rivas, B. L. and *Geckeler, K. E.*: Synthesis and Metal Complexation of Poly(ethyleneimine) and Derivatives. Vol. 102, pp. 171–188.
Robin, J.J.: The Use of Ozone in the Synthesis of New Polymers and the Modification of Polymers. Vol. 167, pp. 35–79.

Robin, J. J. see Boutevin, B.: Vol. 102, pp. 105–132.
Roe, R.-J.: MD Simulation Study of Glass Transition and Short Time Dynamics in Polymer Liquids. Vol. 116, pp. 111–114.
Roovers, J., Comanita, B.: Dendrimers and Dendrimer-Polymer Hybrids. Vol. 142, pp 179–228.
Rothon, R. N.: Mineral Fillers in Thermoplastics: Filler Manufacture and Characterisation. Vol. 139, pp. 67–108.
Rozenberg, B. A. see Williams, R. J. J.: Vol. 128, pp. 95–156.
Rühe, J., Ballauff, M., Biesalski, M., Dziezok, P., Gröhn, F., Johannsmann, D., Houbenov, N., Hugenberg, N., Konradi, R., Minko, S., Motornov, M., Netz, R. R., Schmidt, M., Seidel, C., Stamm, M., Stephan, T., Usov, D. and *Zhang, H.*: Polyelectrolyte Brushes. Vol. 165, pp. 79–150.
Ruckenstein, E.: Concentrated Emulsion Polymerization. Vol. 127, pp. 1–58.
Ruiz-Taylor, L. see Mathieu, H. J.: Vol. 162, pp. 1–35.
Rusanov, A. L.: Novel Bis (Naphtalic Anhydrides) and Their Polyheteroarylenes with Improved Processability. Vol. 111, pp. 115–176.
Russel, T. P. see Hedrick, J. L.: Vol. 141, pp. 1–44.
Rychly, J. see Lazár, M.: Vol. 102, pp 189–222.
Ryner, M. see Stridsberg, K. M.: Vol. 157, pp. 2751.
Ryzhov, V. A. see Bershtein, V. A.: Vol. 114, pp. 43–122.

Sabsai, O. Y. see Barshtein, G. R.: Vol. 101, pp. 1–28.
Saburov, V. V. see Zubov, V. P.: Vol. 104, pp. 135–176.
Saito, S., Konno, M. and *Inomata, H.*: Volume Phase Transition of N-Alkylacrylamide Gels. Vol. 109, pp. 207–232.
Samsonov, G. V. and *Kuznetsova, N. P.*: Crosslinked Polyelectrolytes in Biology. Vol. 104, pp. 1–50.
Santa Cruz, C. see Baltá-Calleja, F. J.: Vol. 108, pp. 1–48.
Santos, S. see Baschnagel, J.: Vol. 152, p. 41–156.
Sato, T. and *Teramoto, A.*: Concentrated Solutions of Liquid-Christalline Polymers. Vol. 126, pp. 85–162.
Schaller, C. see Bohrisch, J.: Vol. 165, pp. 1–41.
Schäfer R. see Köhler, W.: Vol. 151, pp. 1–59.
Scherf, U. and *Müllen, K.*: The Synthesis of Ladder Polymers. Vol. 123, pp. 1–40.
Schmidt, M. see Förster, S.: Vol. 120, pp. 51–134.
Schmidt, M. see Rühe, J.: Vol. 165, pp. 79–150.
Schmidt, M. see Volk, N.: Vol. 166, pp. 29–65.
Scholz, M.: Effects of Ion Radiation on Cells and Tissues. Vol. 162, pp. 97–158.
Schopf, G. and *Koßmehl, G.*: Polythiophenes – Electrically Conductive Polymers. Vol. 129, pp. 1–145.
Schulz, E. see Munz, M.: Vol. 164, pp. 97–210.
Seppälä, J. see Löfgren, B.: Vol. 169, pp. 1–12.
Sturm, H. see Munz, M.: Vol. 164, pp. 87–210.
Schweizer, K. S.: Prism Theory of the Structure, Thermodynamics, and Phase Transitions of Polymer Liquids and Alloys. Vol. 116, pp. 319–378.
Scranton, A. B., Rangarajan, B. and *Klier, J.*: Biomedical Applications of Polyelectrolytes. Vol. 122, pp. 1–54.
Sefton, M. V. and *Stevenson, W. T. K.*: Microencapsulation of Live Animal Cells Using Polycrylates. Vol. 107, pp. 143–198.
Seidel, C. see Holm, C.: Vol. 166, pp 67–111.
Seidel, C. see Rühe, J.: Vol. 165, pp. 79–150.

Shamanin, V. V.: Bases of the Axiomatic Theory of Addition Polymerization. Vol. 112, pp. 135–180.
Sheiko, S. S.: Imaging of Polymers Using Scanning Force Microscopy: From Superstructures to Individual Molecules. Vol. 151, pp. 61–174.
Sherrington, D. C. see Cameron, N. R.,Vol. 126, pp. 163–214.
Sherrington, D. C. see Lin, J.: Vol. 111, pp. 177–220.
Sherrington, D. C. see Steinke, J.: Vol. 123, pp. 81–126.
Shibayama, M. see Tanaka, T.: Vol. 109, pp. 1–62.
Shiga, T.: Deformation and Viscoelastic Behavior of Polymer Gels in Electric Fields. Vol. 134, pp. 131–164.
Shim, H.-K., Jin, J.: Light-Emitting Characteristics of Conjugated Polymers. Vol. 158, pp. 191–241.
Shoda, S. see Kobayashi, S.: Vol. 121, pp. 1–30.
Siegel, R. A.: Hydrophobic Weak Polyelectrolyte Gels: Studies of Swelling Equilibria and Kinetics. Vol. 109, pp. 233–268.
Silvestre, F. see Calmon-Decriaud, A.: Vol. 207, pp. 207–226.
Sillion, B. see Mison, P.: Vol. 140, pp. 137–180.
Simon, F. see Spange, S.: Vol. 165, pp. 43–78.
Singh, R. P. see Sivaram, S.: Vol. 101, pp. 169–216.
Singh, R. P. see Desai, S. M.: Vol. 169, pp. 231–293.
Sinha Ray, S. see Biswas, M: Vol. 155, pp. 167–221.
Sivaram, S. and *Singh, R. P.*: Degradation and Stabilization of Ethylene-Propylene Copolymers and Their Blends: A Critical Review. Vol. 101, pp. 169–216.
Söderqvist Lindblad, M., Liu, Y., Albertsson, A.-C., Ranucci, E., Karlsson, S.: Polymer from Renewable Resources.Vol. 157, pp. 139–161
Spange, S., Meyer, T., Voigt, I., Eschner, M., Estel, K., Pleul, D. and *Simon, F.*: Poly(Vinylformamide-co-Vinylamine)/Inorganic Oxid Hybrid Materials. Vol. 165, pp. 43–78.
Stamm, M. see Möhwald, H.: Vol. 165, pp. 151–175.
Stamm, M. see Rühe, J.: Vol. 165, pp. 79–150.
Starodybtzev, S. see Khokhlov, A.: Vol. 109, pp. 121–172.
Stegeman, G. I. see Canva, M.: Vol. 158, pp. 87–121.
Steinke, J., Sherrington, D. C. and *Dunkin, I. R.*: Imprinting of Synthetic Polymers Using Molecular Templates. Vol. 123, pp. 81–126.
Stenberg, B. see Jacobson, K.: Vol. 169, pp. 151–176.
Stenzenberger, H. D.: Addition Polyimides. Vol. 117, pp. 165–220.
Stephan, T. see Rühe, J.: Vol. 165, pp. 79–150.
Stevenson,W. T. K. see Sefton, M. V.: Vol. 107, pp. 143–198.
Stridsberg, K. M., Ryner, M., Albertsson, A.-C.: Controlled Ring-Opening Polymerization: Polymers with Designed Macromoleculars Architecture. Vol. 157, pp. 27–51.
Sturm, H. see Munz, M.: Vol. 164, pp. 87–210.
Suematsu, K.: Recent Progress of Gel Theory: Ring, Excluded Volume, and Dimension. Vol. 156, pp. 136–214.
Sugimoto, H. and *Inoue, S.*: Polymerization by Metalloporphyrin and Related Complexes. Vol. 146, pp. 39–120.
Suginome, M., Ito, Y.: Transition Metal-Mediated Polymerization of Isocyanides. Vol. 171, pp. 77–136.
Sumpter, B. G., Noid, D. W., Liang, G. L. and *Wunderlich, B.*: Atomistic Dynamics of Macromolecular Crystals. Vol. 116, pp. 27–72.
Sumpter, B. G. see Otaigbe, J. U.: Vol. 154, pp. 1–86.
Sun, H.-B., Kawata, S.: Two-Photon Photopolymerization and 3D Lithographic Microfabrication. Vol. 170, pp. 169–273.

Suter, U. W. see Gusev, A. A.: Vol. 116, pp. 207–248.
Suter, U. W. see Leontidis, E.: Vol. 115, pp. 283–318.
Suter, U. W. see Rehahn, M.: Vol. 131/132, pp. 1–475.
Suter, U. W. see Baschnagel, J.: Vol. 152, p. 41–156.
Suzuki, A.: Phase Transition in Gels of Sub-Millimeter Size Induced by Interaction with Stimuli. Vol. 110, pp. 199–240.
Suzuki, A. and *Hirasa, O.*: An Approach to Artifical Muscle by Polymer Gels due to Micro-Phase Separation. Vol. 110, pp. 241–262.
Swiatkiewicz, J. see Lin, T.-C.: Vol. 161, pp. 157–193.

Tagawa, S.: Radiation Effects on Ion Beams on Polymers. Vol. 105, pp. 99–116.
Takata, T., Kihara, N., Furusho, Y.: Polyrotaxanes and Polycatenanes: Recent Advances in Syntheses and Applications of Polymers Comprising of Interlocked Structures. Vol. 171, pp. 1–75.
Takeuchi, D. see Osakada, K.: Vol. 171, pp. 137–194.
Tan, K. L. see Kang, E. T.: Vol. 106, pp. 135–190.
Tanaka, H. and *Shibayama, M.*: Phase Transition and Related Phenomena of Polymer Gels. Vol. 109, pp. 1–62.
Tanaka, T. see Penelle, J.: Vol. 102, pp. 73–104.
Tauer, K. see Guyot, A.: Vol. 111, pp. 43–66.
Teramoto, A. see Sato, T.: Vol. 126, pp. 85–162.
Terent'eva, J. P. and *Fridman, M. L.*: Compositions Based on Aminoresins. Vol. 101, pp. 29–64.
Theodorou, D. N. see Dodd, L. R.: Vol. 116, pp. 249–282.
Thomson, R. C., Wake, M. C., Yaszemski, M. J. and *Mikos, A. G.*: Biodegradable Polymer Scaffolds to Regenerate Organs. Vol. 122, pp. 245–274.
Thünemann, A. F., Müller, M., Dautzenberg, H., Joanny, J.-F., Löwen, H.: Polyelectrolyte complexes. Vol. 166, pp. 113–171.
Tieke, B. see v. Klitzing, R.: Vol. 165, pp. 177–210.
Tokita, M.: Friction Between Polymer Networks of Gels and Solvent. Vol. 110, pp. 27–48.
Traser, S. see Bohrisch, J.: Vol. 165, pp. 1–41.
Tries, V. see Baschnagel, J.: Vol. 152, p. 41–156.
Tsuruta, T.: Contemporary Topics in Polymeric Materials for Biomedical Applications. Vol. 126, pp. 1–52.

Uemura, T., Naka, K. and *Chujo, Y.*: Functional Macromolecules with Electron-Donating Dithiafulvene Unit. Vol. 167, pp. 81–106.
Usov, D. see Rühe, J.: Vol. 165, pp. 79–150.
Uyama, H. see Kobayashi, S.: Vol. 121, pp. 1–30.
Uyama, Y: Surface Modification of Polymers by Grafting. Vol. 137, pp. 1–40.

Varma, I. K. see Albertsson, A.-C.: Vol. 157, pp. 99–138.
Vasilevskaya, V. see Khokhlov, A.: Vol. 109, pp. 121–172.
Vaskova, V. see Hunkeler, D.: Vol.: 112, pp. 115–134.
Verdugo, P.: Polymer Gel Phase Transition in Condensation-Decondensation of Secretory Products. Vol. 110, pp. 145–156.
Vettegren, V. I. see Bronnikov, S. V.: Vol. 125, pp. 103–146.
Vilgis, T. A. see Holm, C.: Vol. 166, pp. 67–111.
Viovy, J.-L. and *Lesec, J.*: Separation of Macromolecules in Gels: Permeation Chromatography and Electrophoresis. Vol. 114, pp. 1–42.
Vlahos, C. see Hadjichristidis, N.: Vol. 142, pp. 71–128.

Voigt, I. see Spange, S.: Vol. 165, pp. 43–78.
Volk, N., Vollmer, D., Schmidt, M., Oppermann, W., Huber, K.: Conformation and Phase Diagrams of Flexible Polyelectrolytes. Vol. 166, pp. 29–65.
Volksen, W.: Condensation Polyimides: Synthesis, Solution Behavior, and Imidization Characteristics. Vol. 117, pp. 111–164.
Volksen, W. see Hedrick, J. L.: Vol. 141, pp. 1–44.
Volksen, W. see Hedrick, J. L.: Vol. 147, pp. 61–112.
Vollmer, D. see Volk, N.: Vol. 166, pp. 29–65.

Wake, M. C. see Thomson, R. C.: Vol. 122, pp. 245–274.
Wandrey C., Hernández-Barajas, J. and *Hunkeler, D.*: Diallyldimethylammonium Chloride and its Polymers. Vol. 145, pp. 123–182.
Wang, K. L. see Cussler, E. L.: Vol. 110, pp. 67–80.
Wang, S.-Q.: Molecular Transitions and Dynamics at Polymer/Wall Interfaces: Origins of Flow Instabilities and Wall Slip. Vol. 138, pp. 227–276.
Wang, S.-Q. see Bhargava, R.: Vol. 163, pp. 137–191.
Wang, T. G. see Prokop, A.: Vol. 136, pp. 1–52; 53–74.
Wang, X. see Lin, T.-C.: Vol. 161, pp. 157–193.
Webster, O. W.: Group Transfer Polymerization: Mechanism and Comparison with Other Methods of Controlled Polymerization of Acrylic Monomers. Vol. 167, pp. 1–34.
Whitesell, R. R. see Prokop, A.: Vol. 136, pp. 53–74.
Williams, R. J. J., Rozenberg, B. A., Pascault, J.-P.: Reaction Induced Phase Separation in Modified Thermosetting Polymers. Vol. 128, pp. 95–156.
Winkler, R. G. see Holm, C.: Vol. 166, pp. 67–111.
Winter, H. H., Mours, M.: Rheology of Polymers Near Liquid-Solid Transitions. Vol. 134, pp. 165–234.
Wittmeyer, P. see Bohrisch, J.: Vol. 165, pp. 1–41.
Wu, C.: Laser Light Scattering Characterization of Special Intractable Macromolecules in Solution. Vol 137, pp. 103–134.
Wunderlich, B. see Sumpter, B. G.: Vol. 116, pp. 27–72.

Xiang, M. see Jiang, M.: Vol. 146, pp. 121–194.
Xie, T. Y. see Hunkeler, D.: Vol. 112, pp. 115–134.
Xu, Z., Hadjichristidis, N., Fetters, L. J. and *Mays, J. W.*: Structure/Chain-Flexibility Relationships of Polymers. Vol. 120, pp. 1–50.

Yagci, Y. and *Endo, T.*: N-Benzyl and N-Alkoxy Pyridium Salts as Thermal and Photochemical Initiators for Cationic Polymerization. Vol. 127, pp. 59–86.
Yannas, I. V.: Tissue Regeneration Templates Based on Collagen-Glycosaminoglycan Copolymers. Vol. 122, pp. 219–244.
Yang, J. S. see Jo, W. H.: Vol. 156, pp. 1–52.
Yamaoka, H.: Polymer Materials for Fusion Reactors. Vol. 105, pp. 117–144.
Yasuda, H. and *Ihara, E.*: Rare Earth Metal-Initiated Living Polymerizations of Polar and Nonpolar Monomers. Vol. 133, pp. 53–102.
Yaszemski, M. J. see Thomson, R. C.: Vol. 122, pp. 245–274.
Yoo, T. see Quirk, R. P.: Vol. 153, pp. 67–162.
Yoon, D. Y. see Hedrick, J. L.: Vol. 141, pp. 1–44.
Yoshida, H. and *Ichikawa, T.*: Electron Spin Studies of Free Radicals in Irradiated Polymers. Vol. 105, pp. 3–36.

Zhang, H. see Rühe, J.: Vol. 165, pp. 79–150.
Zhang, Y.: Synchrotron Radiation Direct Photo Etching of Polymers. Vol. 168, pp. 291–340.
Zhou, H. see Jiang, M.: Vol. 146, pp. 121–194.
Zubov, V. P., Ivanov, A. E. and *Saburov, V. V.*: Polymer-Coated Adsorbents for the Separation of Biopolymers and Particles. Vol. 104, pp. 135–176.

Subject Index

Ablative photodecomposition 98
Abzymes 3
Acetal 64, 74, 81, 92, 97, 119, 123, 140
Acetal crosslink 76, 165, 170
4-Acetoxystyrene 60, 67, 70, 72, 77, 142, 149, 158, 170
Acid amplified imaging 221
Acid amplifier 53
Acid-breakable resin 76
Acid concentration 205
Acid diffusion 52, 91, 206, 217–221
– –, anisotropic 158
Acid gradient 119
Acid-hardening resist 159, 193, 194
Acid-labile end group 144
Acid sensitive dye 52, 205
Acidolysis, A_{AL}-1 55, 62
Acrylamide 5
Acrylate 5
Activated carbon filtration 90
Activation energy 64, 66, 78, 81, 91, 92, 106, 137, 142
Adamantyl methacrylate 99, 191
Adamantylmethyl methacrylate 101
Aerial image 220
Aerobic autooxidation 102
Airborne base contamination 89, 91, 92, 198, 217
Aldehyde 157
Aldol condensation 21
Alicyclic methacrylate 99
Alkyl vinylsulfonate 136
Alkylation 59
–, C- 58, 66, 82, 148, 153, 156, 158
–, O- 58, 82, 153, 156, 158
All-dry bilayer lithography 178, 188
All-dry development 188
Alternating copolymer 70, 71, 105, 108, 116, 129

Alternating phase shift 133
Amine gradient process 119
Amino acids 5
Amorphous molecule 87
Androstane 101
Androsterone 161
Anhydride 62, 114, 146–148, 161, 198
Anionic polymerization 64, 127, 129, 131
– –, living 59, 64, 67, 68, 72, 126, 140, 143, 144, 185
Anisotropy 27
Annealing 93, 95, 198
Annular illumination 180, 214
Antibody 2, 3, 19
Anti-reflection coating (ARC) 93
APEX resist 89, 95, 219
Aqueous base soluble polymer 59, 61
Aqueous system 14, 23
Argon fluoride (ArF) excimer laser 47, 70, 97, 98, 101, 161, 162, 193
Artificial receptor 19
Aspect ratio 174, 176
Associating PAG 91
Atomic absorption 137
Atomic force microscope (AFM) 174, 207, 216
Auger spectroscopy 207
2,2-Azobis(isobutyronitrile) (AIBN) 59, 66, 135, 149, 173, 181, 182, 198

Backbone cleavage 140, 143
Base hydrolysis 67, 84, 110, 157, 158
Beads 11
Benzopinacole 149, 150
Benzoyl peroxide (BPO) 59, 66
Benzyl acetate 155
Benzyl methacrylate 71, 187
Benzylic carbocation 142, 151, 155, 156

Benzylic stabilization 150, 160
Bias 197, 201, 216, 217
Bilayer 52, 138, 141, 159, 175–185, 187, 198, 201, 214
Bilayer resist 139
Bile acid 101
Bis(4-*tert*-butylphenyl)iodonium cyclamate 106
1,4-Bis(2-hydroxyhexafluoroisopropyl)-benzene 126, 131
1,3-Bis(trimethylsilyl)isopropyl methacrylate 182
Bisazide 199
Blend 72, 81, 83, 129, 130, 141, 164, 183, 207
Block copolymers 71, 72, 119, 124, 126, 173, 182, 202
*t*BOC 55, 59–61, 64, 70–74, 82, 84, 89, 116, 123, 130, 131, 135, 143, 179, 184, 188, 191, 198
– resist 55–57, 71, 89, 90, 188, 190, 191, 208, 217–219
Bulk polymer 11
tert-Butyl acrylate (TBA) 77, 95, 106, 109, 131, 133, 191
tert-Butyl metacrylate (TBMA) 71, 82, 99, 109, 119, 126, 131, 133, 173, 181, 191, 198, 202
tert-Butyl 2-trifluoromethylacrylate (TBTFMA) 129, 130, 131, 132, 133
tert-Butyl 4-vinylbenzoate 64
4-*tert*-Butoxycarbonyloxystyrene (BOCST) 66, 70, 156, 157, 165
4-*tert*-Butyl(dimethyl)silyloxy-α-methylstyrene 143
4-*tert*-Butyl(dimethyl)silyloxystyrene 59, 67
4-*tert*-Butoxystyrene 64, 68
Butyllithium 127

C-alkylation 58
^{13}C NMR 52, 57, 161, 205
–, inverse gated 57, 108, 205–206
–, solid state 106, 206
^{14}C labeling 93, 208
Cage effect 49, 173
Calixarene 84, 87, 156
Calix[4]resocinarene 84
CAMP 59, 89
Capacity 13–16, 28
Capillary electrophoresis 18

Capillary force 175
Carbobenzoxy-L-phenylalanine 27, 28
Carbon dioxide 55, 59, 85, 172–175
Carbon monoxide 130, 133
Casting solvent 43
CASUAL 183
Catalyst 4
Catalytic antibody 20, 21
Catalytic chain length 55, 219
Cationic photoinitiator 164, 199
Cationic polymerization 58, 66, 75, 79, 114, 140–143, 199
– –, ring-opening 164
Ceiling temperature (T_c) 139–144
Cell membrane 23
Chain conformation 208
Chain dynamics 106, 206
Chain entanglement 86
Chain relaxation 10, 14
Chain transfer agent 112
Charge transfer 107, 109
Chemical amplification of resist lines (CARL) 197, 198
Chemical contrast 212
Chiral recognition 4
Chirality 5, 16
Chloromethyltriazine 153
Cholate 101, 106
–, *tert*-butyl 106, 113, 114
Cholesterol 9
Chromatographic stationary phase 16
Chromatographic support 4, 8
Chrome-on-glass (COG) 138, 180
Claisen rearrangement 148
COBRA 112–114, 118, 119
COMA 107, 110, 116, 119, 182, 206
Condensation 76, 84, 152, 156, 157, 159, 179, 193
Contact angle 207
Contact hole 96, 97, 154, 197, 214
Contrast 42, 44, 69, 172, 191, 193–195, 202, 211, 212
Contrast boosted resist 159
Contrast enhancement 159
Copolymerization 108
–, kinetics 206
Corona 202
Covalent interactions 4, 6
Cross polarization 106
Crosslinker 10, 23, 27, 155, 161, 193, 204

Subject Index

Crosslinking 55, 69, 74, 75, 119, 152, 157–160, 164, 166, 192, 193, 196–203
–, thermal 170, 178, 192
18-Crown-6 127, 131
Crown ether 3
Cryptate 3
Cupric ion 26
Curing 196
–, photochemical 164
Cyclodextrine 3
Cyclohexenyl 148
Cyclohexylmercuric chloride 109
Cyclophane 3
Cyclopolymerization 80, 118, 119, 133, 140–146, 161, 185
Cyclopropyl 146

2D-NOESY 206
2D-WISE 206
Decarboxylation 172
De-crosslink 165, 170
Deep UV resist 139
Deesterification 146
Degree of crosslinking 10, 11
Dehydration 151, 157, 159, 166
Dehydration, intermolecular 158
Dehydration, intramolecular 151, 152
Dehydrofluoration 20
Dehydrogenation 65
Delayed acid generation 91
Dendrimer 86, 156, 207
Depolymerization 53, 59, 92, 139–145, 185
Depolymerization, thermodynamically-driven 185
Deprotection 54–139, 172, 173, 179, 181, 188, 190, 198–206, 212, 218, 219
Depth-of-focus (DOF) 176, 214–215
Desilylation 59, 67, 72, 143, 187, 198
DESIRE 192–194, 201
Detection 20
Developer selectivity 78, 128, 16172, 209, 212
Development 42
– contrast 85, 110, 123, 130, 180, 133, 212
Di-*tert*-butyl dicarbonate 61
Di(*tert*-butylphenyl)iodonium pentafluorobutanesulfonate 51, 69, 219
Di(*tert*-butylphenyl)iodonium trifluoromethanesulfonate 99
1,8-Diaminonaphtalene 27
Diazepam 19

Diazonaphthoquinone (DNQ) 45, 47, 55, 68, 133, 198
Dicyclopentyloxy methacrylate 165
Diels–Alder reaction 21, 61, 130
Differential scanning calorimetry (DSC) 204
Diffraction 220
Diffusion, Fickian/Case II 194
Diffusion coefficient 218, 219
Diffusion length 218
Diffusion promoter 194
Dihydrofuran 130
Dimerization 152
α,α-Dimethylbenzyl methacrylate 71
N-(3,5-Dinitrobenzoyl)-α-methylbenzyl-amine 17
vic-Diol 149, 150, 169
Diphenyliodonium salt 81
Diphenylsilanediol 159
Dissolution behavior 46, 102, 118, 208, 209
Dissolution contrast 96, 132
Dissolution inhibition 46, 52, 68, 84, 113, 133, 137, 150, 198
Dissolution inhibitor 46, 80, 81, 83, 101, 105, 106, 123, 124, 130, 137, 149, 159, 204
– –, multi-functional 84
Dissolution inhibitor precursor 158
Dissolution kinetics 69, 209
Dissolution modifying agent (DMA) 113
Dissolution promoter 80, 150, 159
Dissolution rate 48, 58, 68–71, 110, 116
DNA replication 2
Dose latitude 215
Drug 5
Dry development 175, 201
Dry etch 141, 176
– – resistance 42, 71, 98, 99, 101, 118, 124, 141, 172, 217
Dual solubility change 169
Dual tone imaging 53, 55, 62, 152, 165, 188
Dye 5, 190, 205
Dynamic random access memory (DRAM) 47, 56, 89, 97

Electron beam projection lithography (EPL) 137
Electron beam resist 126, 139, 144
Electron deficient monomer 129
Electron transfer 50

Electrophilic substitution 153, 155
Ellipsometry 204, 211
Emulsion 12
Enantiomer 5, 10, 17
End cap 140, 185
Endo group 59
Ene reaction 126
Enthalpy/entropy of polymerization 142
Environmental stabilization 93
Enzyme 2–5, 9, 22
Epi-Rez SU-8 164
Epoxy resin 164
Equilibrium monomer concentration 142
ESCA 207
ESCAP 77, 78, 95–97, 122, 137, 181, 206–209, 218, 220
Ester, tertiary 62, 64, 96, 112, 123
Esterification 161, 162
Etching 42
Ethoxymethyl 126, 131, 132
Exposure latitude (EL) 194, 215
Extinction coefficient 211
Extraction, solid phase 18

^{19}F NMR 206
F$_2$ excimer lasers 47, 61, 122
Fittig bis lactone 85
FLEX 214
Flood exposure 148, 188, 190, 192
Fluorescence 207
Fluoroalcohol 61, 184
Fluoroalkyl methacrylate 119
Fluorocarbon plasma 177, 178, 192
Fluoropolymer 80, 121, 122, 135
Focus latitude 214
Focused ion beam 198
Foot 216
Free volume 91–95, 204
Fries rearrangement 67
Functional monomer 4, 5, 8, 11, 23
Functional polymers 43
4-(3-Furyl-3-hydroxypropyl)styrene 157

G$_s$ 139, 212
G$_x$ 212
Gas chromatography 109
Gel dose 212
Gel layer 113, 211
Gel permeation chromatography (GPC) 74, 77, 203
Geometry optimization 136

Glass transition temperature (T$_g$) 59, 64, 67–70, 87, 93, 95, 99, 101, 105, 115, 119, 143, 184, 191, 196, 198, 201–204
Glicoluril 161
Graft polymerization lithography 199
Grignard reagent 127
Group transfer polymerization 71, 119, 173, 202

^{1}H NMR 60, 160
– analysis, in situ 108, 127, 206
Half-wavelength lithography 96
Hard bake 178, 185, 194, 196
Hexafluoroacetone 61, 126
Hexafluoroisopropanol (HFA) 61, 80, 86, 116, 118, 122, 123, 126, 131, 136, 191
Hexafluoroisopropyl vinylsulfonate 135
Hexamethoxymelamine 159, 167
Host-guest ratio 8
meso-Hydrobenzoin 149
Hydrogel 24
Hydrogen bonding 8, 27, 68, 72, 80, 149, 150, 167, 206
Hydrogenation 66, 79, 114, 126
Hydrolysis 60, 76, 77, 119, 124, 132, 135, 138, 149, 167, 169, 187, 198
–, of anhydride 110, 114, 119, 198
–, of lactone 85, 104, 114, 162
Hydrolytic stability 48, 49, 62, 65, 95, 96, 204
Hydrophilic polymer 55, 112
Hydrophilicity 100–102, 123, 129, 151
Hydrophobic interactions 8
Hydrophobicity 101, 122, 202, 207
Hydrosilylation 184
γ-Hydroxy acid 161
4-Hydroxy-2,3,5,6-tetrafluorostyrene 132
4-(2-Hydroxyhexafluoroisopropyl)styrene (STHFA) 131, 132, 135
Hydroxyphenyl methacrylate 61
4-Hydroxystyrene (HOST) 61, 65, 70–72, 77, 83, 95, 169, 181, 206
Hyperbranched polymer 86, 87

Image blur 217–221
Image spread function model 220
Image spreading 219
Immersion lithography 121, 137, 223
Impedance 113
Imprinted site 9, 11
In situ materials 11

Indenecarboxylic acid 46, 153
Industry roadmap 137, 222
Interactions, non-covalent 4, 8, 9
–, pi-pi 8
Interfacial layer 113, 207, 209
Interfacial mixing 91, 178
Interferometric lithography 220
Interferometry 113, 208, 209
Ion conductivity 52, 188
Ionic interactions 8
IR microscopy 206
IR spectroscopy 56, 110, 149, 161, 162, 188, 191, 194, 206, 211, 212
–, gas phase 206
–, in situ 109
–, reflectance-absorbance 138, 206
Isobornyl methacrylate 101, 131, 191
Isobutene 55, 59, 63
Isocyanate 190
Isomerisation 21
2-Isopropenyl-2-oxazoline 168, 169

Ketal 64, 72, 92, 137, 206
β-Keto carboxylic acid 172
KRS 97, 137, 138, 206
Krypton fluoride (KrF) excimer laser 46, 47

Lactone 85, 114
δ-Lactone 161
γ-Lactone 161
β-Lactone 85
Ladder polymer 178
Langmuir–Blodgett 24
Laser ablationmicroprobe mass spectroscopy (LAMMS) 207
Laser confocal microscope 52
Laser end-point detection 209
Latent electrophile 152, 155–157, 160, 167
Latent image 208, 220
Length scale of mixing 106, 206
Leu-enkephaline 19
Light scattering 203
Line collapse 138, 174, 175, 194, 216
Line edge roughness (LER) 87, 88, 97, 137, 140, 146, 184, 190–194, 201, 202, 217, 220
Linear resolution 213
Linewidth slimming 93
Linewidth variation 176
Lipophilic polymer 55, 88, 129

Lipophilicity 80, 141, 151
Liquid crystallyne imprinted network 27
Liquid-crystal network 26
Lisocholate 101
Lithographic process, environmentally friendly 166
Lithography, 157-nm 80
–, 193-nm 79, 161, 181, 201
–, graft polymerization 199
–, photo/electron beam/X-ray 45
Lock-and-key 3
Low energy electron beam projection lithography (LEEPL) 138
Lyotropic phase 24

Magic angle spinning (MAS) 106
Main chain scission 47, 59, 60, 63, 139–143, 146, 203
MALDI-TOF 184, 207
Maleic anhydride 61, 107, 109, 115, 118, 119, 161, 166, 198, 206
Malonic acid 172
Mass-persistent resist 85, 86
Mass spectroscopy 58, 207
Maxium dissolution rate 96
Mayo–Lewis model 142
Mechanistic study 156
Melamine 179, 193
Membrane 12
Membrane osmometry 203
Membrane receptor 2
Mercury method 108, 109
Mesogenic 27
Mesomorphic state 27
Mesomorphous polymer 27, 29
Metal content 203
Metal-ions 5
Metal-template bond 8
Methacrylate 79, 80, 126, 127, 187, 191, 201, 206
Methacrylate terpolymer resist 83, 101, 161, 183
Methacrylic acid (MAA) 82, 99, 101, 182
Methacrylonitrile 135
Methacryloxyethyltrimethylsilane 181
Methacryloxyethyltris(trimethylsilyl)silane 181
Methoxymethyl 126
N-Methoxymethylated melamine 153
Methyl methacrylate (MMA) 82, 99, 127, 182, 187

Methyl styrenesulfonate 169
2-Methyladamantanol 129
2-Methyladamantyl 2-trifluoromethylacrylate 131
2-Methyladamantyl metacrylate 126
Methylcyclopentyl (MCP) 114
α-Methylstyrene 71, 135
Mevalonic lactone 100
Micelle 202
Micelle 24
Michaelis–Menten 21
Microbridge 154
Micro-electromechanical system (MEMS) 164, 166
Microelectronic devices 41
Minimum dissolution rate 96
Molecular dynamics 68
Molecular mechanics 68
Molecular orbital 104, 136, 140
Molecular recognition 2, 3
Molecular resist 88
Monomer reactivity ratio 77, 142
Monomer-template complex 6
Monomer-template interactions 7
Moore's law 41
Morphine 19
Morphology of the network 13
Multi-functional phenolic compound 84
Multilayer 175, 176
Multiple anion nonvolatile acetal (MANA) 139
Multiple exposure 214

Naphthalene 105
Near edge X-ray absorption fine structure (NEXAFS) 207
Neutron reflectometry 208
Next generation lithography (NGL) 47, 137, 223
2-Nitrobenzyl benzenesulfonate 92
NMP uptake 93–95
NMR analysis 67, 156, 181, 203–206
Norbornene 61, 79, 80, 108, 127, 161, 206
Norbornene hexafluoroalcohol (NBHFA) 116–124, 127–130, 133, 150
Norbornene-maleic anhydride 79, 105–110
Norbornene-sulfur dioxide 79
Notching 216
Novolac 45, 55, 59, 61, 64, 65, 80–82, 84, 99, 106, 113, 141, 156, 164, 185, 194–196

– / DNQ positive photoresist 56, 72, 80, 89, 96, 153, 192, 195
Nucleotides 5
Numerical aperture (NA) 45, 96, 176, 213

Off-axis illumination 96, 191, 214
Onium salt 49, 57, 164
Organic media 14
Organometallic monomer 199
Organometallic polymer 176, 199
Organosilicon polymer 176
Oscillator strength 136
Outgas 137, 138, 146, 184, 206–209
Oxygen plasma 176, 178, 185, 187, 194, 201
Oxygen reactive ion etching (RIE) 97, 141, 176, 187, 188–193, 195, 198–202
Oxygen RIE pattern transfer 178, 185, 195, 196

PAG decomposition 206
PAG distribution 52, 207
Palladium (II) complex 112
Palladium catalyst 112, 123, 130
Pattern size bias 197
PEB temperature stability 217
Penultimate model 129, 132
$1H,1H$-Perfluorooctyl methacrylate 173
Pesticide 5
PF-APEX 131, 132
PF-ESCAP/PF^2-ESCAP 131, 132
Phase separation 72, 130
Phase shift mask 96, 162, 214
Photoactive compound (PAC) 45, 47
Photobase 91
Photochemical ablation 97
Photochemical acid generator (PAG) 48–53, 70, 204
Photochemical base generator 48
Photochemical curing 164
Photodecomposable base 83, 91, 106
Photodecomposition, ablative 98
Photolysis product 50
Photomask 138, 166
Photo-oxidation 193
Photosensitive polymer 199
Piezoelectric quartz crystal oscillator 209
Pinacol-pinacolone rearrangement 149–151, 158, 169
Planarizing layer 176, 178, 185, 195, 198

Subject Index

Plasma-developable photoresist (PDP) 176
Plasticization 48, 94, 101, 146, 198, 204
Polarity change 55, 62, 148, 157, 159, 188
Polarity reversal 146, 166
Polyaldehyde 140
Polycarbonate 145
Polyclonality 8
Polyether 146
Polyformal 145
Polyhydroxybenzylsilsesquioxane 179
Polyimidazoles 5
Polyisoprene 199
Polymer-bound sulfonic acid 201
Polymeric acid generator 51
Polymeric dissolution inhibitor 82, 133, 141
Polymerization 53
–, kinetics 127
–, rate 142
Polymethacrylate 47, 62, 63, 98, 113
Polynortricyclene 118
Polyphenylacetylene 142
Polyphenylmethylsilsesquioxane 178
Polyphthalaldehyde 92, 140, 184
Polypyrroles 5
Polysiloxane 159, 178, 201
Polysiloxane 5
Polysilsesquioxane 122, 182, 184
Polyurethane 5
Poly(α-acetoxystyrene) 85, 141
Poly(adamantly methacrylate) 99
Poly[4,5-bis(trimethylsilyl)phthalaldehyde] 186
Poly(4-tert-butoxycarbonyloxy-α-methylstyrene) 59
Poly(4-tert-butoxycarbonyloxystyrene sulfone) 59, 60, 70, 89, 118
Poly(4-tert-butoxycarbonyloxystyrene) (PBOCST) 55–62, 66, 72, 85, 89, 94, 205, 218, 219
Poly(4-tert-butoxystyrene) 148
Poly(tert-butyl 4-vinylbenzoate) 85, 146, 191
Poly(tert-butyl methacrylate) (PTBMA) 63, 147, 148, 218
Poly(4-chlorophthalaldehyde) 141
Poly(2-cyclopropyl-2-propyl 4-vinylbenzoate) 146
Poly(di-tert-butoxysiloxane) 179
Poly(2,3-dichloropropyl acrylate) 176

Poly(4-hydroxy-α-methylstyrene) 59, 143
Poly[4-(1-hydroxyethyl)styrene] 157
Poly[4-(2-hydroxyhexafluoroisopropyl)-styrene (PSTHFA) 60, 133
Poly[4-2-(hydroxyl-2-propyl)styrene] 151
Poly(4-hydroxystyrene sulfone) 59, 60
Poly(4-hydroxystyrene) (PHOST) 55, 59, 63–75, 82, 84, 95, 104, 141, 148, 153, 156–159, 164, 188–198, 205, 207, 208, 219
Poly(methacrylic acid) 62, 91, 146
Poly(3-O-methacryloyl-D-glucopyranose) 168
Poly(methyl 2-trifluoromethylacrylate) (PMTFMA) 126
Poly(methyl methacrylate) (PMMA) 98, 139, 217
Poly(methyl vinylsulfonate) 133
Poly(norbornene hexafluoroalcohol) (PNBHFA) 129, 130, 133, 211
Poly(norbornene sulfone) 118, 122, 191
Poly(3-methyl-2-(4-vinylphenyl)-2,3-butanediol) 149
Poly(4-methylene-4H-1,3-benzodioxin-2-one) 85
Poly(2,3-norbornene) 63, 111, 112
Poly(4-phenoxymethylstyrene) 148
Poly(4-trimethylsilyloxystyrene) 186, 187
Poly(4-trimethylsilylphthalaldehyde) 178, 185
Poly(4-vinylbenzoic acid) 62, 63, 178, 186
Poly(vinyl alcohol) 124, 167
Poly(vinylsulfonyl fluoride) 133
Pores 13
Porous material 12
Positron annihilation spectroscopy (PALS) 204
Postapply bake (PAB) 41, 95, 204
Postexposure bake (PEB) 48, 58, 64, 91, 185
Postexposure delay 90, 111
Postexposure delay, stability 92, 94, 96, 217
PREVAIL 137
PRIME 192
Process latitude 42
Projection electron beam lithography (PEL) 97
Propylene glycol methyl ether acetate (PGMEA) 43
Protecting group 54–139
Protective overcoat 91

Proteins 3, 12, 26
β-Proton elimination 62, 142, 145, 151
Proton spin diffusion 106
Proximity effect 138

Quantification of acid 52
Quantum chemical calculation 50, 57, 133
Quantum yield 46–48, 52, 139, 212
Quartz crystal microbalance (QCM) 69, 113, 204, 208–211
Quencher 91
Quinone 25

Racemic mixture 5, 7
Radiation sensitive polymers 41, 44
Radical polymerization 10, 58–61, 64–67, 77, 79, 85, 105, 114, 116, 126, 127, 132, 142, 149, 157, 198
–, living 67
Radius of gyration 67
Rayleigh's eqation 45
Reactive ion etching (RIE) 59, 98
Real time infrared reflectance-absorbance spectroscopy 138
Rearrangement 146–151, 203
Receptor 3
Recognition sites 14, 23, 29
Reflectance 113, 211
REFLOW 119
Refractive index 94, 121, 137, 209, 211, 223
Regioselective 20, 21
RELACS 119
Relaxation agent 205
Residual acid 203
Residual casting solvent 93, 173, 204, 205
Resisdual base 203
Resistance 211
Resists 41, 159
–, acid-hardening 159, 193, 194
–, bilayer 139
–, contrast boosted resist 159
–, positive/negative 44
–, silicon-added bilayer (SABRE) 195
–, water-processable 166–172
Resolution 42, 44–46, 91, 96, 137, 139, 153, 164, 176, 188, 213, 217–221, 223
Resolution enhancement technique (RET) 45, 96, 213
Resolution factor 17

Reverse polarity change 149–152, 158, 161, 169
Ring parameter 99
Ring-opening 85
Ring-opening metathesis polymerization (ROMP) 79, 85, 114
Rinse 174, 207, 216
Rutherford backscattering spectrometry (RBS) 53, 124, 194, 207

^{29}Si NMR 181, 206
SAFIER 119
Sameridine 18, 19
Sauerbery equation 209
SCALPEL 137
Scanning electron microscope (SEM) 207, 217
–, top-down/cross-sectional 216
Scanning probe microscope (SPM) 207
Scanning tunneling microscope (STM) 52, 207, 222
Scanning viscoelasticity microsope (SVM) 204
Scattering 220
Scintillation 208
Secondary ion mass spectroscopy (SIMS) 53, 207
Segregation 124, 207
Self-condensation 157–159, 166
Self-development 92, 140, 141
Semiconductor 41
Sensitivity 42, 43, 47, 48, 193, 212, 223
Sensitivity curve 44, 211, 212
Sensitivity enhancement 47, 137, 139
Sensitizing molecule 51
Sensors 4, 16, 20
Separation factor 10, 17, 18
Separation performance 17
Shelf life 93, 110, 161
Shot noise 137, 223
Silanol 159, 187
– condensation 178, 179
Silica gel 4, 5
Silicon-added bilayer resist (SABRE) 195
Silsesquioxane 159
β-Silyl carbocation 180
Silylating agent 188–191, 194, 198, 202
Silylation 195, 201
–, after wet development 195–198
–, bilayer 194
–, diffusion-controlled 188–194, 201

Subject Index

–, flood 198
–, gas phase 194, 197, 202, 209
–, liquid phase 194, 198, 202
–, reactivity-controlled 188–191, 201
Silylation process after alkaline wet development (SILYAL) 196
Simulation 154, 219
Size exclusion chromatography (SEC) 203
Skin 89, 90
Small angle neutron scattering (SANS) 67, 208
S_N2' addition-elimination 127
Sol-gel reaction 179
Solubility parameter 95, 102
Solvent, dissociating power 13
–, environmentally benign 166
Solvent-induced stress 143
Spatial evolution 208
Specificity 8–11, 14, 16
Spirobornane 119
Stability 16
Stabilizing additive 91
Standing wave 93
Stationary phase 17
Stereoselective 20, 21
Steroid 5, 22, 101, 106, 111, 124
Stochastic kinetics simulator (CKS) 218
Stress cracking 101
Strip 178, 187
STUPID model 136
SUCCESS 82
Sugar derivatives 5
Sulfonamide 114, 118
Sulfonic acid ester 133
Sulfur dioxide 58, 59, 61, 66, 116, 143, 194
SUPER 193
Supercritical CO_2 43, 119, 150, 166, 172–175
Supercritical fluid (SCF) 43, 173
Supramolecular organisation 23
Surface imprinting technique 12, 23
Surface modification 198, 199
Surface tension 173–175
Surface-insoluble layer 89, 90, 208
Surface-silylated single layer resist (SSS) 198
Suspension 12
Sweet PAG 106

Swelling 55, 113, 114, 161, 164, 167, 169, 172, 194, 202, 211
Symmetry adapted cluster configuration interaction (SAC-CI) 137

T-top 89, 119
Template extraction 12–14
Template molecule 4–6
TEMPO 67, 182
Terminal model 129, 131
Tetracyclododecene 107, 114
Tetrafluoroethylene (TFE) 80, 122–124
Tetrahydrofuranyl (THF) 184
Tetrahydropyranyl (THP) 64, 72, 124, 132
Tetrahydropyranyl methacrylate 71, 101, 119, 126, 173, 181
Theophylline 19, 27, 28
Thermal analysis 203–204
Thermal development 141, 145, 146, 185
Thermal expansion 208
Thermal flow 93, 96, 119, 196, 202
– – resistance 46, 75
Thermal probe 204
Thermal stability 48, 49, 55, 62, 65, 72, 95, 96, 204
Thermogradient plate (TGP) 206
Thermogravimetric analysis (TGA) 62, 72, 204, 144, 206
Thermolysis 66, 207
–, autocatalytic 59, 74
Thin film imaging 138, 204
Thin layer 12, 18
Time-dependent density functional theory (TD-DFT) 133, 136
Tool contamination 137, 141
Topography 176–178, 188
Top-surface imaging (TSI) 97, 188, 190–193, 201, 214
Total internal reflection fluorescence spectroscopy 207
Trace atmospheric gas analyzer (TAGA) 57, 58, 63
Transannular polymerization 118
Transesterification 162, 166
Transetherification 166
Transition energy 136
Transition state 21
Tricyclononene 130
2-[4-(2,2,2-Trifluoro-1-ethoxymethoxy-1-trifluoromethylethyl)cyclohexane]-hexafluoroisopropyl acrylate 132

2-[4-(2,2,2-Trifluoro-1-ethoxymethoxy-1-trifluoromethylethyl)cyclohexane]-hexafluoroisopropyl 2-trifluoromethylacrylate 131
Trifluoroisopropanol 131
2-Trifluoromethylacrylate 80, 126, 127, 131
2-Trifluoromethylacrylic acid (TFMAA) 127, 129
5-Trifluoromethyl-5-hydroxy-2-norbornene 129
2-Trifluoromethylvinyl acetate 124, 132, 133
Trilayer 177
Trimethylsilyl 64
Trimethylsilyl methacrylate 135, 187
4-Trimethylsilyloxy-α-methylstyrene 135
Triphenylsulfonium hexafluoroantimonate 50, 56, 71, 82, 164, 173
Triphenylsulfonium salt 49, 81, 82
Triphenylsulfonium trifluoromethanesulfonate 50, 92, 161, 166
1,3,5-Tris-(2-(2-hydroxypropyl)benzene 159
1,3,5-Tris(bromoacetyl)benzene 159
3-[Tris(trimethylsilyloxy)silyl]propyl methacrylate 181
Two step-swelling 11
Two-dimensional wide line separation NMR 106

Ultraviolet (UV) 45
–, deep 45, 47, 65, 89, 90, 95, 153, 192, 198
–, extreme 45, 75, 97, 137
–, mid 45, 133
–, near 45, 65, 192, 198
–, spectroscopy 204–205
–, vacuum 45, 133, 135
Uril 179
Ursocholate 101

van der Waals volume 126, 136
VEMA 115
Vesicle 24
Vinyl addition polymerization 79, 111, 123, 130
– – –, living 112
Vinyl ether 74, 75, 80, 115, 129, 165, 170, 199, 206
– / maleic anhydride 79
4-Vinylbenzyl acetate 156
Vinylene carbonate 130
4-Vinylphenylboronic acid 6
Volatile organic compound (VOC) 166

Water-castable and water-developable resist 148
Water-processable resist 166–172
Water-repellent compound 159
Weak interactions 23
Wet development 97
Wittig reaction 66, 67
Wolff rearrangement 46, 92

X-ray lithography 45, 81, 137
X-ray photoelectron spectroscopy 207
X-ray reflectivity 204, 208

Printing: Krips bv, Meppel
Binding: Litges & Dopf, Heppenheim